Molecular Magnetism: A Themed Issue in Honor of Professor Dai-Zheng Liao on the Occasion of His 85th Birthday

Molecular Magnetism: A Themed Issue in Honor of Professor Dai-Zheng Liao on the Occasion of His 85th Birthday

Guest Editors

Peng Cheng
You Song
Hui-Zhong Kou
Jinkui Tang

MDPI

Basel • Beijing • Wuhan • Barcelona • Belgrade • Novi Sad • Cluj • Manchester

Guest Editors
Peng Cheng
Department of Chemistry
Nankai University
Tianjin
China

You Song
School of Chemistry and
Chemical Engineering
Nanjing University
Nanjing
China

Hui-Zhong Kou
Department of Chemistry
Tsinghua University
Beijing
China

Jinkui Tang
School of Chemistry and
Chemical Engineering
Beijing Institute of
Technology
Beijing
China

Editorial Office
MDPI AG
Grosspeteranlage 5
4052 Basel, Switzerland

This is a reprint of the Special Issue, published open access by the journal *Magnetochemistry* (ISSN 2312-7481), freely accessible at: https://www.mdpi.com/journal/magnetochemistry/special_issues/U75H935UZ7.

For citation purposes, cite each article independently as indicated on the article page online and as indicated below:

Lastname, A.A.; Lastname, B.B. Article Title. *Journal Name* **Year**, *Volume Number*, Page Range.

ISBN 978-3-7258-7160-5 (Hbk)
ISBN 978-3-7258-7161-2 (PDF)
https://doi.org/10.3390/books978-3-7258-7161-2

Cover image courtesy of Peng Cheng

Contents

About the Editors

Peng Cheng

Peng Cheng received his BSc, MSc, and PhD from Nankai University in 1985, 1991, and 1994, respectively. He started his academic career at the same university and was appointed a full professor in 1996. From 1997 to 1999, he worked at Laboratoire de Chimie de Coordination du Centre National de la Recherche Scientifique (CNRS) as a visiting professor and at Texas A&M University as a postdoctoral research fellow. His research focuses on functional coordination polymers and molecule-based materials, primarily involving the design and synthesis of molecular materials, structural regulation, and studies on optical, electrical, magnetic, and catalytic properties. He has authored "Lanthanide Metal-Organic Frameworks" (Springer, 2015) and "Lanthanides: Fundamentals and Applications" (Elsevier, 2023), in addition to co-authoring more than 500 articles in international journals.

You Song

You Song received his PhD from Nanjing University in 2000. He then completed two postdoctoral fellowships, first at the University of Tokyo as an STA and JSPS fellow (2000–2002) and later at Nanjing University (2003–2005). After completing his postdoctoral position in 2005, he continued at Nanjing University as an associate professor and was promoted to full professor in 2009. In 2005, he also served as a short-term visiting professor at Academia Sinica and National Taiwan University. Since 2021, Prof. Song has been a member of the Editorial Board of *Magnetochemistry*. His primary research interests include molecular magnetism, spin frustration, and electron spin qubits. Prof. Song has co-authored over 300 articles in prestigious international journals. He has delivered keynote speeches at several prestigious conferences, including but not limited to the 8th International Symposium of Chinese Inorganic Chemists (2010), the 14th Asian Chemical Congress (2011), and the 3rd International Conference on Functional Molecular Materials (2017).

Hui-Zhong Kou

Hui-Zhong Kou received his PhD from Nankai University in 1999. Thereafter, he completed a postdoctoral position at Peking University (1999–2001) before joining Tsinghua University as a lecturer. He was promoted to associate professor in 2003 and to full professor in 2007. Between 2004 and 2005, he worked as a JSPS fellow at Kyushu University. In 2007, he was awarded the "New Century Excellent Talents" research grant by the Ministry of Education of China. Since 2020, Prof. Kou has served as a member of the Editorial Board for the journal *Magnetochemistry*. His primary research interests include rare-earth-based single-molecule magnets, spin crossover complexes, and multifunctional molecular materials. Prof. Kou has co-authored more than 200 articles in international journals.

Jinkui Tang

Jinkui Tang received his PhD from Nankai University in 2001. He then undertook two postdoctoral positions at Universität Karlsruhe (TH) (2001–2006) and Leiden University (2006–2007) before initiating his independent career at Changchun Institute of Applied Chemistry, CAS. Since 2025, he has been a distinguished professor at the Beijing Institute of Technology. Prof. Tang was awarded a "National Science Fund for Distinguished Young Scholars" research grant in 2015 and the "Newton Advanced Fellowship" in 2016. Since 2020, he has served as a member of the

Editorial Board of the journal *Magnetochemistry*. His primary research interests include lanthanide coordination chemistry and molecular magnetism. Prof. Tang has co-authored more than 200 articles in international journals.

magnetochemistry

MDPI

Editorial

Molecular Magnetism: A Themed Issue in Honor of Professor Dai-Zheng Liao on the Occasion of His 85th Birthday

Peng Cheng [1,*], You Song [2,*], Hui-Zhong Kou [3,*] and Jinkui Tang [4,*]

1 Department of Chemistry, Key Laboratory of Advanced Energy Materia Chemistry, and Frontiers Science Center for New Organic Matter, College of Chemistry, Nankai University, Tianjin 300071, China
2 State Key Laboratory of Coordination Chemistry, School of Chemistry and Chemical Engineering, Nanjing University, Nanjing 210023, China
3 Department of Chemistry, Tsinghua University, Beijing 100084, China
4 School of Chemistry and Chemical Engineering, Beijing Institute of Technology, Beijing 102488, China
* Correspondence: pcheng@nankai.edu.cn (P.C.); yousong@nju.edu.cn (Y.S.); kouhz@mail.tsinghua.edu.cn (H.-Z.K.); jtang@bit.edu.cn (J.T.)

This Special Issue of *Magnetochemistry* is dedicated to Professor Dai-Zheng Liao on the occasion of his 85th birthday. Professor Liao was born in 1939 in Fujian, China. He graduated from the Chemistry Department of Fudan University in 1962. In the same year, he began his academic career at the Chemistry Department (now the College of Chemistry) at Nankai University, where he continued to work until his retirement in 2011. As both a mentor and a scientist, he guided eighteen master's and twenty-five doctoral students and published over 700 scientific papers in peer-reviewed journals throughout his distinguished career.

Professor Liao has long been dedicated to teaching *Advanced Inorganic Chemistry* and *Quantum Chemistry*. During his teaching career, Professor Liao focused on delivering high-quality education, relying on creative solutions to engage his students. At a time when multimedia tools were not widely available, he put great effort into designing effective teaching materials. One notable example was his creation of a large, one-square-meter poster filled with key formulas, which became a distinctive feature of his lectures. In an era when academic resources were limited in China and the literature was in print, Professor Liao strove to make the latest research accessible to his students. He diligently sought out the few available resources, often relying on physical copies of papers to stay connected with cutting-edge developments. His efforts helped bring global advancements to his students, encouraging them to stay informed and motivated in pursuing their own research. The photographs in Figure 1 capture moments from Professor Liao's research career, including his visit to Academia Sinica in Taipei and his mentorship of Dr. Hui-Zhong Kou at Nankai University. In his free time, Professor Liao loved playing table tennis, which helped him stay active and maintain the same energy and enthusiasm he brought to his work and mentorship.

Professor Liao was a pioneer in the development of molecular magnetism research in China. Between 1985 and 1986, he served as a visiting professor at Kyushu University in Japan, where he was first introduced to the emerging field of molecular magnetism, which was then a rapidly growing area of coordination chemistry. Upon returning to China, Professor Liao was one of the first to lead a research team in the field of molecular magnetism. He made significant contributions to spin-crossover complexes, irregular spin-state structures, magnetic ordering, radical complexes, and single-molecule magnets. In recognition of his work, he was awarded the Second Prize of the National Natural Science

Award of China in 2003 for the project on "Foundational Studies in Molecular Magnetism". His pioneering work played a key role in the growth of molecular magnetism in China, inspiring scholars from Nanjing University, Peking University, and Sun Yat-sen University to initiate research in the field, resulting in many significant academic contributions.

Figure 1. (**a**) Professor Liao visiting Academia Sinica in Taipei, 2000. (**b**) Professor Liao (right) supervising Dr. Hui-Zhong Kou (left) during an experiment at Nankai University, 1997.

To honor Professor Liao's profound contributions to the field of molecular magnetism, *Magnetochemistry* is pleased to present this Special Issue, entitled "Molecular Magnetism: A Tribute to Professor Dai-Zheng Liao on the Occasion of His 85th Birthday", featuring fifteen contributions from leading researchers in the field of molecular magnetism in China.

In chronological order, the first contribution in this Special Issue—by Y.-X. Wang, Y.-F. Wang, and B. Yin from Northwest University [1]—is a theoretical study, based on ab initio electronic structure calculation, on a group of 16 pentacoordinate Dy(III) single-ion magnets. They emphasize the contribution from the equatorial coordinating atoms might be even more important than that from the axial coordinating atoms for good Dy(III) SMMs and reveal that widening the axial ∠O–Dy–O might be a probable way of improving the SMM performance of penta-coordinated Dy-SIMs.

Y. Qian, Y. Gao, L. Xu, R. K. Kremer, J. Zhang, and X.-M. Ren from Nanjing Tech University, the Max-Planck-Institute für Festkörperforschung, and Nanjing University [2] report the variable-temperature crystal structures of two isomorphous salts: [1-benzyl-4-aminopyridinium][$M(mnt)_2$] (M = Ni or Cu; mnt^{2-} = maleonitriledithiolate). A magnetostructural phase transition occurred at T_C ~ 190 K in the S = ½ Ni complex at ambient pressure, converting paramagnetic tetramers into nonmagnetic spin-paired dimers. No such transition was observed in the nonmagnetic Cu complex in the same temperature range. The magnetic and phase transition properties of the Ni complex under varying pressures show that T_C increases linearly with pressure (0.003–0.88 GPa) at a rate of 90 K GPa^{-1}.

Y. Zhou, Z. Yao, N. Li, F. Liu, J. Zhao, and X. Bu from Tianjin University of Technology and Nankai University [3] present mixed-valence metal formates $[CH_3NH_2CH_3]_n$ $[Cr^{III}_{1-x}Fe^{III}_xFe^{II}(HCO_2)_6]_n$ that exhibit the *N*-, *P*-, and *Q*-type Néel's ferrimagnetism. The positive and negative exchange bias in *N*- and *P*-type ferrimagnets originates from the antiferromagnetic exchange interaction between the uncompensated spin of the host ferrimagnetic lattice and the pinned compensated spin of the antiferromagnetic clusters as a guest, which is rooted in the valence disorder of the iron ions.

H.-H. Cui, D.-Z. Wang, S. Li, L. Wang, X. Yu, X. Liu, J. Wang, M. Wang, and Y. Tang from Nantong University [4] report two mononuclear octahedral Co(II) complexes: $[Co(L)X_2]$ (L = 1-(prop-2-en-1-yl)-1H-imidazole, X = NCS^- and $NCSe^-$). Magnetic data revealed

https://doi.org/10.3390/magnetochemistry12020025

large easy-plane magnetic anisotropy in both complexes, and they display field-induced slow magnetic relaxation. For both complexes, the Raman mechanism was found to be the dominant process in the whole temperature range. The magnetic relaxation of $[Co(L)(NCSe)_2]$ is faster, likely due to the presence of the hydrogen-bonding system.

B. Zhang, Y. Zhang, D. Wang, Z. Wang, G. Chang, Z. Gao, Y. Guo, F. Liu, Z. Zhao, X. Zhang, B. Qu, P. Xu, J. Wang, F. Dong, T. Liang, Y. Sun, D. Yang, Q. Li, X. Luo, and R. Feng from the Institute of Chemistry (CAS), Peking University, the National Center for Nanoscience and Technology, and the Institute of High-Energy Physics (CAS) [5] present an organic–inorganic hybrid $(BEDT\text{-}TTF)_3[Cu_2(\mu\text{-}C_2O_4)_3]\cdot CH_3CH_2OH\cdot 1.2H_2O$ prepared via electrocrystallization. The three donor organic molecules have a total formal charge of +2, consistent with TTF core bond lengths and Raman spectra. It is a semiconductor with room-temperature conductivity: σ_{rt} = 0.04 S/cm and $E\alpha$ = 40 meV. The antiferromagnetic copper-oxalate framework exhibits spin frustration, making it a quantum spin liquid candidate.

Y. Zhou, J. Xie, C. Jin, Y. Ma, and L. Li from Nankai University [6] report a heterolanthanide complex $[Dy_{0.56}La_{1.44}(hfac)_7(NITPhMeImbisH)]$ derived from a nitronyl nitroxide biradical NITPhMeImbis with $Dy(hfac)_3$ and $La(hfac)_3$ ($hfac^-$ = hexafluoroacetylacetonate). In the complex, the protonated $NITPhMeImbisH^+$ ligand chelates one Ln(III) via two neighboring NO units and binds La(III) through another NO group, forming a dinuclear structure. Magnetic measurements show dominant ferromagnetic couplings in the complex. Spin dynamics studies reveal frequency-dependent χ'' peaks under a dc field (field-induced relaxation), a combination of Orbach and QTM processes, with U_{eff} = 15.14 K, $\tau_0 = 3.04 \times 10^{-7}$ s, and $\tau_{QTM} = 3.61 \times 10^{-4}$ s.

Z. Pan, Y.-S. Ding, L. Li, and Z. Zheng from Southern University of Science and Technology [7] present trinuclear rare-earth chloride-bridged clusters $[(Li(THF)(Et_2O))(Cp^*RE)_3(\mu\text{-}Cl)_4(\mu_3\text{-}Cl)_2(\mu_4\text{-}Cl)]$ (RE = Y, Gd, Tb, Dy, etc.; Cp* = pentamethylcyclopentadienide) synthesized by reacting LiCp* with $RECl_3$ (1:1 molar ratio) in $THF{:}Et_2O$ (1:9). These clusters are highly soluble in common organic solvents. Magnetic studies show Gd_3 has coexistent ferromagnetic and antiferromagnetic superexchange interactions; Dy_3 exhibits intramolecular dipolar interactions and slow magnetic relaxation below 23 K with an energy barrier of 125(6) cm^{-1} under a zero DC field.

A contribution by X.-F. Chen, T. Wang, D. Liao, N. Wu, Y. Peng, S.-Y. Zhang, and Z.-B. Hu from the Jiangxi University of Science and Technology and Gannan Normal University [8] report two organic–inorganic materials: $(TMAA)_2[CoCl_4]$ and $(TMAA)_2[MnCl_4]$ ($TMAA^+$ = N,N,N-trimethyl-1-adamantylammonium). Both exhibit first-order structural phase transition at high temperatures; near the transition point, their dielectric properties and $\chi_M T$ values show significant abnormal changes, indicating dielectric and spin bistability. Moreover, an external magnetic field induces obvious changes in their dielectric constants, providing strong evidence for magnetic–dielectric coupling effects.

C.-M. Liu from the Institute of Chemistry (CAS) [9] reports two new Dy_2 complexes: $[Dy_2(L_{Schiff}\text{-}1)_2(DMF)_2(dpp)_2]\cdot 0.5DMF$ and $[Dy_2(L_{Schiff}\text{-}2)_2(DMF)_2(dpp)_2]\cdot 2DMF$, based on two compartmental Schiff-base ligands. Magnetic studies show ferromagnetic interactions between Dy^{3+} ions in both complexes, which act as zero-field SMMs with U_{eff}/k values of 49.7 K for $[Dy_2(L_{Schiff}\text{-}1)_2(DMF)_2(dpp)_2]\cdot 0.5DMF$ and 151.8 K for $[Dy_2(L_{Schiff}\text{-}2)_2(DMF)_2(dpp)_2]\cdot 2DMF$. This work indicates that heterocycle groups (pyrazine vs. pyridine) on the ligands affect the SMM properties of the reported complexes.

L. Miao, D.-M. Zhu, C.-M. Liu, Y.-Q. Zhang, and H.-Z. Kou from Tsinghua University, Nanjing Normal University, and the Institute of Chemistry (CAS) [10] report three lanthanide complexes, $[Dy(HL^1\text{-}o)(NO_3)_2(CH_3OH)_2]NO_3\cdot CH_3OH$, and tetranuclear complexes, $[Ln_4(L^1\text{-}c)_2(L^2)_2(\mu_3\text{-}OH)_2(NO_3)_2(CH_3OH)_4](NO_3)_2\cdot 2CH_3CN\cdot 5CH_3OH\cdot 2H_2O$

https://doi.org/10.3390/magnetochemistry12020025

(Ln = Dy or Gd). Magnetic measurements reveal the mononuclear Dy(III) complex is an SMM, the tetranuclear Dy(III) complex shows slow magnetic relaxation, and the tetranuclear Gd(III) complex is a magnetic cooling material with a $-\Delta S_m$ of 9.81 J kg^{-1} K^{-1} at 2 K and 5 T. Theoretical calculations on the mononuclear Dy(III) complex indicate good magnetic anisotropy with a calculated energy barrier of 194.6 cm^{-1}.

Y.-W. Geng, T. Guo, X.-Q. Wang, and T. Han from Xi'an Jiaotong University [11] used the sulfur-containing ligand of 2-pyridinethiol 1-oxide (HL) and successfully synthesized a series of hourglass-like Ln_9 clusters: $[Ln_9(L)_{17}(\mu_3\text{-}OH)_9(\mu_4\text{-}OH)]\cdot nH_2O$ (Ln = Gd, n = 3; Ln = Tb, n = 3; Ln = Dy, n = 1). Magnetic data analysis reveals that Gd_9 shows a significant MCE, with the entropy change ($-\Delta S_m$) reaching a maximum of 34.41 J kg^{-1} K^{-1} at 2 K under ΔH = 7 T, and Dy_9 exhibits SMM characteristics. Gd_9 possesses the highest molar mass among the reported Gd_9 clusters with MCE and a competitive $-\Delta S_m$ value.

L. Zhou, H. Lv, Y. Liang, D. Liu, Z. Yao, S. Luo, and Z. Chen from Guangxi Normal University and the Guizhou University of Engineering Science [12] report a dinuclear dysprosium complex, $[Dy_2(HOBQ)_4Cl_6]$, prepared by reacting $DyCl_3$ with benzo[*h*]quinolin-10-ol (HOBQ). Magnetic measurements revealed that the complex is a field-induced single-molecule magnet with an energy barrier of 71(2) K and an obvious magnetic hysteresis loop. The magneto-structural correlation is corroborated by ab initio CASSCF calculations, demonstrating that a compressed octahedral geometry with two axial phenoxy groups can achieve Dy(III) single-molecule magnets.

X. Wang, S. Qin, X. Li, W. Zuo, Q. Wang, L. Li, Y. Ma, J. Tang, and B. Zhao from Nankai University and the Changchun Institute of Applied Chemistry [13] present a centrosymmetric dinuclear dysprosium complex, $[Dy_2(H_2dapp)_2(\mu\text{-}OH)_2(H_2O)_2]\cdot 4TCNQ\cdot 2CH_3OH$, which was synthesized using a pentadentate Schiff-base ligand and the $TCNQ^{\cdot -}$ radical anion. The complex exhibits ferromagnetic Dy–Dy magnetic interactions and shows weak magnetic relaxation under zero field, with an effective energy barrier of approximately 2.82 K and a pre-exponential factor τ_0 of 6.88×10^{-6} s. This behavior is proposed to arise from the short intermolecular Dy···Dy distance of 7.97 Å, which facilitates enhanced intermolecular dipolar interactions and quantum tunneling of magnetization.

A contribution by X. Gou, X. Sun, P. Cheng, and W. Shi from Nankai University and the Suzhou Institute of Nano-Tech and Nano-Bionics (SINANO) [14] present a new review on the recent advances in photo-responsive molecular nanomagnets, which show great potential in optical switches, smart sensors, and data storage devices. The design strategies and the photo-responsive mechanism based on spin transition, photocyclization, and photogenerated radicals are discussed in detail.

S.-S. Bao and L.-M. Zheng of Nanjing University [15] report two kinds of dysprosium(III) single-molecule magnets (Ln-SMMs), which show stimulus response changes in photoluminescence and magnetic properties. They also investigated the effect of pressure on magnetic properties and observed narrowing of the butterfly-shaped hysteresis loop and acceleration of magnetic relaxation. These changes are mainly caused by intramolecular π–π interaction and external stimuli (such as grinding and pressure), which provide potential value for information storage and molecular device applications.

As a true mentor, Professor Liao guided his students not only in research but also in shaping their characters and professional development. He took a personal interest in their growth, engaging with them both academically and personally. Even after retirement, he continued to inspire young scientists, fostering a passion for education and maintaining close connections with his students. His dedication as an educator has made a lasting impact on the field of chemistry in China.

This Special Issue reflects Professor Liao's lasting influence, bringing together contributions from scholars who have been directly or indirectly influenced by his guidance and

https://doi.org/10.3390/magnetochemistry12020025

research. We sincerely thank all the contributors, whose work is a testament to the ongoing impact of Professor Liao's mentorship and legacy.

Author Contributions: P.C.; Y.S.; H.-Z.K. and J.T. contributed to the writing of this Editorial. All authors have read and agreed to the published version of the manuscript.

Conflicts of Interest: The authors declare no conflicts of interest.

References

1. Wang, Y.-X.; Wang, Y.-F.; Yin, B. Theoretical Study of Pentacoordinated Lanthanide Single-Ion Magnets via Ab Initio Electronic Structure Calculation. *Magnetochemistry* **2025**, *11*, 3. [CrossRef]
2. Qian, Y.; Gao, Y.; Xu, L.; Kremer, R.K.; Zhang, J.; Ren, X.-M. Comprehensively Understanding the Transformation of Paramagnetic Tetramer to Spin-Paired Dimer in an S = ½ Molecular Crystal. *Magnetochemistry* **2025**, *11*, 8. [CrossRef]
3. Zhou, Y.; Yao, Z.; Li, N.; Liu, F.; Zhao, J.; Bu, X. Positive and Negative Exchange Bias in *N*-, *P*- and *Q*-Type Ferri-Magnets of Niccolite Metal Formates $[CH_3NH_2CH_3]_n[Cr^{III}{}_{1-x}Fe^{III}{}_xFe^{II}(HCO_2)_6]_n$. *Magnetochemistry* **2025**, *11*, 10. [CrossRef]
4. Cui, H.-H.; Wang, D.-Z.; Li, S.; Wang, L.; Yu, X.; Liu, X.; Wang, J.; Wang, M.; Tang, Y. Magnetic Anisotropy and Slow Magnetic Relaxation in Two Mononuclear Octahedral Cobalt(II) Complexes. *Magnetochemistry* **2025**, *11*, 11. [CrossRef]
5. Zhang, B.; Zhang, Y.; Wang, D.; Wang, Z.; Chang, G.; Gao, Z.; Guo, Y.; Liu, F.; Zhao, Z.; Zhang, X.; et al. An Organic–Inorganic Hybrid Semiconducting Quantum Spin Liquid Candidate: $(BEDT\text{-}TTF)_3[Cu_2(\mu\text{-}C_2O_4)_3\cdot CH_3CH_2OH\cdot 1.2H_2O]$. *Magnetochemistry* **2025**, *11*, 12. [CrossRef]
6. Zhou, Y.; Xie, J.; Jin, C.; Ma, Y.; Li, L. Magnetic Relaxation in a Heterolanthanide Binuclear Complex Involving a Nitronyl Nitroxide Biradical. *Magnetochemistry* **2025**, *11*, 26. [CrossRef]
7. Pan, Y.; Ding, Y.-S.; Li, L.; Zheng, Z. Synthesis, X-Ray Crystal Structures, and Magnetic Properties of a Series of Trinuclear Rare-Earth Hepta-Chloride Clusters. *Magnetochemistry* **2025**, *11*, 38. [CrossRef]
8. Chen, X.-F.; Wang, T.; Liao, D.; Wu, N.; Peng, Y.; Zhang, S.-Y.; Hu, Z.-B. Multifunctional Synergistic Response Induced by Phase Transition in Molecular Compounds. *Magnetochemistry* **2025**, *11*, 41. [CrossRef]
9. Liu, C.-M. Two Dy_2 Zero-Field Single-Molecule Magnets Derived from Hydrazone Schiff Base-Bridging Ligands and 1,3-Di(2-pyridyl)-1,3-propanedione. *Magnetochemistry* **2025**, *11*, 58. [CrossRef]
10. Miao, L.; Zhu, D.-M.; Liu, C.-M.; Zhang, Y.-Q.; Kou, H.-Z. Synthesis, Crystal Structures and Magnetic Properties of Lanthanide Complexes with Rhodamine Benzoyl Hydrazone Ligands. *Magnetochemistry* **2025**, *11*, 68. [CrossRef]
11. Geng, Y.-W.; Guo, T.; Wang, X.-Q.; Han, T. Enhanced Magnetocaloric Effect and Single-Molecule Magnet Behavior in a Series of Sulfur-Containing Ligand-Based Ln_9 Clusters (Ln = Gd, Tb, and Dy). *Magnetochemistry* **2025**, *11*, 70. [CrossRef]
12. Zhou, L.; Lv, H.; Liang, Y.; Liu, D.; Yao, Z.; Luo, S.; Chen, Z. A Dinuclear Dysprosium(III) Single Molecule Magnet of Benzo[*h*]quinolin-10-ol. *Magnetochemistry* **2025**, *11*, 73. [CrossRef]
13. Wang, X.; Qin, S.; Li, X.; Zuo, W.; Wang, Q.; Li, L.; Ma, Y.; Tang, J.; Zhao, B. A Dy_2 Complex Constructed by $TCNQ^{\cdot -}$ Radical Anions with Slow Magnetic Relaxation Behavior. *Magnetochemistry* **2025**, *11*, 75. [CrossRef]
14. Gou, X.; Sun, X.; Cheng, P.; Shi, W. Molecular Nanomagnets with Photomagnetic Properties: Design Strategies and Recent Advances. *Magnetochemistry* **2025**, *11*, 77. [CrossRef]
15. Qin, Y.-H.; Su, Q.-Q.; Bao, S.-S.; Zheng, L.-M. Dysprosium Complexes Incorporating Halogen-Substituted Anthracene: Piezochromism and Single-Molecule Magnet Properties. *Magnetochemistry* **2025**, *11*, 102. [CrossRef]

magnetochemistry

MDPI

Article

Theoretical Study of Pentacoordinated Lanthanide Single-Ion Magnets via Ab Initio Electronic Structure Calculation

Yu-Xi Wang, Yu-Fei Wang and Bing Yin *

Key Laboratory of Synthetic and Natural Functional Molecule of the Ministry of Education, Lab of Theoretical Molecular Magnetism, College of Chemistry and Materials Science, Northwest University, Xi'an 710127, China
* Correspondence: rayinyin@nwu.edu.cn

Abstract: A theoretical study, based on ab initio electronic structure calculation, is performed in a group of 16 pentacoordinate Dy-SIMs. Theoretical results provide a reasonable explanation of the observed SMM performance based on a concise criterion, i.e., the co-existence of long τ_{QTM} and high U_{eff}. To have the desired electronic structure favoring good SMM performance, the contribution from the equatorial coordinating atoms might be even more important than that from the axial coordinating atoms. Widening the axial ∠O–Dy–O might be a probable way to improve the SMM performance of pentacoordinated Dy-SIMs. Starting from existing systems, a rigid-scan type exploration indicates the possibility of U_{eff} higher than 1600 K.

Keywords: single-molecule magnets; ab initio calculation; crystal field analysis; axial bond angle

1. Introduction

Single-molecule magnets (SMMs) refer to a type of molecular systems displaying magnet behavior at the unimolecular level [1–11]. Because of their potential to retain magnetic information in one single molecule, SMMs have gained widespread attention from researchers, especially in the field of ultra-high-density storage of data [4–9]. Compared to hard-disk drives, the information storage density using SMMs can reach to 300 Tbit in^{-2}, which is up to 100 times higher [4,7]. Thus SMM could be the components in future revolutionary micro-electronic devices.

As research progresses, people have discovered the importance of lanthanide single-molecule magnets (Ln-SMMs) especially the mononuclear structures which are also called as lanthanide single-ion magnets (Ln-SIMs). Actually, many recent breakthroughs in the field of SMM are provided by Ln-SIMs [12–24]. However, a practical application of SMMs is still out of reach. As one important figure of merit of SMM, the blocking temperature (T_B) means the highest temperature to observe blocked magnetization of a sample. Until now, the record T_B values of SMMs have been within 60–80 K [10,18–21]. Although they exceed the boiling point of liquid nitrogen, these records still remain far below room temperature. Thus, enhancing the performance, e.g., increasing T_B, is one central task in the current stage of SMM.

It is well known that the coordination environment dictates the performance of SMMs, especially in the case of Ln-SIMs. Recently reducing the coordination number (CN) has been suggested to be an effective strategy for enhancing the performance of Ln-SIMs [25]. This is because a low coordination environment can lead to high magnetic axiality, which is crucial for maintaining a preferred orientation of magnetic moment to achieve a long relaxation time. Actually, the record-holding dysprosocenium SIMs just adopt sandwiched structures presenting effective/pseudo two coordination of the central ion.

However, the SMM performances of most synthesized low CN Ln-SIMs are apparently inferior to those of dysprosocenium systems. Thus, the potential of low CN structures as high-performance SMMs has not been amply realized. The search for low CN Ln-SIMs with good SMM performance needs rational guidance, especially from theoretical studies.

Recently, the SMM performance of a group of typical low CN Ln-SIMs, i.e., tetracoordinated structures, has been well interpreted via the ab initio electronic structure calculation [26–44]. A theoretical study also indicated possible routes to improve the SMM performance and then pointed out the structures holding the possibility of T_B higher than 50 K for tetracoordinated Ln-SIMs [39].

Encouraged by its capability, ab initio electronic structure calculation is carried out here in another typical type of low CN Ln-SIMs, i.e., the pentacoordinate complexes synthesized in recent years [45–49]. Here, we selected 16 structures as the objects of this study. For the symmetry of these 16 structures, **1Dy** to **12Dy** are close to D_{4d}, and **13Dy** to **16Dy** are close to D_{3d}. In addition to interpreting the reported systems, this work also tries to give suggestions for improving the SMM performance and to predict possible improvements (Figure 1).

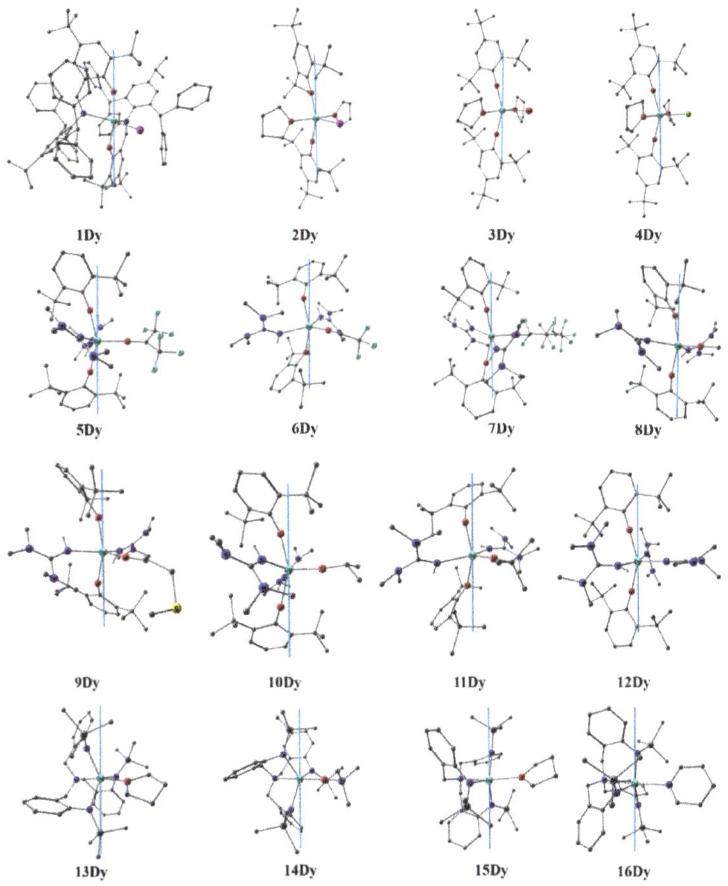

Figure 1. The structures of all SMMs involved in the study.

2. Theoretical Background and Computational Details

The microscopic mechanism underlying the SMM behavior is the magnetic relaxation at the molecular level. Due to the co-existence of several relaxation pathways, a comprehensive magnetic relaxation treatment needs to include the characteristics of the electronic and vibrational states, the coupling between them and the interaction between the system and the environment. Sophisticated methods to accomplish this task have been proposed by several groups [50–52]. However, for this type of method, huge computational cost is hardly avoidable.

Recently, a concise criterion for good SMM has been proposed to be the co-existence of long quantum tunneling of magnetization (QTM) time τ_{QTM} and a high effective barrier of magnetic reversal U_{eff} [39–43]. For Ln-SIMs, both τ_{QTM} and U_{eff} can be obtained from one single ab initio calculation for a given system. Thus, the application of this criterion is easy for a large number of Ln-SIMs due to its low computational cost [53–55]. The reliability of this criterion has been verified in structurally similar systems, e.g., tetracoordinated Ln-SIMs and square antiprism ones [39,40].

$$\tau_{QTM}^{-1} = \frac{\beta B_{ave}}{h} \cdot \frac{g_{XY}^2}{2(g_{XY}^2 + g_Z^2)^{\frac{1}{2}}}, \; g_{XY} = \sqrt{g_X^2 + g_Y^2} \tag{1}$$

$$\tau_{QTM}^{-1} = \frac{\beta B_{ave}}{h} \cdot \frac{1}{2} \cdot \frac{x_{aniso}^2 g_{XY}^2}{\sqrt{x_{aniso}^2 g_{XY}^2 + (3 - 2x_{aniso}^2) g_Z^2}} \tag{2}$$

$$x_{aniso} = \frac{B_{trans}}{B_{ave}}, \; B_X = B_Y = B_{trans} \tag{3}$$

The systems studied here are all Kramers SIMs of which the zero-field QTM rate τ_{QTM}^{-1}, i.e., the reciprocal of τ_{QTM}, is calculated by Equation (1). β and h in Equation (1) are the Bohr magneton and Planck constant, respectively. The principal g-factors, i.e., $g_{X\text{-}Z}$, of the ground Kramers doublet (KD) in Equation (1) are obtained from ab initio calculation. B_{ave} in Equation (1) is the averaged strength of the internal magnetic field which is empirically estimated as 20 milli Tesla (mT) here. The reliability of this selection has been proved in our previous works [39,41,42]. For field-induced systems, another parameter x_{aniso}, describing the anisotropy of the magnetic field, is needed as shown in Equations (2) and (3). Here, the applied direct current (DC) field is assumed to be added in the Z principal direction. In this case, B_X = B_Y = 20 mT and B_Z is the sum of 20 mT and the applied DC field in experiment.

$$U_{eff}(T) = \sum_i \frac{\tau_{QTM,i}^{-1,eff}(T)}{N} E_i \tag{4}$$

$$N = \sum_i \tau_{QTM,i}^{-1,eff}(T) \tag{5}$$

U_{eff} could be calculated as a weighted sum of energies of both ground and excited KDs (Equation (4)) [41,42]. The normalization factor N is the sum of the effective QTM rates of all the involved KDs (Equation (5)). The effective rate of the ith KD (Equation (6)) depends on its principal g-factors (Equation (7)) and the Boltzmann population (Equation (8)).

In this work, theoretical predictions of QTM time via Equations (1)–(3) and effective barrier via Equations (4)–(5) are labeled as τ_{QTM}^{Zee} and U_{eff}^{Zee} respectively. The corresponding experimental results are labeled as τ_{QTM}^{exp} and U_{eff}^{exp} respectively.

$$\tau_{QTM,i}{}^{-1}(T) \propto \frac{\exp(-E_i/k_B T)}{Z} \tau_{QTM,i}^{-1,eff} \tag{6}$$

$$\tau_{QTM,i}^{-1,eff} = \frac{g_{XY,i}^2}{2(g_{XY,i}^2 + g_{Z,i}^2)^{\frac{1}{2}}} \quad (7)$$

$$Z = \sum_i \exp(-E_i/k_BT) \quad (8)$$

The ab initio calculations in this work consist of two steps: first, a set of spin eigenstates are obtained via the state-averaged complete active space self-consistent field method (SA-CASSCF) [56]; then, the final states, i.e., KDs here, are obtained via state interaction which diagonalizes the spin-orbit-coupling (SOC) matrix under the basis of spin eigenstates from the first step.

A free academic version of MOLCAS 8.0 was used to perform ab initio calculation [57]. The active space consisted of 9 electrons in 7 orbitals and 21 spin sextets were included in the SA-CASSCF step [39–43]. The scalar relativistic effect was accounted via DKH2 transformation. Then, state interaction was performed via the RASSI-SO module [58], with the SOC integrals from the AMFI method [59]. We choose the ANO-RCC relativistic basis set [60,61], including VQZP for Dy, VDZP for C, VDZ for H and VTZP for the others. The SINGLE_ANISO module was used to gain the g-factors and other magnetic parameters [62,63]. The numerical results of the ab initio calculation are included in Tables S1~S16.

3. Results and Discussion

3.1. The Comparison Between Theoretical Predictions and Experimental Results

The most quoted evidence for SMM characteristics is the temperature- or frequency-dependence of the imaginary part of the alternating-current (AC) magnetic susceptibility, especially a peak. The highest temperature to observe such a peak is labeled as T_{AC} here. T_{AC} only represents the short-term magnetic memory effect and thus is less conceptually sound than a magnetic hysteresis loop or zero-field cooled susceptibility. However, T_{AC} is available for most SMMs, including all the Dy-SIMs studied here. In comparison, neither magnetic hysteresis nor zero-field cooled susceptibility holds such an availability. T_{AC} has been used to quantify T_B for a lot of SMMs [9]. Our recent works have also verified the capability of T_{AC} as a common measurement of SMM performance [42,64]. Thus, T_{AC} at 1000 Hz is utilized here.

As shown in Figure 2, higher T_{AC} is generally located at a position closer to the lower right corner of the plane defined by τ_{QTM} and U_{eff}. This corner just corresponds to both the longest τ_{QTM} and the highest U_{eff}. **1Dy**, which holds the highest T_{AC} here, has both the longest τ_{QTM} and the highest U_{eff} according to either theoretical prediction or experimental results (Table 1). Thus, the co-existence of longer τ_{QTM} and higher U_{eff} does correspond to better SMM performance in the pentacoordinated Ln-SIMs. The proposed criterion also works here.

Formally, this criterion only includes the QTM and thermally-activated (TA) pathways, i.e., Orbach and TA-QTM, but neglects Raman pathway. Thus, in principle, it is only necessary rather than sufficient. However, one recent in-depth statistical analysis has indicated the existence of a correlation between U_{eff} and Raman parameters, especially in high U_{eff} cases [9]. Thus, when applying the criterion, the effect of the Raman pathway might be accounted implicitly. Also, in structurally similar systems, the effect of the Raman pathway might not vary sharply [39,40].

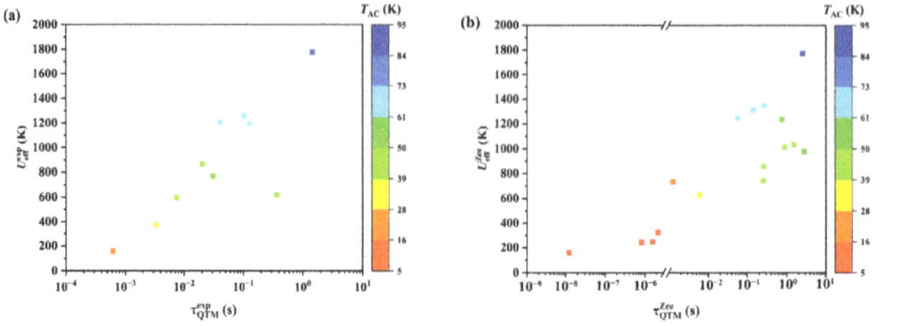

Figure 2. The relation among T_{AC}, τ_{QTM} and U_{eff}. (**a**) T_{AC} vs. experimental τ_{QTM} and U_{eff}. (**b**) T_{AC} vs. theoretical τ_{QTM} and U_{eff}.

Table 1. The experiment values and theoretical predictions of τ_{QTM} (s) and U_{eff} (K).

	Refcode	τ^{exp}_{QTM}	τ^{Zee}_{QTM}	U^{exp}_{eff}	$U^{Zee}_{eff}/U^{KD}_{eff}$	T_{AC} [a]
1Dy	LEVLEH	1.43	2.49	1780	1776/1856	95
2Dy	XUWDAX	1.26×10^{-1}	2.62×10^{-1}	1201	1350/1282	64
3Dy	XUWCUQ	3.98×10^{-2}	1.35×10^{-1}	1210	1315/1239	64
4Dy	XUWCOK	1.00×10^{-1}	5.42×10^{-2}	1262	1249/1173	64
5Dy	ENACOO	N/A [b]	7.41×10^{-1}	1176	1236/999	60
6Dy	ENACII	N/A	1.51	905	1035/964	49
7Dy	ENACUU	2.02×10^{-2}	8.77×10^{-1}	872	1014/951	41
8Dy	ENABON	3.02×10^{-2}	2.77	773	981/801	50
9Dy	ZESGAJ	3.68×10^{-1}	2.56×10^{-1}	622	859/731	45
10Dy	E.NACAA	7.45×10^{-3}	2.51×10^{-1}	601	745/684	40
11Dy	ENACEE	3.35×10^{-3}	5.75×10^{-3}	378	633/374	33
12Dy	ENABIH	6.11×10^{-4}	1.25×10^{-3}	160	738/422	22
13Dy	DEYRIO	N/A	2.12×10^{-6}	36	328/230	6
14Dy	FEYREK	N/A	1.53×10^{-6}	19	250/124	5
15Dy	DEYRIO	N/A	8.00×10^{-7}	N/A	245/0	6
16Dy	FEYRAG	4.03×10^{-4}	1.19×10^{-8}	N/A	160/0	6

[a] The highest AC frequency is usually 1000 Hz, the exception is **9Dy** (1488 Hz). [b] Data were not provided by experiments.

As shown in Figure 3, the order of either τ^{Zee}_{QTM} or U^{Zee}_{eff} is generally consistent to that of the corresponding experimental τ^{exp}_{QTM} and U^{exp}_{eff} respectively. With only one exception of **16Dy**, the deviations in τ^{Zee}_{QTM} are all less than two orders of magnitude (Figure 3a). For the SIMs holding the first four longest τ^{exp}_{QTM} here, i.e., **1Dy**–**4Dy**, the deviations in τ^{Zee}_{QTM} are all less than one order of magnitude. It needs to indicate that even the theoretical results from those sophisticated methods may still bear deviations of one or two orders of magnitude [53]. The large deviation in τ^{Zee}_{QTM} of field-induced **16Dy** might be attributed to the difficulty of obtaining accurate x_{aniso} [43]. As a vector field, magnetic fields have different sources, e.g., dipolar interaction with other electronic magnetic moment in the sample and hyperfine interaction due to nuclear spin [43]. Thus, the anisotropy of a magnetic field depends on both the electronic and nuclear contributions.

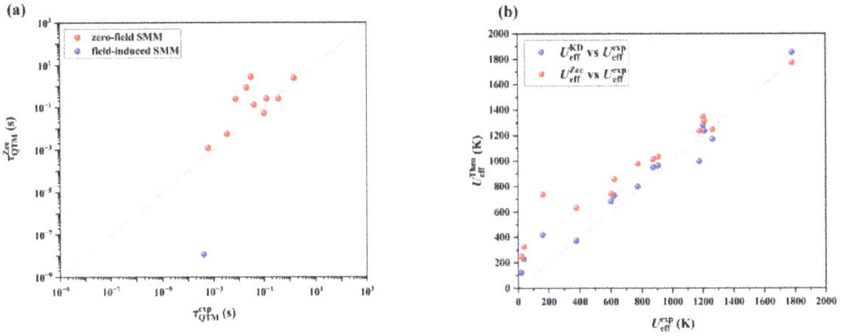

Figure 3. Comparison between theoretical prediction and experimental fitting. (**a**) τ_{QTM}^{Zee} vs. τ_{QTM}^{exp}. (**b**) U_{eff}^{Zee} vs. U_{eff}^{exp}.

In the aspect of effective barrier, we do find some large discrepancies between theoretical U_{eff}^{Zee} and experimental U_{eff}^{exp}, i.e., **11Dy–16Dy**. Their T_{AC} values are around 20 K or apparently lower (5~6 K) while other Ln-SIMs have T_{AC} values lying within the range of 40~95 K. Therefore, those large discrepancies mainly occur in systems holding inferior SMM performance as represented by low T_{AC}. In these inferior systems, pure dominance of the Orbach pathway is hardly possible and other pathways strongly drive the fitted U_{eff}^{exp} away from having a solid physical meaning [40]. Thus, in these systems, U_{eff}^{exp} becomes closer to a purely phenomenological parameter and U_{eff}^{Zee} is not obligated to be close to it [40].

Besides U_{eff}^{Zee}, we can also calculate U_{eff} as the energy of a given excited KD, identified as the most probable one wherein magnetic reversal takes place. This theoretical U_{eff} is labeled as U_{eff}^{KD} here after (Table S17 and Figure S1). As shown in Figure 4b, these two types of theoretical barrier are generally consistent with each other. They both identify that the barrier of **1Dy** is clearly higher than those of all the others.

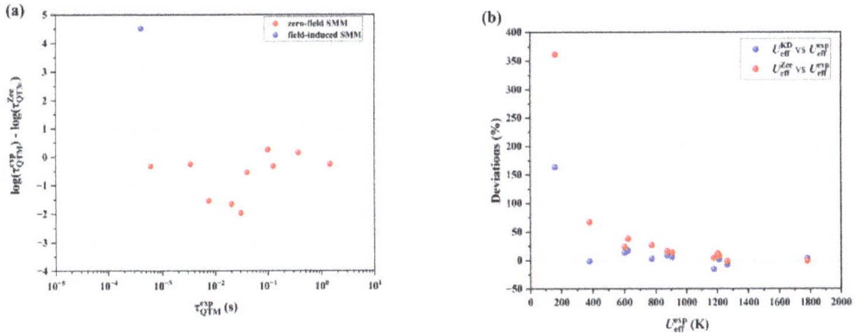

Figure 4. The deviations in τ_{QTM}^{Zee} and U_{eff}^{Zee} compared to experimental results. (**a**) Order-of-magnitude deviation in τ_{QTM}^{Zee}. (**b**) percentage deviation in U_{eff}^{Zee}.

3.2. Mechanisms of Magnetic Relaxation

According to Equation (4), U_{eff}^{Zee} can be decomposed into contributions from various KDs(Figure 5). This decomposition is capable of providing some mechanistic information about the magnetic relaxation. Since U_{eff}^{Zee} varies with the temperature value used in the calculation, two cases are analyzed for one Ln-SIM: the saturated case and the reproducing case [64]. The saturated case refers to the result when U_{eff}^{Zee} becomes a saturated value which is obtained by using 300 K in Equation (4). The reproducing case refers to the result when U_{eff}^{Zee} exactly reproduces U_{eff}^{exp} with a special temperature value T_{rep} used in Equation (4). It needs to be clarified that there is no direct relationship between T_{rep} and T_B in principle.

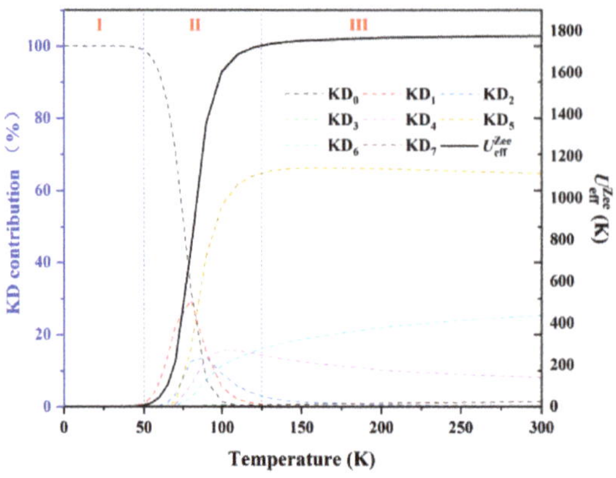

Figure 5. U_{eff}^{Zee} and the contributions from various KDs of **1Dy**.

As shown in Table 2, the most important KDs of the saturated U_{eff}^{Zee} of **1Dy** are KD_5 and KD_6, of which the sum contribution is 91%. These two KDs remain to be the most important ones (68%) in the reproducing case while KD_4 (16%) also plays a role. Thus, the most important KDs of U_{eff} of **1Dy** are some highly excited ones which do not change heavily between the saturated case and the reproducing case. The contributions from the ground and other lower excited KDs are negligible. These results suggest that, in **1Dy**, the slow thermally-activated relaxation, proceeding mainly via KD_5 and KD_6, is important while the ground KD QTM is not efficient. This is consistent to the long τ_{QTM} and high U_{eff} and T_{AC} of **1Dy**. Similar results occur in **2Dy–5Dy**.

Table 2. The contributions of various KDs to U_{eff} in the saturated and reproducing cases.

	Saturated Case	Reproducing Case	T_{rep}
1Dy	KD5 + KD6 (91%)	KD5 + KD6 (68%), KD4 (16%)	100 [a]
2Dy	KD3 + KD4 (97%)	KD4 + KD3 (82%)	79
3Dy	KD3 + KD4 (95%)	KD3 + KD4 (77%), KD2 (21%)	86
4Dy	KD3 + KD4 (96%)	KD3 + KD4 (86%)	90 [a]
5Dy	KD4 + KD5 + KD6 (79%)	KD4 + KD5 + KD6 (75%), KD3 (15%)	160
6Dy	KD3 (38%) + KD5 (36%) + KD4 (18%)	KD3 (47%) + KD0 (28%) + KD2 (11%)	49
7Dy	KD4 (53%) + KD5 (28%) + KD6 (16%)	KD4 (62%) + KD1 (9%) + KD5 (8%)	56
8Dy	KD3 (42%) + KD5 (38%) + KD4 (12%)	KD3 (52%) + KD2 (19%) + KD0 (13%)	48
9Dy	KD5 (40%) + KD3 (38%) + KD2 (15%)	KD2 (48%) + KD3 (25%) + KD0 (18%)	44
10Dy	KD2 (46%) + KD3 (35%) + KD4 (18%)	KD2 (71%) + KD0 (15%) + KD3 (12%)	38
11Dy	KD3 (53%) + KD2 (43%)	KD2 (54%) + KD0 (25%) + KD3 (12%)	40
12Dy	KD3 (52%) + KD2 (16%) + KD7 (12%)	KD0 (74%) + KD3 (13%)	42
13Dy	KD2 (48%) + KD1 (39%) + KD3 (10%)	KD0 (83%) + KD1 (16%)	31
14Dy	KD3 (66%) + KD4 (14%)	KD0 (86%) + KD1 (13%)	23
15Dy	KD1 (70%) + KD2 (21%)	N/A [b]	N/A
16Dy	KD1 (44%) + KD2 (30%) + KD0 (16%)	N/A	N/A

[a] In these cases, the saturated theoretical barrier is still a little bit lower than the experimentally fitted one. Thus, T_{rep} in these cases is taken as the temperature at 2/3 position of region II. [b] U_{eff}^{exp} were not provided by experiments.

For other Ln-SIMs' U_{eff}, the contributions from the lower excited KDs and even the ground KD_0 are more important. In some reproducing cases, i.e., **12Dy–14Dy**, KD_0 even becomes the most important one. These results support a smaller contribution from slow thermally activated relaxation and a larger contribution from fast ground KD QTM. This is consistent with their inferior SMM performance as represented by lower T_{AC} when compared to **1Dy–5Dy**.

3.3. Crystal-Field Analysis and Theoretical Magneto-Structural Correlation

Since 20 mT B_{ave} is used for most Ln-SIMs here, their SMM performance ought to be mainly dictated by the principal g-factors and energies of the KDs. These are all determined by the electronic structure. The crystal field (CF) Hamiltonian (Equation (9)) is a suitable theoretical tool for interpreting the electronic structure of Ln-SIMs. Since O_k^q are common operators, the electronic structure characteristics arise solely from the CF parameter (CFP) B_k^q [11,30,41,65].

Accurate CFPs could be extracted from ab initio results via irreducible tensor operator (ITO). The point charge electrostatic model (PCEM) can also help to give a rough estimate of CFPs as a sum of contributions from the coordinating atoms (Equation (10)) [65]. Z_{eff} and R_j in Equation (10) are the effective charge and distance to the central ion of a given coordinating atom j respectively. Y_k^q is the spherical harmonic function of which the variables are azimuthal coordinates θ_j and ϕ_j of the coordinating atoms.

$$\hat{H}_{CF} = \sum_{k=2,4,6} \sum_{q=-k}^{k} B_k^q \hat{O}_k^q(\vec{J}) \tag{9}$$

$$\left| B_k^q \right| \sim \sum_{j}^{\text{ligand}} \left\{ \frac{Z_{\mathrm{eff},j}}{R_j^{k+1}} Y_k^q(\theta_j, \phi_j) \right\} \tag{10}$$

It must be clarified that Equation (10) only provides a rough estimate of the relative magnitude of CFPs rather than the actual value. The PCEM estimate does not even have the unit of energy. It also neglects the covalent contribution which has been shown to be capable of playing an important role recently [66,67]. However, there are some advantages of the PCEM estimate. First, as it is related to the position and charge of the coordinating atoms, it is chemically intuitive [39–41]. Second, it can help measuring the importance of different coordinating atoms since it is a sum of atomic contributions.

In principle, CFPs up to at least $k = 6$ need to be included. However, recent works have indicated that 2nd-rank CFPs, especially the diagonal term B_2^0, can be the leading ones to give a reasonable explanation by themselves [64]. All the ITO B_2^0 are negative and hence we only need to discuss their magnitude, as indicated by their absolute values $|B_2^0|$.

As shown in Figure 6, a larger $|B_2^0|$ usually corresponds to higher T_{AC} or U_{eff}. The best SMM here, **1Dy** (T_{AC} = 95 K), holds the largest $|B_2^0|$ (Table S18). **2Dy–4Dy** also have a high T_{AC} (86 K) and their $|B_2^0|$ values, are slightly smaller than that of **1Dy** but clearly larger than all the others. Although being less accurate, the PCEM estimate also captures the general trend (Figure 6 and Table S18) and thus subsequent PCEM analysis should be reliable.

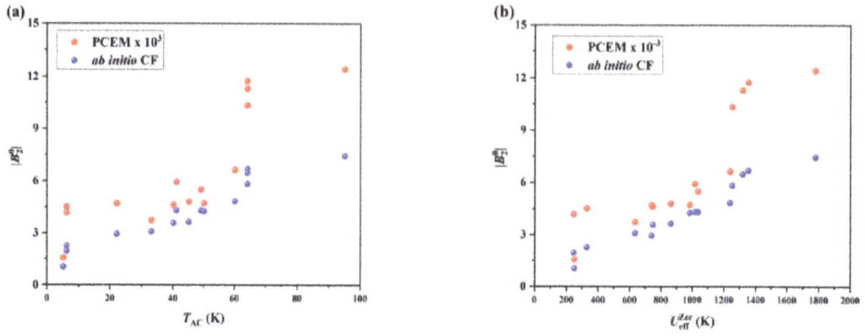

Figure 6. Crystal field analysis based on both ab initio ITO and PCEM. (**a**) T_{AC} vs. $|B_2^0|$ in the studied systems. (**b**) U_{eff}^{Zee} vs. $|B_2^0|$ in the studied systems.

Based on their positions with respect to the magnetic easy axis, the coordinating atoms could be divided into two types: the axial ones lying close to the easy axis and the equatorial ones perpendicular to the easy axis (Table S20) [65]. PCEM suggests that the contribution from axial atoms tends to increase $|B_2^0|$ while that from equatorial atoms is destructive.

1Dy holds the largest PCEM estimate of $|B_2^0|$ but its contribution from the two axial atoms is even smaller than that of **2Dy** (Table S20). Interestingly the amount of destructive contribution from the equatorial atoms of **1Dy** (-4.45×10^{-3}) is clearly smaller than that of **2Dy** (-5.39×10^{-3}). This smaller destructive contribution is one important reason for the largest PCEM estimate of $|B_2^0|$ of **1Dy**. Although **3Dy** has the largest constructive contribution to $|B_2^0|$ from the axial coordinating atoms (Table S20), its destructive contribution from equatorial atoms is also larger than that of **1Dy**. Thus, the eventual $|B_2^0|$ of 3Dy is smaller than that of **1Dy**.

For the Dy-SIMs studied here, the bond angle between the central Ln ion and two axial atoms is around 150°. A previous study of tetracoordinated Ln-SIMs has indicated that widening this angle can improve the SMM performance significantly [39]. Linear or quasi-linear O–Dy–O arrangement will facilitate the generation of high magnetic axiality due to the oblate shape of the electron density of the ground state of the central Dy^{III} ion. This widening might be made possible by applying pressure [68]. The two axial O atoms of **1Dy** come from the same bidentate ligand HL (2-((2,6-dibenzhydryl-4-isopropylphenylimino)methyl)-4,6-di-*tret*-butylphenol) and thus widening of the axial bond angle is probably restricted there. Consequently, we chose to explore **2Dy** and **5Dy** wherein only monodentate ligands were involved. This exploration was performed in a rigid-scan way wherein only the axial bond angle is varied. Clearly this rigid-scan exploration has limited precision, and the results only provide some possibility. However, the capability of rigid-scan has already been verified in our recent work on Ln-SIMs [39].

As show in Figure 7, widening ∠O–Dy–O up to 180° can lead to an increase in τ_{QTM} by one order of magnitude, i.e., from around 0.25 s to around 2.75 s. Meanwhile, U_{eff} also experiences a sharp increase by about 600 K, from ~1300 K (150°) to ~1900 K (180°). For **5Dy**, there is no significant change in τ_{QTM} with widening ∠O–Dy–O. Meanwhile, U_{eff} increases by about 400 K from around 1200 K (150°) to around 1600 K (180°). Thus, starting from the reported pentacoordiated Ln-SIMs, widening the axial bond angle might be a probable way to improve the SMM performance.

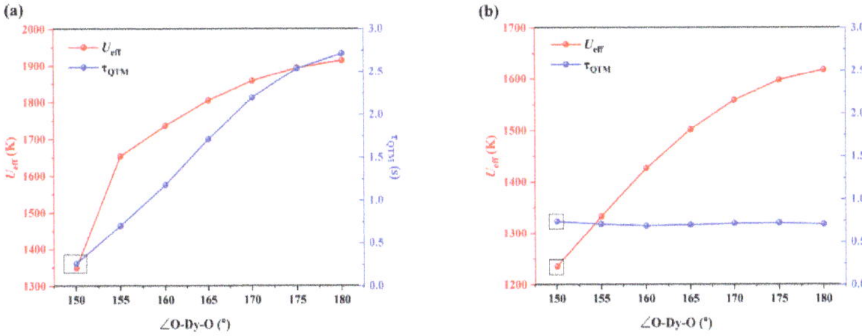

Figure 7. Theoretical prediction of two structures in a rigid-scan way. (**a**) τ_{QTM} and U_{eff} along ∠O–Dy–O of **2Dy**. (**b**) τ_{QTM} and U_{eff} along ∠O–Dy–O of **5Dy**.

4. Conclusions

A theoretical study, based on ab initio electronic structure calculation, was carried out in a group of 16 pentacoordinate Dy-SIMs. The experimentally observed SMM performance is well explained by a concise criterion, i.e., the co-existence of τ_{QTM} and U_{eff}.

Crystal field analysis indicated that 2nd-rank CFPs, especially the diagonal term B_2^0, are the leading ones to generate the desired electronic structure favoring good SMM performance. To have the needed CFPs, the contribution from the equatorial coordinating atoms might be even more important than that from the axial coordinating atoms.

Widening the axial bond angle between the central Ln ion and two axial atoms might be a probable way to improve the SMM performance of pentacoordinated Ln-SIMs. Starting from existing systems, a rigid-scan type exploration indicates the possibility of U_{eff} higher than 1600 K.

Supplementary Materials: The following supporting information can be downloaded at: https://www.mdpi.com/article/10.3390/magnetochemistry11010003/s1, Table S1. ab initio results of 1Dy (LEVLEH); Table S2. ab initio results of 2Dy (XUWDAX); Table S3. ab initio results of **3Dy** (XUWCUQ); Table S4. ab initio results of **4Dy** (XUWCOK), Table S5. ab initio results of **5Dy** (ENACII); Table S6. ab initio results of **6Dy** (ENACII); Table S7. ab initio results of **7Dy** (ENACUU); Table S8. ab initio results of **8Dy** (ENABON); Table S9. ab initio results of **9Dy** (ZESGAJ); Table S10. ab initio results of **10Dy** (ENACAA); Table S11. ab initio results of **11Dy** (ENACEE); Table S12. ab initio results of **12Dy** (ENABIH); Table S13. ab initio results of **13Dy** (DEYRIO); Table S14. ab initio results of **14Dy** (FEYREK); Table S15. ab initio results of **15Dy** (DEYRIO); Table S16. ab initio results of **16Dy** (FEYRAG); Table S17. Various U_{eff}^{KD} values according to transition magnetic moment $|\mu|$, angle of easy axis of excited KD with respect to that of ground KD (θ) and crystal field wave function composition; Table S18. ab initio and PCEM crystal field parameters (in cm^{-1} and a.u. respectively) of Dy-SIMs [a]; Table S19. The distance R to Dy^{3+} (in Å), atomic charge Z_{eff} (in $|e|$) and angle θ, ϕ with respect to the ab initio magnetic easy axis (in °) of atoms in the first sphere; Table S20. The average distance R to Dy^{3+} (in Å), average atomic charge Z_{eff} (in $|e|$), average angle θ with respect to the ab initio magnetic easy axis (in °) and contribution to $|B_2^0|$ from the atoms in the first sphere; Table S21. Theoretical prediction of 2Dy τ_{QTM} and U_{eff} with different ∠O–Dy–O; Table S22. Theoretical prediction of **5Dy** τ_{QTM} and U_{eff} with different ∠O–Dy–O; Table S23. Cartesian coordinates of complexes; Figure S1. Possible relaxation mechanism based on transition magnetic moment for the Ln-SIMs studied here; Figure S2. U_{eff}^{Zee} and the contributions from various KDs of other SIMs.

Author Contributions: Conceptualization, B.Y.; Formal analysis, Y.-X.W. and B.Y.; Investigation, Y.-X.W. and Y.-F.W.; Resources, B.Y.; Writing—original draft, Y.-X.W.; Writing—review & editing, B.Y.; Supervision, B.Y.; Project administration, B.Y. All authors have read and agreed to the published version of the manuscript.

Funding: This work is supported by the National Natural Science Foundation of China (No. 22373076).

Data Availability Statement: The original contributions presented in this study are included in the article/Supplementary Material. Further inquiries can be directed to the corresponding author.

Acknowledgments: We thank Yuanhe Huang and De-Cai Fang (Beijing Normal University) for their long-term support and encouragement.

Conflicts of Interest: The authors declare no conflicts of interest.

References

1. Sessoli, R.; Gatteschi, D.; Caneschi, A.; Novak, M.A. Magnetic bistability in a metal-ion cluster. *Nature* **1993**, *365*, 141–143. [CrossRef]
2. Sessoli, R.; Tsai, H.L.; Schake, A.R.; Wang, S.; Vincent, J.B.; Folting, K.; Gatteschi, D.; Christou, G.; Hendrickson, D.N. High-spin molecules: $[Mn_{12}O_{12}(O_2CR)_{16}(H_2O)_4]$. *J. Am. Chem. Soc.* **1993**, *115*, 1804–1816. [CrossRef]
3. Gatteschi, D.; Sessoli, R.; Villain, J. *Molecular Nanomagnets*; Oxford University Press: New York, NY, USA, 2006. [CrossRef]
4. Liddle, S.T.; van Slageren, J. Improving f-element single molecule magnets. *Chem. Soc. Rev.* **2015**, *44*, 6655–6669. [CrossRef] [PubMed]
5. Liu, J.-L.; Chen, Y.-C.; Tong, M.-L. Symmetry strategies for high performance lanthanide-based single-molecule magnets. *Chem. Soc. Rev.* **2018**, *47*, 2431–2453. [CrossRef]
6. Parmar, V.S.; Mills, D.P.; Winpenny, R.E.P. Mononuclear Dysprosium Alkoxide and Aryloxide Single-Molecule Magnets. *Chem. Eur. J.* **2021**, *27*, 7625–7645. [CrossRef] [PubMed]
7. Zabala-Lekuona, A.; Seco, J.M.; Colacio, E. Single-Molecule Magnets: From Mn12-ac to dysprosium metallocenes, a travel in time. *Coord. Chem. Rev.* **2021**, *441*, 213984. [CrossRef]
8. Chilton, N.F. Molecular Magnetism. *Annu. Rev. Mater. Res.* **2022**, *52*, 79–101. [CrossRef]
9. Duan, Y.; Rosaleny, L.E.; Coutinho, J.T.; Giménez-Santamarina, S.; Scheie, A.; Baldoví, J.J.; Cardona-Serra, S.; Gaita-Ariño, A. Data-driven design of molecular nanomagnets. *Nat. Commun.* **2022**, *13*, 7626. [CrossRef] [PubMed]
10. Gould, C.A.; McClain, K.R.; Reta, D.; Kragskow, J.G.C.; Marchiori, D.A.; Lachman, E.; Choi, E.-S.; Analytis, J.G.; Britt, R.D.; Chilton, N.F.; et al. Ultrahard magnetism from mixed-valence dilanthanide complexes with metal-metal bonding. *Science* **2022**, *375*, 198–202. [CrossRef] [PubMed]
11. Ungur, L.; Chibotaru, L.F. Strategies toward High-Temperature Lanthanide-Based Single-Molecule Magnets. *Inorg. Chem.* **2016**, *55*, 10043–10056. [CrossRef]
12. Jiang, S.-D.; Wang, B.-W.; Su, G.; Wang, Z.-M.; Gao, S. A Mononuclear Dysprosium Complex Featuring Single-Molecule-Magnet Behavior. *Angew. Chem. Int. Ed.* **2010**, *49*, 7448–7451. [CrossRef]
13. Ungur, L.; Le Roy, J.J.; Korobkov, I.; Murugesu, M.; Chibotaru, L.F. Fine-tuning the Local Symmetry to Attain Record Blocking Temperature and Magnetic Remanence in a Single-Ion Magnet. *Angew. Chem. Int. Ed.* **2014**, *53*, 4413–4417. [CrossRef]
14. Gupta, S.K.; Rajeshkumar, T.; Rajaraman, G.; Murugavel, R. An air-stable Dy(III) single-ion magnet with high anisotropy barrier and blocking temperature. *Chem. Sci.* **2016**, *7*, 5181–5191. [CrossRef]
15. Liu, J.; Chen, Y.-C.; Liu, J.-L.; Vieru, V.; Ungur, L.; Jia, J.-H.; Chibotaru, L.F.; Lan, Y.; Wernsdorfer, W.; Gao, S.; et al. A Stable Pentagonal Bipyramidal Dy(III) Single-Ion Magnet with a Record Magnetization Reversal Barrier over 1000 K. *J. Am. Chem. Soc.* **2016**, *138*, 5441–5450. [CrossRef] [PubMed]
16. Chen, Y.-C.; Liu, J.-L.; Ungur, L.; Liu, J.; Li, Q.-W.; Wang, L.-F.; Ni, Z.-P.; Chibotaru, L.F.; Chen, X.-M.; Tong, M.-L. Symmetry-Supported Magnetic Blocking at 20 K in Pentagonal Bipyramidal Dy(III) Single-Ion Magnets. *J. Am. Chem. Soc.* **2016**, *138*, 2829–2837. [CrossRef] [PubMed]
17. Feng, M.; Tong, M.-L. Single Ion Magnets from 3d to 5f: Developments and Strategies. *Chem. Eur. J.* **2018**, *24*, 7574–7594. [CrossRef]
18. Goodwin, C.A.P.; Ortu, F.; Reta, D.; Chilton, N.F.; Mills, D.P. Molecular magnetic hysteresis at 60 kelvin in dysprosocenium. *Nature* **2017**, *548*, 439–442. [CrossRef] [PubMed]
19. Guo, F.-S.; Day, B.M.; Chen, Y.-C.; Tong, M.-L.; Mansikkamäki, A.; Layfield, R.A. Magnetic hysteresis up to 80 kelvin in a dysprosium metallocene single-molecule magnet. *Science* **2018**, *362*, 1400–1403. [CrossRef]
20. Vanjak, J.C.; Wilkins, B.O.; Vieru, V.; Bhuvanesh, N.S.; Reibenspies, J.H.; Martin, C.D.; Chibotaru, L.F.; Nippe, M. A High-Performance Single-Molecule Magnet Utilizing Dianionic Aminoborolide Ligands. *J. Am. Chem. Soc.* **2022**, *144*, 17743–17747. [CrossRef] [PubMed]
21. Vincent, A.H.; Whyatt, Y.L.; Chilton, N.F.; Long, J.R. Strong Axiality in a Dysprosium(III) Bis(borolide) Complex Leads to Magnetic Blocking at 65 K. *J. Am. Chem. Soc.* **2023**, *145*, 1572–1579. [CrossRef]
22. Zhu, Z.; Zhao, C.; Feng, T.; Liu, X.; Ying, X.; Li, X.-L.; Zhang, Y.-Q.; Tang, J. Air-Stable Chiral Single-Molecule Magnets with Record Anisotropy Barrier Exceeding 1800 K. *J. Am. Chem. Soc.* **2021**, *143*, 10077–10082. [CrossRef] [PubMed]

23. Wang, H.-S.; Zhang, K.; Wang, J.; Hu, Z.; Song, Y.; Zhang, Z.; Pan, Z.-Q. Regulating the distortion degree of the square antiprism coordination geometry in Dy–Na single ion magnets. *CrystEngComm* **2021**, *23*, 3175–3184. [CrossRef]
24. Wang, H.-S.; Zhang, K.; Wang, J.; Hu, Z.-B.; Zhang, Z.; Song, Y.; Zhang, Y.-Q. Influence of the Different Types of Auxiliary Noncarboxylate Organic Ligands on the Topologies and Magnetic Relaxation Behavior of Zn–Dy Heterometallic Single Molecule Magnets. *Inorg. Chem.* **2021**, *60*, 9941–9955. [CrossRef] [PubMed]
25. Ungur, L.; Chibotaru, L.F. Magnetic anisotropy in the excited states of low symmetry lanthanide complexes. *Phys. Chem. Chem. Phys.* **2011**, *13*, 20086–20090. [CrossRef] [PubMed]
26. Chilton, N.F.; Goodwin, C.A.P.; Mills, D.P.; Winpenny, R.E.P. The first near-linear bis(amide) f-block complex: A blueprint for a high temperature single molecule magnet. *Chem. Commun.* **2015**, *51*, 101–103. [CrossRef] [PubMed]
27. Chilton, N.F. Design Criteria for High-Temperature Single-Molecule Magnets. *Inorg. Chem.* **2015**, *54*, 2097–2099. [CrossRef]
28. Luzon, J.; Sessoli, R. Lanthanides in molecular magnetism: So fascinating, so challenging. *Dalton Trans.* **2012**, *41*, 13556–13567. [CrossRef] [PubMed]
29. Ungur, L.; Thewissen, M.; Costes, J.-P.; Wernsdorfer, W.; Chibotaru, L.F. Interplay of Strongly Anisotropic Metal Ions in Magnetic Blocking of Complexes. *Inorg. Chem.* **2013**, *52*, 6328–6337. [CrossRef] [PubMed]
30. Chibotaru, L.F. Theoretical Understanding of Anisotropy in Molecular Nanomagnets. In *Molecular Nanomagnets and Related Phenomena*; Gao, S., Ed.; Springer: Berlin/Heidelberg, Germany, 2015; pp. 185–229.
31. Ungur, L.; Chibotaru, L.F. Computational Modelling of the Magnetic Properties of Lanthanide Compounds. In *Lanthanides and Actinides in Molecular Magnetism*; Wiley: Hoboken, New Jersey, USA, 2015. [CrossRef]
32. Gómez-Coca, S.; Aravena, D.; Morales, R.; Ruiz, E. Large magnetic anisotropy in mononuclear metal complexes. *Coord. Chem. Rev.* **2015**, *289–290*, 379–392. [CrossRef]
33. Baldoví, J.J.; Duan, Y.; Morales, R.; Gaita-Ariño, A.; Ruiz, E.; Coronado, E. Rational Design of Lanthanoid Single-Ion Magnets: Predictive Power of the Theoretical Models. *Chem. Eur. J.* **2016**, *22*, 13532–13539. [CrossRef]
34. Gupta, T.; Rajaraman, G. Modelling spin Hamiltonian parameters of molecular nanomagnets. *Chem. Commun.* **2016**, *52*, 8972–9008. [CrossRef] [PubMed]
35. Ungur, L. 1—Introduction to the electronic structure, luminescence, and magnetism of lanthanides. In *Lanthanide-Based Multifunctional Materials*; Martín-Ramos, P., Ramos Silva, M., Eds.; Elsevier: Amsterdam, The Netherlands, 2018; pp. 1–58.
36. Kotrle, K.; Herchel, R. Are Inorganic Single-Molecule Magnets a Possibility? A Theoretical Insight into Dysprosium Double-Deckers with Inorganic Ring Systems. *Inorg. Chem.* **2019**, *58*, 14046–14057. [CrossRef]
37. Castro-Alvarez, A.; Gil, Y.; Llanos, L.; Aravena, D. High performance single-molecule magnets, Orbach or Raman relaxation suppression? *Inorg. Chem. Front.* **2020**, *7*, 2478–2486. [CrossRef]
38. Aravena, D.; Ruiz, E. Spin dynamics in single-molecule magnets and molecular qubits. *Dalton Trans.* **2020**, *49*, 9916–9928. [CrossRef]
39. Yang, Q.-Q.; Wang, Y.-F.; Wang, Y.-X.; Tang, M.-J.; Yin, B. Ab initio prediction of key parameters and magneto-structural correlation of tetracoordinated lanthanide single-ion magnets. *Phys. Chem. Chem. Phys.* **2023**, *25*, 18387–18399. [CrossRef] [PubMed]
40. Liu, H.; Li, J.-F.; Yin, B. The coexistence of long τ_{QTM} and high U_{eff} as a concise criterion for a good single-molecule magnet: A theoretical case study of square antiprism dysprosium single-ion magnets. *Phys. Chem. Chem. Phys.* **2022**, *24*, 11729–11742. [CrossRef] [PubMed]
41. Wu, X.; Li, J.-F.; Yin, B. The interpretation and prediction of lanthanide single-ion magnets from ab initio electronic structure calculation: The capability and limit. *Dalton Trans.* **2022**, *51*, 14793–14816. [CrossRef] [PubMed]
42. Yin, B.; Li, C.-C. A method to predict both the relaxation time of quantum tunneling of magnetization and the effective barrier of magnetic reversal for a Kramers single-ion magnet. *Phys. Chem. Chem. Phys.* **2020**, *22*, 9923–9933. [CrossRef] [PubMed]
43. Yin, B.; Luo, L. The anisotropy of the internal magnetic field on the central ion is capable of imposing great impact on the quantum tunneling of magnetization of Kramers single-ion magnets. *Phys. Chem. Chem. Phys.* **2021**, *23*, 3093–3105. [CrossRef] [PubMed]
44. Dergachev, V.D.; Nakritskaia, D.D.; Varganov, S.A. Strong Relativistic Effects in Lanthanide-Based Single-Molecule Magnets. *J. Phys. Chem. Lett.* **2022**, *13*, 6749–6754. [CrossRef] [PubMed]
45. Zhu, Z.; Zhang, Y.-Q.; Li, X.-L.; Guo, M.; Lu, J.; Liu, S.; Layfield Richard, A.; Tang, J. Tuning Magnetic Relaxation in Square-Pyramidal Dysprosium Single-Molecule Magnets Using Apical Alkoxide Ligands. *CCS Chem.* **2021**, *3*, 388–398. [CrossRef]
46. Guo, Y.; Liu, K.; Qin, Y.; Wu, Q.; Hu, K.; Mei, L.; Chai, Z.; Liu, X.; Yu, J.; Shi, W. Role of molecular symmetry in the magnetic relaxation dynamics of five-coordinate Dy(iii) complexes. *Dalton Trans.* **2023**, *52*, 2703–2711. [CrossRef]
47. Zhu, Z.; Ying, X.; Zhao, C.; Zhang, Y.-Q.; Tang, J. A new breakthrough in low-coordinate Dy(iii) single-molecule magnets. *Inorg. Chem. Front.* **2022**, *9*, 6061–6066. [CrossRef]
48. Parmar, V.S.; Ortu, F.; Ma, X.; Chilton, N.F.; Clérac, R.; Mills, D.P.; Winpenny, R.E.P. Probing Relaxation Dynamics in Five-Coordinate Dysprosium Single-Molecule Magnets. *Chem. Eur. J.* **2020**, *26*, 7774–7778. [CrossRef]
49. Ying, X.; Zhu, Z.; Zhao, C.; Zhang, Y.-Q.; Tang, J. Five-Coordinated Dysprosium Single-Molecule Magnet Functionalized by the SMe Group. *Inorg. Chem.* **2022**, *61*, 20547–20551. [CrossRef] [PubMed]

50. Bartolomé, E.; Arauzo, A.; Luzón, J.; Bartolomé, J.; Bartolomé, F. Chapter 1—Magnetic Relaxation of Lanthanide-Based Molecular Magnets. In *Handbook of Magnetic Materials*; Brück, E., Ed.; Elsevier: Amsterdam, The Netherlands, 2017; Volume 26, pp. 1–289.
51. Gatteschi, D.; Sessoli, R. Quantum Tunneling of Magnetization and Related Phenomena in Molecular Materials. *Angew. Chem. Int. Ed.* **2003**, *42*, 268–297. [CrossRef] [PubMed]
52. Garanin, D.A.; Chudnovsky, E.M. Thermally activated resonant magnetization tunneling in molecular magnets: $Mn_{12}Ac$ and others. *Physical Review B* **1997**, *56*, 11102–11118. [CrossRef]
53. Lunghi, A.; Sanvito, S. Computational design of magnetic molecules and their environment using quantum chemistry, machine learning and multiscale simulations. *Nat. Rev. Chem.* **2022**, *6*, 761–781. [CrossRef] [PubMed]
54. Reta, D.; Kragskow, J.G.C.; Chilton, N.F. Ab Initio Prediction of High-Temperature Magnetic Relaxation Rates in Single-Molecule Magnets. *J. Am. Chem. Soc.* **2021**, *143*, 5943–5950. [CrossRef] [PubMed]
55. Briganti, M.; Santanni, F.; Tesi, L.; Totti, F.; Sessoli, R.; Lunghi, A. A Complete Ab Initio View of Orbach and Raman Spin–Lattice Relaxation in a Dysprosium Coordination Compound. *J. Am. Chem. Soc.* **2021**, *143*, 13633–13645. [CrossRef] [PubMed]
56. Roos, B.O.; Taylor, P.R.; Sigbahn, P.E.M. A complete active space SCF method (CASSCF) using a density matrix formulated super-CI approach. *Chem. Phys.* **1980**, *48*, 157–173. [CrossRef]
57. Aquilante, F.; Autschbach, J.; Carlson, R.K.; Chibotaru, L.F.; Delcey, M.G.; De Vico, L.; Fdez Galván, I.; Ferré, N.; Frutos, L.M.; Gagliardi, L.; et al. Molcas 8: New capabilities for multiconfigurational quantum chemical calculations across the periodic table. *J. Comput. Chem.* **2016**, *37*, 506–541. [CrossRef]
58. Malmqvist, P.Å.; Roos, B.O.; Schimmelpfennig, B. The restricted active space (RAS) state interaction approach with spin–orbit coupling. *Chem. Phys. Lett.* **2002**, *357*, 230–240. [CrossRef]
59. Heß, B.A.; Marian, C.M.; Wahlgren, U.; Gropen, O. A mean-field spin-orbit method applicable to correlated wavefunctions. *Chem. Phys. Lett.* **1996**, *251*, 365–371. [CrossRef]
60. Roos, B.O.; Lindh, R.; Malmqvist, P.-Å.; Veryazov, V.; Widmark, P.-O. Main Group Atoms and Dimers Studied with a New Relativistic ANO Basis Set. *J. Phys. Chem. A* **2004**, *108*, 2851–2858. [CrossRef]
61. Roos, B.O.; Lindh, R.; Malmqvist, P.-Å.; Veryazov, V.; Widmark, P.-O.; Borin, A.C. New Relativistic Atomic Natural Orbital Basis Sets for Lanthanide Atoms with Applications to the Ce Diatom and LuF_3. *J. Phys. Chem. A* **2008**, *112*, 11431–11435. [CrossRef]
62. Chibotaru, L.F.; Ungur, L. Ab initio calculation of anisotropic magnetic properties of complexes. I. Unique definition of pseudospin Hamiltonians and their derivation. *J. Chem. Phys.* **2012**, *137*, 064112. [CrossRef]
63. Chibotaru, L.F. Ab Initio Methodology for Pseudospin Hamiltonians of Anisotropic Magnetic Complexes. *Adv. Chem. Phys.* **2013**, *153*, 397–519.
64. Wang, Y.-F.; Wang, Y.-X.; Yang, Q.-Q.; Yin, B. Auxiliary Rather Than Dominant. The Role of Direct Dy–S Coordination in Single-Molecule Magnet Unveiled via ab initio Study. *J. Phys. Chem. A* **2024**, *128*, 5285–5297. [CrossRef] [PubMed]
65. Ungur, L.; Chibotaru, L.F. Ab Initio Crystal Field for Lanthanides. *Chem. Eur. J.* **2017**, *23*, 3708–3718. [CrossRef] [PubMed]
66. Briganti, M.; Garcia, G.F.; Jung, J.; Sessoli, R.; Le Guennic, B.; Totti, F. Covalency and magnetic anisotropy in lanthanide single molecule magnets: The DyDOTA archetype. *Chem. Sci.* **2019**, *10*, 7233–7245. [CrossRef] [PubMed]
67. Manvell, A.S.; Pfleger, R.; Bonde, N.A.; Briganti, M.; Mattei, C.A.; Nannestad, T.B.; Weihe, H.; Powell, A.K.; Ollivier, J.; Bendix, J.; et al. LnDOTA puppeteering: Removing the water molecule and imposing tetragonal symmetry. *Chem. Sci.* **2024**, *15*, 113–123. [CrossRef] [PubMed]
68. Pointillart, F.; Le Guennic, B.; Cador, O. Pressure-Induced Structural, Optical and Magnetic Modifications in Lanthanide Single-Molecule Magnets. *Chem. Eur. J.* **2024**, *30*, e202400610. [CrossRef] [PubMed]

magnetochemistry

MDPI

Article

Comprehensively Understanding the Transformation of Paramagnetic Tetramer to Spin-Paired Dimer in an S = ½ Molecular Crystal

Yin Qian [1], Yan Gao [1], Lei Xu [1], Reinhard K. Kremer [2,*], Jin Zhang [1] and Xiao-Ming Ren [1,2,3,*]

1 State Key Laboratory of Materials-Oriented Chemical Engineering and College of Chemistry and Molecular Engineering, Nanjing Tech University, Nanjing 211816, China; yinqian@njtech.edu.cn (Y.Q.); 202261105089@njtech.edu.cn (Y.G.); xl8686666@163.com (L.X.); Zhangjin@ecust.edu.cn (J.Z.)
2 Max-Planck-Institute für Festkörperforschung, Heisenbergstrasse 1, 70569 Stuttgart, Germany
3 State Key Laboratory of Coordination Chemistry, Nanjing University, Nanjing 210023, China
* Correspondence: rekre@fkf.mpg.de (R.K.K.); xmren@njtech.edu.cn (X.-M.R.)

Abstract: In this study, we comparatively analyzed the variable-temperature crystal structures for two isomorphous salts, [1-benzyl-4-aminopyridinium][$M(mnt)_2$] (M = Ni or Cu; mnt^{2-} = maleonitriledithiolate; labeled as **APy-Ni** or **APy-Cu**). Both salts crystallize in the triclinic *P*–1 space group at 296 K, comprising linear $[M(mnt)_2]^-$ (M = Ni or Cu) tetramers. A magnetostructural phase transition occurs at T_C~190 K in S = ½ **APy-Ni** at ambient pressure, with a conversion of paramagnetic tetramers into nonmagnetic spin-paired dimers. The discontinuous alteration of cell parameters at T_C signifies the characteristic of first-order phase transition in **APy-Ni**. No such transition appears in the nonmagnetic **APy-Cu** within the same temperature vicinity, demonstrating the magnetic interactions promoting the structural phase transition in **APy-Ni**, which is further reinforced through a comparison of the lattice formation energy between **APy-Ni** and **APy-Cu**. The phase transition may bear a resemblance to the mechanisms typically observed in spin-Peierls systems. We further explored the magnetic and phase transition properties of **APy-Ni** under varying pressures. Significantly, T_C shows a linear increase with rising pressure within the range of 0.003–0.88 GPa, with a rate of 90 K GPa^{-1}, manifesting that the applied pressure promotes the transition from tetramer to dimer.

Keywords: metal-bis-dithiolene salt; thermotropic phase transition; piezotropic phase transition; linear tetramer; spin-paired dimer

1. Introduction

A fascinating phenomenon, crystal-to-crystal phase transition, has been observed in numerous materials. Such transitions can be categorized into two main types based on structural changes, i.e., displacement-type [1–4] and order–disorder-type [5–9] phase transitions. The former pertains to the relative displacement of atoms or molecules/ions, while the latter involves the rotational motion of groups or molecules/ions within the lattice. Displacement-type phase transitions generally occur with the assistance of phonons. For instance, the Jahn–Teller effect can trigger a displacement-type phase transition in polyatomic molecules with degenerate electronic states [10]. Similarly, a Peierls/spin-Peierls transition can occur in a one-dimensional half-filled band structure/S = ½ magnetic chain system [11–14]. Moreover, a displacement-type phase transition can take place in a ferroelectric/antiferroelectric crystal [1–4]. The order–disorder-type phase transition is typically characterized by the presence of one or more components comprising globular-shaped molecules or ions within the lattice. Such molecules or ions exhibit minimal rotational energy barriers and considerable rotational freedom [5–9].

In a prior study, we documented the salt [1-benzyl-4-aminopyridinium][Ni(mnt)$_2$] (**APy-Ni**). This salt comprises linear [Ni(mnt)$_2$]$^-$ tetramers at ambient conditions, where the anion bears S = ½ spin, and exhibits a magnetostructural phase transition at T_C~190 K. It adopts a paramagnetic tetramer configuration in the high-temperature phase (HTP) and transitions to a spin-paired dimer configuration in the low-temperature phase (LTP) [15]. To gain a comprehensive understanding of the origin of the phase transition in **APy-Ni**, we synthesized the nonmagnetic crystals of [1-benzyl-4-aminopyridinium][Cu(mnt)$_2$] (**APy-Cu**). Notably, not only is **APy-Cu** isomorphous to **APy-Ni** at room temperature, but these two salts also exhibit remarkably analogous cell parameters and packing manner. In this study, we conducted a comparative investigation into the temperature-dependent crystal structures and lattice formation energies of **APy-Ni** and **APy-Cu**. Additionally, we explored the pressure-dependent magnetic and phase transition behaviors of **APy-Ni**. Our results revealed that the T_C exhibits ultra-sensitivity to external pressure (below 0.003 GPa), and the magnetic couplings play a crucial role in promoting the magnetostructural phase transition in **APy-Ni** as well.

2. Materials and Methods

2.1. Preparation of [1-Benzyl-4-aminopyridinium][M(mnt)$_2$] (M = Ni or Cu)

The microcrystalline sample of [1-benzyl-4-aminopyridinium][Ni(mnt)$_2$] (mnt^{2-} = maleonitriledithiolate; labeled as **APy-Ni**) was prepared following the published procedure [15]. The isomorphous analog [1-benzyl-4-aminopyridinium][Cu(mnt)$_2$] (abbr. as **APy-Cu**) was synthesized using the process that is similar to that of **APy-Ni**, only replacing the reactant $NiCl_2{\cdot}6H_2O$ by $CuCl_2{\cdot}2H_2O$. The yields of both salts are more than 60% calculated on the reactant $NiCl_2{\cdot}6H_2O$ or $CuCl_2{\cdot}2H_2O$.

Microanalysis: Calc. for $C_{20}H_{13}N_6NiS_4$ (**APy-Ni**): C, 45.81; H, 2.50; N, 16.03%. Found: C, 45.66; H, 2.62; N, 15.99%.

Microanalysis: Calc. for $C_{20}H_{13}N_6CuS_4$ (**APy-Cu**): C, 45.40; H, 2.48; N, 15.88%. Found: C, 45.28; H, 2.52; N, 15.64%.

2.2. X-Ray Single Crystallography

Diffraction data of selected crystals of **APy-Ni** were obtained on a Stoe IPDS diffractometer with an image plate detector (Mo-Kα, Ag-Kα radiation). Sections and projections of the reciprocal space were calculated from the raw data using diffractometer software [16]. The crystal data and structure refinements for **APy-Ni** at 296 and 100 K are represented in Table S1. The single-crystal X-ray diffraction data of **APy-Cu** at 296 and 100 K were performed by using Graphite monochromated Mo-Kα (λ = 0.71073 Å) on a Bruker D8 QUEST Apex III CCD area detector diffractometer. Data reduction and absorption correction were performed with the SAINT [17] and SADABS [18] software packages, respectively. The crystal structures were solved by direct method and refined on F^2 using the full-matrix least-squares method with the SHELXTL package [19]. All hydrogen atoms were geometrically fixed and placed in an ideal position. The crystal data and structure refinements for **APy-Cu** at 296 and 100 K are summarized in Table S2.

The details of other instruments and methods used in this study are provided in Supplementary Materials.

3. Results and Discussion

3.1. Crystal Structures

The single-crystal X-ray diffraction data were acquired at 296 and 100 K for both **APy-Ni** and **APy-Cu**. The crystallographic details are summarized in Tables S1 and S2 and Table 1, and the packing diagrams are shown in Figures S2–S7. Both **APy-Ni** and **APy-Cu** are isomorphous at room temperature, exhibiting closely comparable cell parameters and packing arrangements. While maintaining a comparable packing mode, a structural phase transition was observed in **APy-Ni** around 190 K. Conversely, no corresponding phase transition was noted in **APy-Cu** within the same temperature vicinity (Figure S1b). As a result, the two salts display slightly distinct packing structures at 100 K.

Table 1. The typical crystallographic data for **APy-Ni** and **APy-Cu** at 296 and 100 K.

Salt	APy-Ni		APy-Cu	
Temp./K	296	100	296	100
Crystal system	Triclinic	Triclinic	Triclinic	Triclinic
Space group	*P*–1	*P*–1	*P*–1	*P*–1
a/Å	11.2264 (8)	11.322 (2)	11.2559 (5)	11.156 (4)
b/Å	14.2879 (10)	13.933 (3)	14.2885 (7)	14.099 (4)
c/Å	15.6204 (11)	15.458 (3)	15.5951 (8)	15.445 (5)
α/°	73.879 (6)	73.07 (3)	74.420 (5)	75.166 (10)
β/°	77.289 (6)	74.32 (3)	77.931 (4)	77.667 (10)
γ/°	67.915 (6)	67.63 (3)	68.227 (4)	67.527 (9)
V/Å^3	2211.5 (3)	2121.9 (7)	2226.8 (2)	2151.8 (12)

Given the analogous molecular and packing structures of **APy-Ni** and **APy-Cu**, we shall focus our comprehensive crystal structure elucidation solely on **APy-Ni**. In both the HTP and LTP, the crystal structure of **APy-Ni** is categorized within the triclinic *P*–1 space group, with nearly identical cell parameters. The asymmetric unit has two pairs of crystallographically independent anions and cations, illustrated in Figure 1a (where the hydrogen atoms are omitted for clarity). Differentiating anions and cations are distinguished as Ni1/Ni2 and N10/N12, respectively. Two distinct planar $[Ni(mnt)_2]^-$ anions show an approximate parallel alignment, with their mean molecular planes composed of Ni and four S atoms, forming a dihedral angle of 2.44° in HTP and 1.12° in LTP. The two different cations present a Λ-shaped configuration. Notably, the bond lengths and bond angles within both anions and cations are in excellent accordance with values previously reported in the literature [15].

As illustrated in Figures 1b and S2–S4, the cations are arranged in a wave-shaped layer, aligning parallel to the (010) crystallographic plane. Meanwhile, the $[Ni(mnt)_2]^-$ anions, situated within channels formed by adjacent cationic layers, form tetrameric stacks along the a-axis direction. Within each tetramer, neighboring anions adopt a parallel face-to-face arrangement, connected by intermolecular S. . .S interactions with adjacent anion tetramers. In addition to the electrostatic interactions between anions and cations, the structure reveals charge-assisted hydrogen bonds. These hydrogen bonds occur between the NH_2 groups in cations and the CN groups in mnt ligands, as well as between hydrogen atoms within phenyl/pyridyl rings and the CN groups of mnt ligands (Figures 2 and S8).

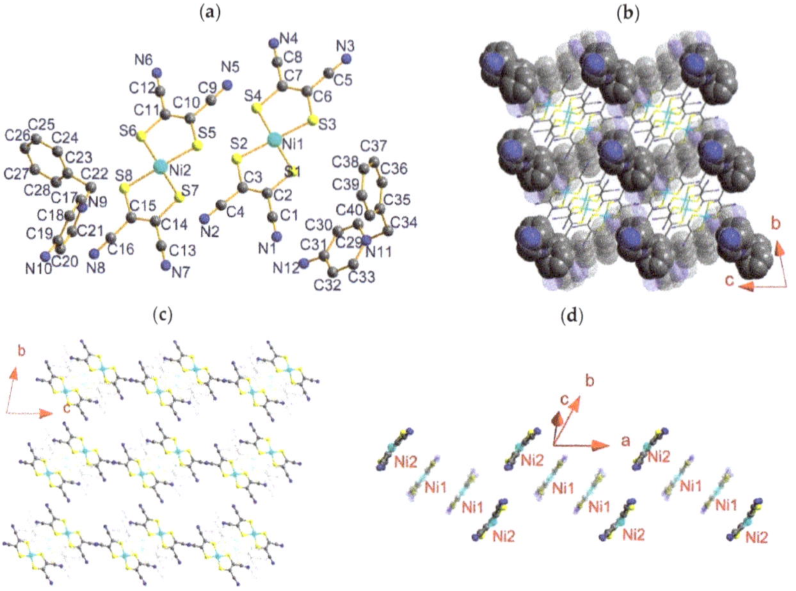

Figure 1. (**a**) An asymmetric unit (here, all hydrogen atoms omitted for clarity), (**b**) packing diagram viewed along a-axis wherein the cations with N10 and N12 are represented in the translucent and opaque modes, (**c**) arranging manner of anions viewed along a-axis wherein the anions with Ni1 and Ni2 are represented in the translucent and opaque modes, (**d**) tetrameric chain of anions along a-axis in **APy-Ni**.

For the sake of facilitating a comprehensive comparison between the crystal structure alterations of **APy-Ni** from its HTP to LTP, we have delineated pertinent distances. These include intermolecular Ni. . .Ni, Ni. . .S, and S. . .S distances, as well as the mean molecular plane of anions within a tetramer and the S. . .S distances between adjacent anion tetramers. These intermolecular contacts are visually highlighted in Figure 2a–c. The distances associated with charge-assisted hydrogen bonds between anions and cations, along with the shortest C. . .C distance between phenyl and pyridyl rings of neighboring cations, are depicted in Figure 2d,e. A summarized overview of these distances can be found in Tables 2 and 3. The phase transition results in discernible changes in the crystal structure, with the primary alterations involving the following aspects: (1) Within a tetramer, the distances concerning the closest neighboring Ni1. . .Ni2 separations (labeled as a1) and Ni1. . .Ni1 separations (labeled as a2) have undergone unequal reductions. This phenomenon arises due to the contraction of two Ni1 anions along the longer molecular axis of $[Ni(mnt)_2]^-$ anion (Figure 2f,g). (2) Concomitantly, the distances between the mean molecular planes of adjacent anions (designated as a7 and a8 in Figure 2b) have also experienced non-uniform reductions. This effect is due to the contraction of the tetramer along its stacking direction. The above-mentioned structural modifications are a direct consequence of the phase transition. The $\Delta = a2 - a1$ value is 0.15 Å in the HTP versus 0.351 Å in the LTP, indicating that the terminal and central $[Ni(mnt)_2]^-$ anions in a tetramer are strongly dimerized in the LTP regarding that in the HTP. From the viewpoint of structural change, the phase transition of tetramer–dimer belongs to the displacement-type structural phase transition. The S. . .S contacts between inter-tetramers (a9–a12) decrease, suggesting that the adjacent tetramers get close from the HTP to the LTP. (3) With respect to those in the HTP, the N. . .H distances between NH_2 groups and CN groups (b5, b7, and b9) are decreased and others expended, and the shortest C. . .C distance between phenyl and pyridyl rings of neighboring cations is increased in the LTP.

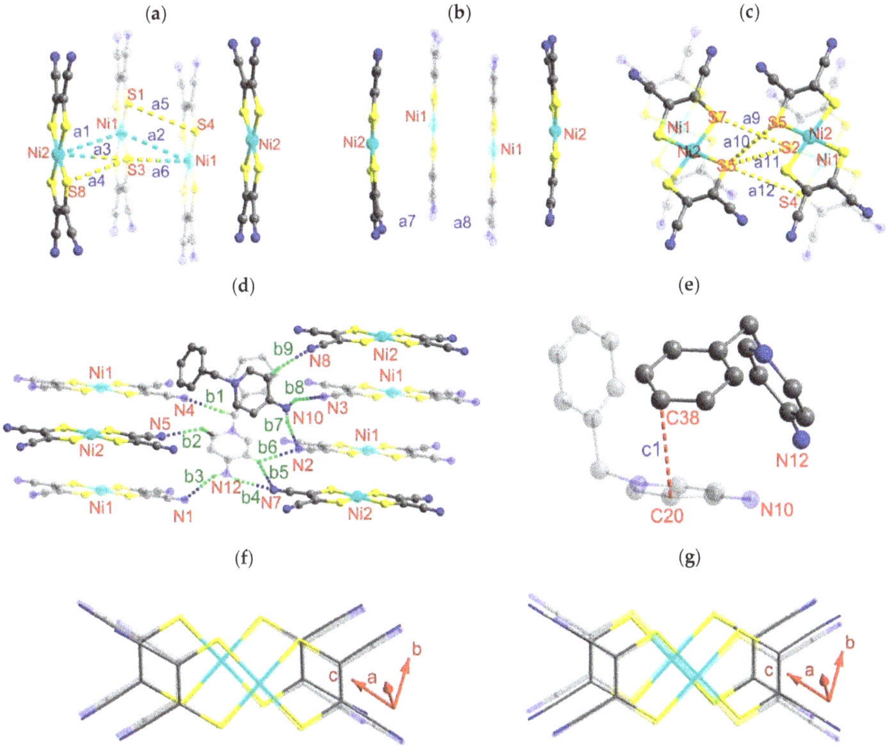

Figure 2. Schematic illustration of (**a**) typical intermolecular Ni. . .Ni, Ni. . .S, and S. . .S distances in an anion tetramer, (**b**) the separation of mean molecule plane between neighboring anions in an anion tetramer, (**c**) the S. . .S contacts between adjacent anion tetramers, (**d**) the charge-assisted H-bonds between adjacent anions and cations, (**e**) the shortest C. . .C distance between phenyl and pyridyl rings of neighboring cations, and top view of an anion tetramer in (**f**) HTP and (**g**) LTP in **APy-Ni**.

Table 2. The typical intermolecular distances (Å) within an anion tetramer of **APy-Ni** and **APy-Cu** at 296 and 100 K.

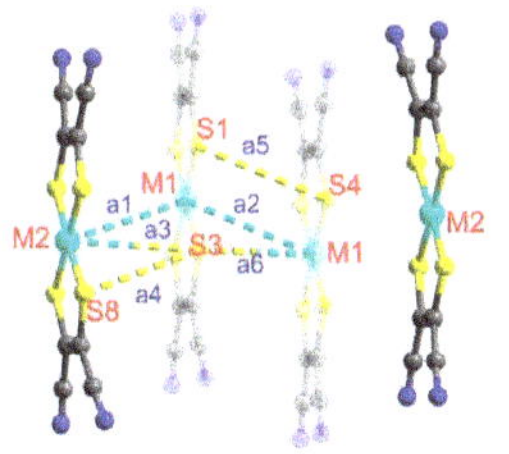

Distance/Å	APy-Ni		APy-Cu	
	296 K	100 K	296 K	100 K
a1	3.953	3.897	4.004	3.947
a2	3.944	3.546	3.984	3.919
a3	3.815	3.801	3.906	3.950
a4	3.867	3.519	3.998	3.953
a5	3.991	3.972	4.098	4.095
a6	3.585	3.549	3.506	3.368

Table 3. The typical intermolecular distances between anions and between anion and cation as well as between cations in both HTP (296 K) and LTP (100 K) of **APy-Ni**.

Distance/Å	296 K	100 K	Distance/Å	296 K	100 K
a7	3.561	3.455	b6	2.692	2.541
a8	3.517	3.542	b7	2.171	2.252
a9	3.886	3.841	b8	2.436	2.289
a10	3.891	3.721	b9	2.691	3.116
a11	3.591	3.589	c1	3.376	3.441

3.2. *Temperature-Dependent Cell Parameters and PXRD Patterns of* ***APy-Ni***

The temperature-dependent cell parameters of **APy-Ni** were assessed through single-crystal X-ray diffraction analysis. The variation in relative cell parameters [$P_{cell}(T)/P_{cell}(293\ K)$] vs. T is presented in Figures 3a and S9, revealing significant anisotropic tendencies. In the HTP, the a-axis and the α angle demonstrate a distinct negative thermal expansion behavior. Conversely, the remaining unit cell parameters decrease as the temperature decreases. Around 190 K, it is noteworthy that all cell parameters undergo a sudden discontinuous alteration, exhibiting the characteristic of the first-order phase transition. In the LTP, the c-axis exhibits negative thermal expansion behavior, while the other cell parameters display positive thermal expansion tendencies.

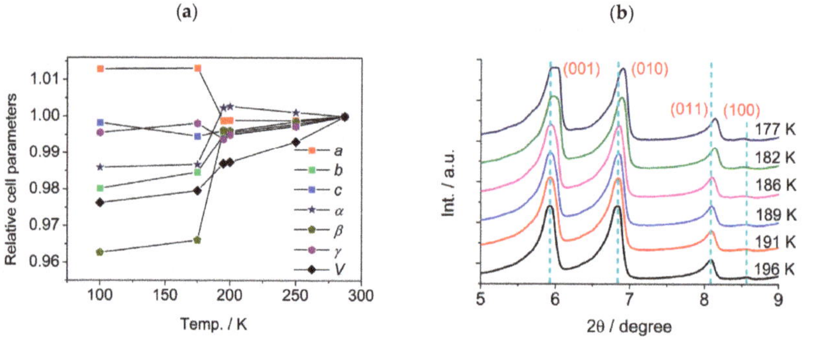

Figure 3. (**a**) Temperature-dependent unit cell parameters; (**b**) variable-temperature powder X-ray patterns around phase transition for **APy-Ni**.

The varying temperature PXRD patterns of **APy-Ni** are illustrated in Figures 3b and S9a,b. These patterns show remarkable similarity within the temperature range of 164–286 K. Transitioning from HTP to LTP, subtle distinctions in the PXRD patterns become apparent. Remarkably, the (001), (010), and (011) diffraction peaks shift toward the higher angle side, while the (100) diffraction peak shifts toward the lower angle side. This observation is consistent with the expansion of the a-axis and the contraction of both b- and c-axes, as shown in Figure 3a. Furthermore, the observable changes appear in the diffraction peaks at $2\theta \approx 9.2$ and 11.9°, which shift toward the lower and higher angle sides, respectively. The relative intensities of double diffraction peaks, such as those at $2\theta \approx 13.5$ and 13.7° as well as 15.6 and 16.0°, exhibit a contrasting trend of alteration (Figure S10b). These notable observations indicated the presence of two-phase equilibria within a temperature range of approximately ±10 K around the critical temperature (T_C). This is evidenced by the superposition of PXRD patterns in the LTP and HTP regions during this temperature interval.

3.3. Magnetic Properties at Ambient Pressure and DSC Features

The plot of χ_m vs. T is depicted in Figure 4a for **APy-Ni**; herein, the molar magnetic susceptibility, χ_m, corresponds to one $[Ni(mnt)_2]^-$ anion per formula. On the basis of the crystal structure analyses in both LTP and HTP, the magnetic behaviors of **APy-Ni** may be described by an S = ½ AFM dimer in the LTP and a linear S = ½ tetramer in the HTP, respectively [20]. An isolated dimer involving two S = ½ centers, which are coupled by a magnetic exchange constant J in line with the spin Hamiltonian $\hat{H} = -2\hat{S}_1 \cdot \hat{S}_2$ [20] the temperature dependences of magnetic susceptibility follow the Bleaney–Bowers expression, i.e., the first term in Equation (1) [21]. In consideration of the Curie–Weiss-type paramagnetism manifested within the low-temperature regime, it becomes essential to address the magnetic impurities associated with this phenomenon. These impurities, often arising from lattice defects, stem from structural imperfections that result in the presence of uncoupled monomers. Additionally, the diamagnetism observed in closed shells and the potential occurrence of van Vleck paramagnetism probably contribute to the whole experimental magnetic susceptibility; therefore, the whole experimental χ_m(T) of **APy-Ni** may be written as the form in Equation (1) in the LTP:

$$\chi_m = \frac{Ng^2\mu_B^2}{k_B T} \cdot \frac{1}{\left[3 + \exp\left(-\frac{2J}{k_B T}\right)\right]} + \frac{C}{T} + \chi_0 \quad (1)$$

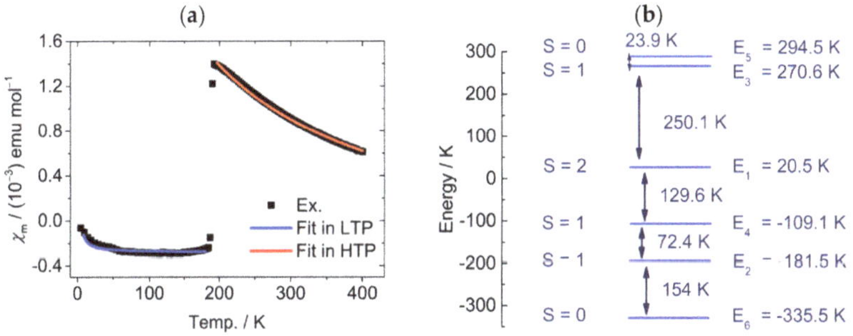

Figure 4. (**a**) Plot of χ_m vs. T in the temperature range of 5–400 K on heating (the solid squares and red/blue lines represent the experimental data and fits, respectively); (**b**) the low-lying state energy level diagram for a linear S = ½ tetramer ($J_1/k_B = -101$ K and J_2/k_B = 161 K) in the HTP for **APy-Ni**.

The best fit produced the parameters of $J/k_B \approx -320$ K, $C = 2.9 \times 10^{-3}$ emu K mol^{-1} and $\chi_0 = -3.17 \times 10^{-4}$ emu mol^{-1} with g = 2.015 (fixed) [20], which is quite close to the average of g-factor value 2.017 obtained from variable-temperature EPR spectra analyses (ref. next section). The amount of magnetic impurity is estimated to be ca. 0.8% based on the fitted Curie constant. The χ_0 value is comparable to the Pascal's constants of **APy-Ni**. The spin gap ($\Delta_{spin}/k_B = 2J/k_B$) in LTP is quantified at ~640 K, and so a big spin gap resulted in **APy-Ni** exhibiting almost diamagnetism at much higher temperature (~180 K; see Figures 4a and 5b). The whole experimental $\chi_m(T)$ of **APy-Ni** in HTP includes three parts, shown in Equations (2)–(4), in which the first term represents the contribution from a linear S = ½ tetramer spin system with two different magnetic exchange constants within a spin-tetramer based on the spin Hamiltonian $\hat{H} = -2J_1(\hat{S}_1 \cdot \hat{S}_2 + \hat{S}_3 \cdot \hat{S}_4) - 2J_2\hat{S}_2 \cdot \hat{S}_3$ [20,22], the J_1 and J_2 correspond to the magnetic exchange constant between the terminal and the neighboring central spins as well as the magnetic exchange constant between two central spins, and the last two terms have the same meanings as those in Equation (1). The best fit for the χ_m–T plot of **APy-Ni** in the HTP (190–400 K) using the Equations (2)–(4) [22]

offered $J_1/k_B \approx -101$ K and $J_2/k_B \approx 161$ K with C and χ_0 fixed to the same values in the LTP. The results obtained from fits suggest that the terminal and central anions feature AFM coupling, while two central anions exhibit FM exchange interactions within an anion tetramer. As illustrated in Figure 4a, both the experimental and the fitted χ_m–T profiles match well with each other, demonstrating that the magnetic coupling model and the fitted results seem reasonable. On the basis of the E_i (i = 1–6) values, the spin energy gap between the singlet ground state (E_6 with S = 0) and the first magnetic excited state (E_1 with S = 1) is ~154 K, and the low-lying state energy level diagram is displayed in Figure 4b.

$$\chi_m = \frac{Ng^2\mu_B^2}{k_B T} \cdot \frac{A}{Z} + \frac{C}{T} + \chi_0 \tag{2}$$

$$A = 10\exp\left(-\frac{E_1}{k_B T}\right) + 2\exp\left(-\frac{E_2}{k_B T}\right) + 2\exp\left(-\frac{E_3}{k_B T}\right) + 2\exp\left(-\frac{E_4}{k_B T}\right) \tag{3}$$

$$Z = 5\exp\left(-\frac{E_1}{k_B T}\right) + 3[\exp\left(-\frac{E_2}{k_B T}\right) + \exp\left(-\frac{E_3}{k_B T}\right) + \exp\left(-\frac{E_4}{k_B T}\right) + \exp\left(-\frac{E_5}{k_B T}\right) + \exp\left(-\frac{E_6}{k_B T}\right)] \tag{4}$$

$$E_1 = -J_1 - \frac{1}{2}J_2 \ (S = 2)$$

$$E_2 = J_1 - \frac{1}{2}J_2 \ (S = 1)$$

$$E_3 = \frac{1}{2}J_2 + \left(J_1^2 + J_2^2\right)^{1/2} \ (S = 1)$$

$$E_4 = \frac{1}{2}J_2 - \left(J_1^2 + J_2^2\right)^{1/2} \ (S = 1)$$

$$E_5 = J_1 + \frac{1}{2}J_2 + \left(4J_1^2 - 2J_1J_2 + J_2^2\right)^{1/2} \ (S = 0)$$

$$E_6 = J_1 + \frac{1}{2}J_2 - \left(4J_1^2 - 2J_1J_2 + J_2^2\right)^{1/2} \ (S = 0)$$

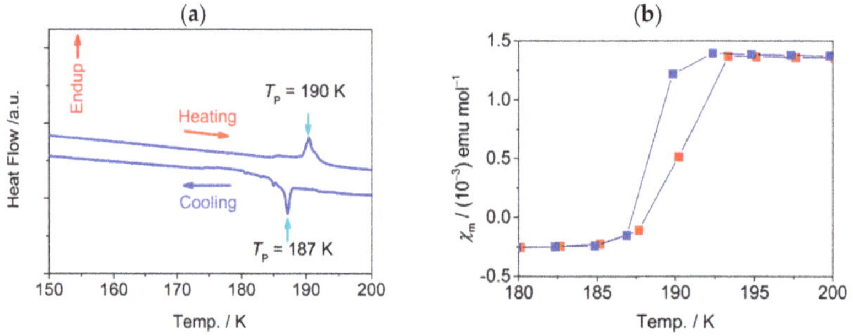

Figure 5. (**a**) DSC plots of **APy-Ni** in the heating–cooling cycle and (**b**) temperature-dependent magnetic susceptibility of **APy-Ni** showing a small hysteresis loop close to the phase transition.

The DSC traces obtained at a heating/cooling rate of 1 K min^{-1} are depicted in Figure 5a. These traces reveal distinct exothermic and endothermic anomalies occurring at ~187 K during cooling and ~190 K during heating, respectively. The noticeable thermal hysteresis observed in both DSC and χ_m vs. T plots (Figure 5) between these temperatures strongly indicates the presence of the first-order phase transition characteristics. This inference finds further support in the abrupt alteration of lattice parameters around the critical temperature (T_C). The enthalpy change (ΔH) associated with the phase transition is quantified at 1.52 kJ mol^{-1}, while an entropy difference (ΔS) of 8.04 J mol^{-1} K^{-1} is determined during heating [15]. In general, the ΔS associated with a phase transition includes the contributions from structure (molecular orientations) entropy changes, vibrational entropy changes, and magnetic entropy changes for a magnetic insulator. The crystal structure analyses for **APy-Ni** revealed analogous packing structures in both the HTP and the LTP. This implies that the entropy change (ΔS) of the phase transition primarily arises from the alterations in spin states. As discussed in the part of magnetic property analysis, **APy-Ni** displays paramagnetism, characterized by the presence of a linear S = ½ tetramer in the HTP, while diamagnetism in the LTP owing to the formation of spin-paired dimers composed of S = ½ $[Ni(mnt)_2]^-$ anions, the spin entropy change (ΔS) corresponding to the magnetostructural phase transition is calculated as ΔS = $R\ln(Z_{HTP}/Z_{LTP})$, with R (=8.314 J mol^{-1} K^{-1}) denoting the gas constant and Z_{HTP}, Z_{LTP} representing the partition functions in the HTP and the LTP sequentially, which are expressed in the form of Equations (5) and (6) correspondingly.

$$Z_{HTP} = 1 + 3\exp\left(-\frac{E_2 - E_6}{k_B T}\right) + 3\exp\left(-\frac{E_4 - E_6}{k_B T}\right) + 5\exp\left(-\frac{E_1 - E_6}{k_B T}\right) + 3\exp\left(-\frac{E_3 - E_6}{k_B T}\right) + \exp\left(-\frac{E_5 - E_6}{k_B T}\right) \quad (5)$$

$$Z_{LTP} = \left[1 + 3\exp\left(-\frac{\Delta}{k_B T}\right)\right]^2 \quad (6)$$

In Equation (5), E_i (i = 1–6) denotes the spin energy levels (Figure 4b) in the HTP. In Equation (6), Δ signifies the spin energy gap between the singlet ground state and the triplet excited state in the LTP. The squared term arises from the transformation of a spin-tetramer in the HTP into two spin-dimers in the LTP. Taking T_C as 190 K, the values of E_i (i = 1 to 6) in the HTP and Δ in the LTP were determined through the fits of variable-temperature magnetic susceptibilities. Subsequently, the spin entropy change (ΔS) was computed to be ~10.24 J mol^{-1} K^{-1}. This result, slightly bigger than the ΔS value of 8.04 J mol^{-1} K^{-1} determined by DSC measurements, provides further support for the conclusion that the observed ΔS at T_C primarily originates from modifications in the spin states.

3.4. EPR Spectra of *APy-Ni* in the Temperature of 110–292 K

Figure S11 displays EPR spectra of **APy-Ni** in 110–292 K. Above 170 K, **APy-Ni** exhibits isotropic broad Lorentzian-shaped signals, which can be well fit by taking the field derivative of both the '+' and '−' absorption signal into consideration. The fits revealed a typical line width of ~1500 Oe, the g-factors about 1% larger than the free electron value, and the averaged g-factor amounts to 2.017 in 170–292 K (Figure S11b), indicating small orbital contributions to the magnetic moment, and this conclusion is consistent with the theoretic analysis, i.e., the singly occupied molecular orbital (SOMO) in the $[Ni(mnt)_2]^-$ anion with D_{2h} point group symmetry is comprised of the d_{x2-y2} orbital of Ni and the group orbitals of $[(mnt)_2]^{4-}$ ligands with a_g symmetry [23], and the spin population on the Ni atom was calculated to be only ~0.21 [24]. As shown in Figure 6a, the EPR spectrum of **APy-Ni** displays the superposition of two signals at 160 K, namely, an isotropic broad signal arising from tetramers together with a narrow and asymmetric signal, as well as only narrow and asymmetric signal below 160 K, and its intensity is enhanced upon cooling (Figure 6b). The inset in Figure 6a shows that the χ_m value decays exponentially in 110–190 K and tends to be unchanged below 160 K, demonstrating that, below 160 K, no magnetic excited state is populated in the thermal equilibrium distribution of a magnetic tetramer. Thus, the narrow and asymmetric resonance signals originate from the uncoupled magnetic impurities, which arise from the defect of the lattice.

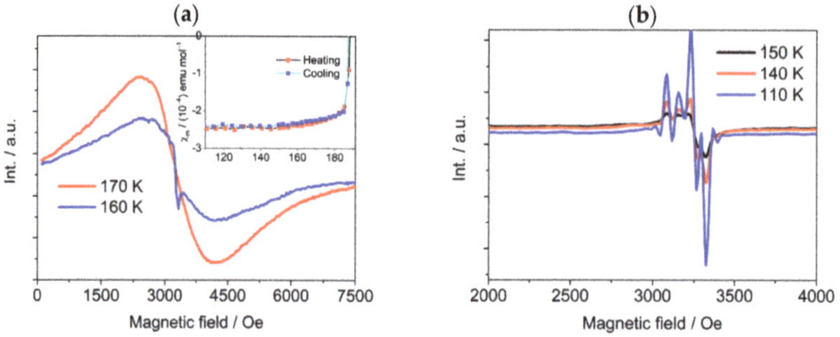

Figure 6. EPR spectra of a powdered polycrystalline sample of **APy-Ni** at (**a**) 160 and 170 K (inset: χ_m–T plots in 110–190 K) and (**b**) 110, 140, and 150 K.

*3.5. Phase Transition Behavior of **APy-Ni** Under Pressures*

With respect to inorganic matters, molecular crystals display greater compressibility and, as a result, exhibit higher sensitivity for pressure detection, and this is verified by some examples, e.g., a Cu^{2+} complex with nitroxide [Cu(hfac)$_2$NN-PzMe] shows a spin crossover-like phenomenon with $T_C \approx 150$ K at ambient pressure, which significantly shifts from 150 to 300 K as the hydrostatic pressure increases from 10^{-4} to 0.14 GPa [25]. Two salts, [R-BzPy][Ni(mnt)$_2$] (R-BzPy$^+$ = 1-(4′-nitrobenzyl)pyridinium with R = NO_2 and 1-(4′-brominobenzyl)pyridinium with R = Br), show spin-Peierls transition at $T_{SP} \approx 184$ and 112 K under ambient pressure, respectively, and T_{SP} rises at an average rate of 1.66 K GPa^{-1} for R = NO_2 and 0.82 K GPa^{-1} for R = Br [26].

From a theoretical standpoint, the Clausius–Clapeyron relation elucidates the connection between pressure (P) and the critical temperature (T_C) of a phase transition within the equilibrium of two phases. This is expressed as follows:

$$\frac{dP}{dT} = \frac{\Delta H}{T_C \Delta V} \quad (7)$$

where P and T represent pressure and temperature, respectively. T_C denotes the critical temperature of the phase transition, while ΔH and ΔV refer to the alterations in enthalpy and volume associated with the phase transition at T_C. This equation can be rewritten as follows:

$$lnT_c = \frac{\Delta V}{\Delta H}P + C \quad (8)$$

where C is a constant. From the LTP to the HTP, **APy-Ni** shows positive expansion in cell volume and endothermal behavior, indicative of the T_C of tetramer–dimer transformation increasing with increasing pressure.

The pressure behavior of the phase transition was further studied by measuring the temperature dependences of magnetic susceptibility at different applied pressures and under the magnetic field of 1.0×10^4 Oe (Supplementary Materials). The T_C was determined reliably from the peak temperature of the quantity $d(\chi_m T)/dT$ (Figure 7a), which increases with an increase in pressure. In a narrow pressure regime (from ambient to 0.003 GPa), the T_C sharply jumps. Subsequently, the T_C almost linearly increases with a rate of 90 K GPa^{-1} from 0.003 to 0.88 GPa (Figures 7b and S12). Obviously, this slope in the plot of T_C vs. P is much less than in the initially applied pressure stage because the Coulomb repulsion between ions increases rapidly as the pressure is elevated, resulting in the compressibility of the crystalline sample decreases. To the best of our knowledge, the averaged $\Delta T_C/\Delta P$ value of **APy-Ni** in the pressure range of 0.003–0.88 GPa is the largest one in the reported molecule-based spin transition compounds [27]. The T_C in **APy-Ni** is remarkably sensitive to pressure, especially under initial lower pressure, demonstrating that the magnetic couplings between adjacent $[Ni(mnt)_2]^-$ anions are highly sensitive to the intermolecular arrangement [28]; moreover, the applied pressure promotes the transformation of tetrameric phase to dimeric phase. It is worth noting that the sample recovered from 0.88 GPa to ambient pressure shows the same T_C as the pristine sample, indicating that the transformation under pressure is reversible for **APy-Ni**. Materials that display a reversible piezotropic phase transition and demonstrate a linear correlation between applied pressure and critical temperature hold the capability to convert mechanical stimuli into various outputs, including optical [29,30], electrical [31,32], or magnetic signals [25,26,33,34]. These materials can be employed for real-time monitoring of pressure fluctuations, offering potential applications in fields such as unsteady flow field measurement [35,36] and artificial intelligence development [37,38].

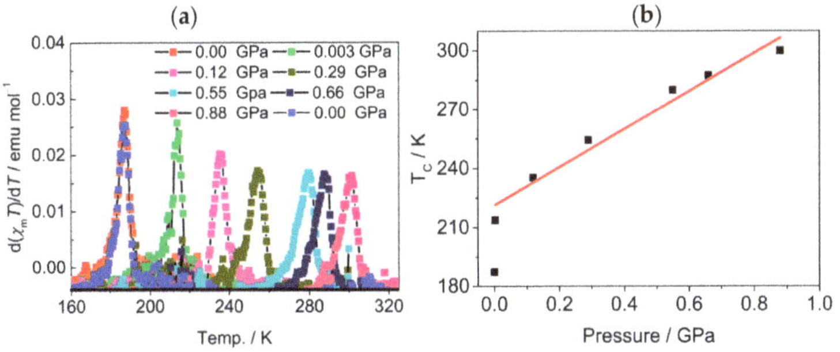

Figure 7. (**a**) Plots of $d(\chi_m T)/dT$ vs. T at different external pressures and (**b**) pressure dependence of T_C for **APy-Ni**.

3.6. *The Formation Energy of Lattice of* ***APy-Ni*** *and* ***APy-Cu***

The formation energy of lattice ($\Delta E_{lattice}$) per molar unit cell was defined as the difference between the cumulative energy of the constituent components (E_{anion} and E_{cation}) and the total energy of the crystal ($E_{crystal}$) per molar unit cell and calculated utilizing the Density Functional Theory (DFT) method via the $Dmol^3$ program [39–41]. In the crystal structures of isomorphous **APy-Ni** and **APy-Cu**, both at 296 K and 100 K, four pairs of anions and cations exist. The computed values of $\Delta E_{lattice}$, $E_{crystal}$, E_{anion}, and E_{cation} for these crystal structures (**APy-Ni** and **APy-Cu**) at both 296 K and 100 K are summarized in Table 4. The findings indicate that, at both 296 K and 100 K, **APy-Ni** exhibits higher formation energy of lattice compared to **APy-Cu**. This discrepancy is attributed to the magnetic interactions among anions present in the paramagnetic **APy-Ni**, which confer supplementary stability energy for its lattice formation when contrasted with the nonmagnetic **APy-Cu**. As the temperature decreases, lattice contraction leads to a notable elevation in the formation energy of lattice for both **APy-Ni** and **APy-Cu**. Specifically, upon cooling from 296 to 100 K, the formation energy of the lattice increased by 51.98 kJ mol^{-1} for **APy-Ni** and 7.35 kJ mol^{-1} for **APy-Cu**. This difference is noticeable, with the former increase being substantially larger than the latter.

Table 4. The calculated $\Delta E_{lattice}$, $E_{crystal}$, E_{anion}, and E_{cation} for **APy-Ni** and **APy-Cu** at 296 and 100 K.

	* E_{anion}/Hartree	* E_{cation}/Hartree	$E_{crystal}$/Hartree	$\Delta E_{lattice}$/Hartree	$\Delta E_{lattice}$/kJ mol^{-1}
APy-Cu (100 K)	−3756.1185	−573.886	−17,320.9663	0.9483	2489.76
APy-Cu (296 K)	−3756.1185	−573.886	−17,320.9635	0.9455	2482.41
APy-Ni (100 K)	−3623.9881	−573.886	−16,792.4758	0.9794	2571.41
APy-Ni (296 K)	−3623.9881	−573.886	−16,792.456	0.9596	2519.43

* E_{anion} and E_{cation} represent the energy of per anion and per cation, respectively.

Both **APy-Cu** and **APy-Ni** are classified as ionic crystals, with their $\Delta E_{lattice}$ primarily arising from Coulomb interactions, specifically Madelung energy. The $\Delta E_{lattice}$ values of **APy-Cu** exhibit a slight increase at 100 K compared to 296 K, attributed to lattice contraction at lower temperatures, leading to an enhancement of Coulomb interactions. In the case of **APy-Ni**, magnetic exchange interactions contribute significantly, in addition to Coulomb interactions. This divergence can be linked to the notably stronger magnetic interactions present in the LTP compared to the HTP for **APy-Ni**. These findings unveil the pivotal role of magnetic interactions in the transformation from a paramagnetic tetramer to a spin-paired dimer within **APy-Ni**.

It is well known that the electron– or spin–vibronic couplings (so-called electron– or spin–phonon interactions) play a crucial role in the phase transition that occurs in one-dimensional half-filled electron band structure or S = ½ spin systems, e.g., the conventional Peierls or spin-Peierls transition systems [11–14]. It is likely that the magnetoelastic interactions contribute to this displacement-type of tetramer–dimer transition, bearing a resemblance to the mechanisms typically observed in spin-Peierls systems.

4. Conclusions

In this study, we have investigated the crystal structures and magnetic properties of an S = ½ linear tetramer salt (**APy-Ni**). **APy-Ni** exhibits a magnetostructural phase transition with a T_C around 190 K at ambient pressure and adopts a paramagnetic tetramer configuration in the HTP, whereas a diamagnetic dimer configuration in the LTP. Investigation into the pressure-dependent magnetic behaviors has unveiled the remarkable response of T_C to varying pressures attributed to the substantial compressibility of van der Waals crystals. A strong linear correlation between T_C and pressure (P) has been observed within the pressure range of 0.003–0.88 GPa, with an average rate of change ($\Delta T_C/\Delta P$) of ~90 K GPa^{-1}, demonstrating the applied pressure promoting the spin transition by compressing the lattice, which enhances the magnetic couplings within the tetramer. The sensitivity of T_C to applied pressure also suggests that **APy-Ni** holds potential as a material for pressure-sensing applications. The isomorphous nonmagnetic salt (**APy-Cu**) shares similar cell parameters and packing arrangements with **APy-Ni** at ambient conditions. Nevertheless, the absence of the tetramer–dimer transformation within the same temperature range implies a pronounced influence of magnetic interactions on the structural phase transition of **APy-Ni**, resembling the spin-Peierls transition observed in S = ½ linear chain systems. This hypothesis is supported by the calculated lattice formation energy. Overall, these findings provide a foundation for the development of novel molecule-based smart materials that demonstrate sensitivity to pressure variations, particularly within moderate pressure ranges.

Supplementary Materials: The following supporting information can be downloaded at https://www.mdpi.com/article/10.3390/magnetochemistry11020008/s1, Table S1: Crystal data for **APy-Ni**, Table S2: Crystal data for **APy-Cu** at 296 and 100 K. Figure S1: (a) Experimental and simulated PXRD patterns of **APy-Cu**, and (b) DSC plot of **APy-Cu.** Figure S2: Packing diagrams of **APy-Ni** viewed along *b*- and *c*-axes, respectively. Figure S3: Packing diagrams in both ball–stick and space-filling modes of cations in *APy-Ni* viewed along *a*-, *b*-, and *c*-axes. The cations with N10 and N12 are represented in translucent and opaque modes, respectively. Figure S4: (a) Top and (b) side views of a cationic layer in **APy-Ni**. The cations with N10 and N12 are represented in translucent and opaque modes, respectively. Figure S5: ORTEP plots of **APy-Cu** at (a) 296 K and (b) 100 K, all of which the thermal ellipsoid was drawn at the 30% probability level. Figure S6: Packing diagrams of **APy-Cu** at 296 K viewed along *a*-axis direction, which showed (a) the tetramers of anions residual in the channels enclosed by cations and (b) the arrangement of anions. Figure S7: Packing diagrams of anions in the crystal structure of **APy-Cu** at 296 K, showing the arranging manner between neighboring tetramers. Figure S8: The Hirshfeld surfaces of APy-Ni for visualizing intermolecular interactions between cations and anions (a–c) in HTP (296 K) and (d–f) in LTP (100 K). The cation contains N10 in (a,d), as well as N12 in (b,c,e,f), in which the red dots represent the H-bonding interactions between H atoms of the −NH2 group or phenyl ring/pyridyl ring and the −CN groups of mnt^{2-} in the anions. Obviously, the number of red dots increases in LTP compared to that in HTP owing to contraction of cell volume. Figure S9: Variable-temperature cell parameters with error bars of **APy-Ni**. Figure S10: (a,b) Variable-temperature powder X-ray patterns of **APy-Ni**. Figure S11: (a–e) Plots of the EPR signals measured for a powder sample of **APy-Ni** at 9.3 GHz at different temperatures in 110–292 K, and (f) plots of line width *versus* T and g-factor *versus* T in 160−292 K. Figure S12: Plot of TC versus applied pressure with error bars for **APy-Ni**. See refs. [39,41–50].

Author Contributions: Y.Q., Y.G. and L.X.: synthesized the samples, performed the power and part of single-crystal X-ray experiments, and data analysis; Y.Q.: performed Harshifeld surface analysis; J.Z.: performed the calculations for overlap integrals and lattice formation energy; R.K.K.: performed collection and analysis of temperature-dependent EPR spectra; R.K.K. and X.-M.R.: supervised the study and revised the manuscript. All authors have read and agreed to the published version of the manuscript.

Funding: This research was funded by the National Natural Science Foundation of China, grant number 22073047. XMR acknowledges the Max-Planck Society for financial support.

Data Availability Statement: All data generated and analyzed during this study are included in the article and its Supplementary Information. The crystal structures generated in this study have been deposited in the Cambridge Crystallographic Data Centre under accession code CCDC: 2177252, 2177253, 2291616, and 2291618, and can be obtained free of charge from the CCDC via https://www.ccdc.cam.ac.uk/structures/ (accessed on 9 January 2025).

Acknowledgments: The authors thank E. Brücher for the magnetic property determinations, G. Siegle for DSC analyses, and W. Röthenbach for variable-temperature X-ray powder diffraction measurements.

Conflicts of Interest: The authors declare no conflicts of interest.

Abbreviations

The following abbreviations are used in this manuscript:

APy-Ni	[1-benzyl-4-aminopyridinium][$Ni(mnt)_2$]
APy-Cu	[1-benzyl-4-aminopyridinium][$Cu(mnt)_2$]
APy^+	1-benzyl-4-aminopyridinium
mnt^{2-}	Maleonitriledithiolate
AFM	Antiferromagnetic
FM	Ferromagnetic

References

1. Maksimov, E.G. Theoretical Investigations of the Ferroelectric Transition. *Physics—Uspekhi* **2009**, *52*, 603–614. [CrossRef]
2. Nakano, A.; Hasegawa, T.; Tamura, S.; Katayama, N.; Tsutsui, S.; Sawa, H. Antiferroelectric Distortion with Anomalous Phonon Softening in the Excitonic Insulator Ta_2NiSe_5. *Phys. Rev. B* **2018**, *98*, 045139. [CrossRef]

3. Luo, Y.; Zhang, Y.Q.; Xu, G.C. Synthesis and Characterization of a Displacement-Type Ferroelectric–Ferroelectric Phase Transition Compound $[(NH_3)(CH_2)_3(NH_3)]_2[InBr_6]Br\cdot H_2O$. *Inorg. Chem.* **2022**, *61*, 13143–13148. [CrossRef] [PubMed]
4. Morosov, A.I.; Sigov, A.S. The Nature of Central Peak in $SrTiO_3$. *Ferroelectrics* **2011**, *412*, 12–14. [CrossRef]
5. Harada, J.; Shimojo, T.; Oyamaguchi, H.; Hasegawa, H.; Takahashi, Y.; Satomi, K.; Suzuki, Y.; Kawamata, J.; Inabe, T. Directionally Tunable and Mechanically Deformable Ferroelectric Crystals from Rotating Polar Globular Ionic Molecules. *Nat. Chem.* **2016**, *8*, 946–952. [CrossRef]
6. Liu, H.Y.; Zhang, H.Y.; Chen, X.G.; Xiong, R.G. Molecular Design Principles for Ferroelectrics: Ferroelectrochemistry. *J. Am. Chem. Soc.* **2020**, *142*, 15205–15218. [CrossRef]
7. Xu, K.; Zhou, Z.N.; Han, X.B.; Yang, Y.W.; Zhang, W.; Ye, Q. Shape Shifting and Locking in Mechanically Responsive Organic-Inorganic Hybrid Materials for Thermoelastic Actuators. *Angew. Chem. Int. Ed.* **2024**, *63*, e202408247. [CrossRef]
8. Zhou, Q.J.; He, L.; Xu, K.; Han, X.B.; Yin, T.J.; Zhang, W.; Ye, Q. Thermally Reversible Luminescence in Ferroelastic Phase Transition Materials. *Adv. Opt. Mater.* **2024**, *12*, 2302292. [CrossRef]
9. Gui, L.A.; Chen, J.; Zhang, Y.F.; Li, L.H.; Li, J.R.; Hu, Z.B.; Zhang, S.Y.; Zhang, J.; Zhang, Z.; Ye, H.Y.; et al. Room-temperature Magnetocapacitance Spanning 97 K Hysteresis in Molecular Material. *Angew. Chem. Int. Ed.* **2024**, e202416380. [CrossRef]
10. Jahn, H.A.; Teller, E. Stability of Polyatomic Molecules in Degenerate Electronic States—I: Orbital Degeneracy. *Proc. R. Soc. Lond. A* **1937**, *161*, 220–235.
11. Peierls, R.E. *Quantum Theory of Solids*; Oxford University Press: London, UK, 1955; pp. 108–114.
12. Bray, J.W.; Hart, H.R., Jr.; Interrante, L.V.; Jacobs, I.S.; Kasper, J.S.; Watkins, G.D.; Wee, S.H.; Bonner, J.C. Observation of a Spin-Peierls Transition in a Heisenberg Antiferromagnetic Linear-Chain System. *Phys. Rev. Lett.* **1975**, *35*, 744–747. [CrossRef]
13. Braden, M.; Hennion, B.; Reichardt, W.; Dhalenne, G.; Revcolevschi, A. Spin-Phonon Coupling in $CuGeO_3$. *Phys. Rev. Lett.* **1998**, *80*, 3634–3637. [CrossRef]
14. Musfeldt, J.L.; Wang, Y.J.; Jandl, S.; Poirier, M.; Revcolevschi, A.; Dhalenne, G. Infrared Investigation of the Broken-Symmetry Ground State in $GeCuO_3$. *Phys. Rev. B* **1996**, *54*, 469–473. [CrossRef] [PubMed]
15. Ren, X.M.; Meng, Q.J.; Song, Y.; Hu, C.J.; Lu, C.S.; Chen, X.Y.; Xue, Z.L. Unusual Magnetic Property Associated with Dimerization within a Nickel Tetramer. *Inorg. Chem.* **2002**, *41*, 5931–5933. [CrossRef]
16. *Stoe IPDS Software Package Version 2.89*; Stoe & Cie. GmbH: Darmstadt, Germany, 1989.
17. Bruker. *APEX 2, SAINT, XPREP*; Bruker AXS Inc.: Madison, WI, USA, 2007.
18. Bruker. *SADABS*; Bruker AXS Inc.: Madison, WI, USA, 2001.
19. Sheldrick, G.M. A Short History of SHELX. *Acta Crystallogr. Sect. A Found. Crystallogr.* **2008**, *64*, 112–122. [CrossRef]
20. Ren, X.M.; Kremera, R.K.; Meng, Q.J. Investigation of the Magneto-Structural Phase Transition in [1-benzyl-4-aminopyridinium][bis(maleonitriledithiolato)-nickelate]. *J. Mag. Mag. Mater.* **2004**, *272–276*, 924–926. [CrossRef]
21. Carlin, R.L. *Magnetochemistry*; Springer: Berlin/Heidelberg, Germany, 1986; Chapter 5.
22. Rubenacker, G.V.; Drumheller, J.E.; Emerson, K.; Willett, R.D. Magnetic Susceptibility of $((CH_3)_3NH)_2Cu_4Br_{10}$, Chains of Stacked Linear Tetramers. *J. Mag. Mag. Mater.* **1986**, *54–57*, 1483–1484. [CrossRef]
23. Shupack, S.I.; Billig, E.; Clark, R.J.H.; Willams, R.; Gray, H.B. The Electronic Structures of Square-Planar Metal Complexes. V. Spectral Properties of the Maleonitriledithiolate Complexes of Nickel, Palladium, and Platinum. *J. Am. Chem. Soc.* **1964**, *86*, 4594–4602. [CrossRef]
24. Ni, Z.; Ren, X.; Ma, J.; Xie, J.; Ni, C.; Chen, Z.; Meng, Q. Theoretical Studies on the Magnetic Switching Controlled by Stacking Patterns of Bis(maleonitriledithiolato)nickelate(III) Dimers. *J. Am. Chem. Soc.* **2005**, *127*, 14330–14338. [CrossRef]
25. Maryunina, K.Y.; Zhang, X.; Nishihara, S.; Inoue, K.; Morozov, V.A.; Romanenko, G.V.; Ovcharenko, V.I. A Heterospin Pressure Sensor. *J. Mater. Chem. C* **2015**, *3*, 7788–7791. [CrossRef]
26. Ren, X.M.; Nishihara, S.; Akutagawa, T.; Fujita, W.; Awaga, K. Pressure Effect on Spin-Peierls-Like Transition in Quasi-1D Spin Systems $[RBzPy][Ni(mnt)_2]$. *Chem. Phys. Lett.* **2007**, *439*, 318–322. [CrossRef]
27. Tanaka, T.; Fujita, W.; Awaga, K. Pressure Effects on Magnetic Bistability in a Heterocyclic Thiazyl Radical TTTA. *Chem. Phys. Lett.* **2004**, *393*, 150–152. [CrossRef]
28. Coomber, A.T.; Beljonne, D.; Friend, R.H.; Brédas, J.L.; Charlton, A.; Robertson, N.; Underhill, A.E.; Kurmoo, M.; Day, P. Intermolecular Interactions in the Molecular Ferromagnetic $NH_4Ni(mnt)_2\cdot H_2O$. *Nature* **1996**, *380*, 144. [CrossRef]
29. Bao, H.; Wang, W.; Li, X.; Liu, X.; Zhang, L.; Yan, X.; Wang, Y.; Wang, C.; Jia, X.; Sun, P.; et al. Interfacial Stress-Modulated Mechanosensitive Upconversion Luminescence of $NaErF_4$ Based Heteroepitaxial Core–Shell Nanoparticles. *Adv. Opt. Mater.* **2022**, *10*, 2101702. [CrossRef]
30. Matsuda, Y.; Orimo, R.; Abe, Y.; Hiraiwa, Y.; Okamura, Y.; Sunami, Y. Pressure-Sensitive Nano-Sheet for Optical Pressure Measurement. *Sensors* **2021**, *21*, 7168. [CrossRef]
31. Chen, Z.; Zhuo, H.; Hu, Y.; Lai, H.; Liu, L.; Zhong, L.; Peng, X. Wood-Derived Lightweight and Elastic Carbon Aerogel for Pressure Sensing and Energy Storage. *Adv. Funct. Mater.* **2020**, *30*, 1910292. [CrossRef]

32. Zhao, S.; Ran, W.; Wang, D.; Yin, R.; Yan, Y.; Jiang, K.; Lou, Z.; Shen, G. 3D Dielectric Layer Enabled Highly Sensitive Capacitive Pressure Sensors for Wearable Electronics. *ACS Appl. Mater. Interfaces* **2020**, *12*, 32023–32030. [CrossRef]
33. Codjovi, E.; Menendez, N.; Jeftic, J.; Varret, F. Pressure and Temperature Hysteresis in the Spin-Transition Solid $Fe(btr)_2(NCS)_2{\cdot}H_2O$, Pure and Diluted in Ni Matrix. *Acad. Sci. Ser. IIC Chem.* **2001**, *4*, 181–188.
34. Linares, J.; Codjovi, E.; Garcia, Y. Pressure and Temperature Spin Crossover Sensors with Optical Detection. *Sensors* **2012**, *12*, 4479–4492. [CrossRef]
35. Gregory, J.W.; Sakaue, H.; Liu, T.; Sullivan, J.P. Fast Pressure-Sensitive Paint for Flow and Acoustic Diagnostics. *Annu. Rev. Fluid Mech.* **2014**, *46*, 303–330. [CrossRef]
36. Peng, D.; Liu, Y. Fast Pressure-Sensitive Paint for Understanding Complex Flows: From Regular to Harsh Environments. *Exp. Fluids* **2020**, *61*, 8. [CrossRef]
37. Zhang, D.; Xu, S.; Zhao, X.; Qian, W.; Bowen, C.R.; Yang, Y. Wireless Monitoring of Small Strains in Intelligent Robots via a Joule Heating Effect in Stretchable Graphene–Polymer Nanocomposites. *Adv. Funct. Mater.* **2020**, *30*, 1910809. [CrossRef]
38. Persano, L.; Dagdeviren, C.; Su, Y.; Zhang, Y.; Girardo, S.; Pisignano, D.; Huang, Y.; Rogers, J.A. High Performance Piezoelectric Devices Based on Aligned Arrays of Nanofibers of Poly(vinylidenefluoride-co-trifluoroethylene). *Nat. Commun.* **2013**, *4*, 1633. [CrossRef]
39. Delley, B. From Molecules to Solids with the $DMol^3$ Approach. *J. Chem. Phys.* **2000**, *113*, 7756–7764. [CrossRef]
40. Perdew, J.P.; Burke, K.; Ernzerhof, M. Generalized Gradient Approximation Made Simple. *Phys. Rev. Lett.* **1996**, *77*, 3865–3868. [CrossRef]
41. Monkhorst, H.J.; Pack, J.D. Special Points for Brillouin-zone Integrations. *Phys. Rev. B* **1976**, *13*, 5188–5192. [CrossRef]
42. Simon, A. Eine Methode zur Untersuchung extrem luftempfindlicher Substanzen mit der Guinier-Methode. *J. Appl. Crystallogr.* **1970**, *3*, 11. [CrossRef]
43. Kremer, R.K.; Loa, I.; Razavi, F.S.; Syassen, K. Effect of pressure on the magnetic phase transition in α'-NaV_2O_5. *Solid State Commun.* **2000**, *113*, 217. [CrossRef]
44. Frisch, M.J.; Trucks, G.W.; Schlegel, H.B.; Scuseria, G.E.; Robb, M.A.; Cheeseman, J.R. *GAUSSIAN 09*; Revision A.02; Gaussian Inc.: Wallingford, CT, USA, 2009.
45. Lu, T.; Chen, F. Multiwfn: A multifunctional wavefunction analyzer. *J. Comput. Chem.* **2012**, *33*, 580–592. [CrossRef]
46. Becke, A.D. Density-functional thermochemistry. I. The effect of the exchange-only gradient correction. *J. Chem. Phys.* **1993**, *98*, 5648–5652. [CrossRef]
47. Lee, C.; Yang, W.; Parr, R.G. Development of the Colle-Salvetti correlation-energy formula into a functional of the electron density. *Phys. Rev. B* **1988**, *37*, 785–789. [CrossRef] [PubMed]
48. Delley, B. Hardness conserving semilocal pseudopotentials. *Phys. Rev. B* **2002**, *66*, 155125. [CrossRef]
49. Delley, B. An all-electron numerical method for solving the local density functional for polyatomic molecules. *J. Chem. Phys.* **1990**, *92*, 508–517. [CrossRef]
50. Tkatchenko, A.; Scheffler, M. Accurate Molecular Van Der Waals Interactions from Ground-State Electron Density and Free-Atom Reference Data. *Phys. Rev. Lett.* **2009**, *102*, 073005. [CrossRef] [PubMed]

magnetochemistry

MDPI

Article

Positive and Negative Exchange Bias in *N*-, *P*- and *Q*-Type Ferri-Magnets of Niccolite Metal Formates $[CH_3NH_2CH_3]_n[Cr^{III}_{1-x}Fe^{III}_xFe^{II}(HCO_2)_6]_n$

Yu Zhou [1], Zhaoquan Yao [1], Na Li [2], Fuchen Liu [1], Jiongpeng Zhao [1,*] and Xianhe Bu [2,3,4,*]

[1] TKL of Organic Solar Cells and Photochemical Conversion, School of Chemistry and Chemical Engineering, Tianjin University of Technology, Tianjin 300384, China; dianthe1999@163.com (Y.Z.); yaozq@email.tjut.edu.cn (Z.Y.); fcliu@tjut.edu.cn (F.L.)
[2] State Key Laboratory of Elemento-Organic Chemistry, College of Chemistry, Nankai University, Tianjin 300071, China; lina@nankai.edu.cn
[3] Frontiers Science Center for New Organic Matter, Nankai University, Tianjin 300071, China
[4] School of Materials Science and Engineering, Tianjin Key Laboratory of Metal and Molecule Based Material Chemistry, Nankai University, Tianjin 300350, China
* Correspondence: zhaojp@tjut.edu.cn (J.Z.); buxh@nankai.edu.cn (X.B.)

Abstract: Exchange bias (EB) materials, whose magnetization curve can shift along the field axis after field cooling, have attracted tremendous attention and play a crucial role in the development of fundamental physics as well as practical applications of magnetization storage. In this work, the *N*-, *P*-, and *Q*-type ferrimagnets of Néel's notation were realized in mixed valence metal formates $[CH_3NH_2CH_3]_n[Cr^{III}_{1-x}Fe^{III}_xFe^{II}(HCO_2)_6]_n$ by altering x, respectively. The positive and negative EB was found in *N*- and *P*-type ferrimagnets. The exchange anisotropy originates from the antiferromagnetic exchange interaction between the uncompensated spin of the host ferrimagnetic lattice and the pinned compensated spin of the antiferromagnetic clusters as a guest, which is rooted in the valence disorder of the iron ions.

Keywords: coordination chemistry; Niccolite metal formate materials; magnetic materials; exchange bias; ferrimagnetic

1. Introduction

The exchange bias (EB) effect, namely a shift of the magnetization curve along the field axis after field cooling, plays a crucial role in the development of fundamental physics as well as practical applications of magnetization storage [1–3]. It is generally believed that the phenomenon of EB originates from the exchange coupling between ferromagnetic (FM) and anti-ferromagnetic (AFM) interfaces, which leads to another magnetic field (H_i) distinct from the applied field (H_a) [4]. Therefore, the attention of the relative researchers has mainly been focused on the construction of the FM/AFM bilayer systems [5]. As a result, in a negative EB field, the magnetization curve shifting in the direction against that of the cooling field is a usual experimental observation despite the nature of the FM–AFM interaction [6].

However, a positive EB phenomenon has been found in some systems, i.e., TM/TMF_2 (TM = transition metal) layers [7], which surpasses the simple picture of mechanisms presented by FM/AFM bilayers [8]. Several obscure theories were developed to comprehend such phenomena [9,10]. However, the complexity of hetero-structural interfaces, such as magnetic anisotropy, roughness, coupling and defect which have an obvious effect on EB, makes it hard to target the origin of the EB in these systems [11]. Moreover, in the conventional materials with TM/TMF_2 layers [7], the positive EB was often found in magnetic materials with negative magnetization, in which the magnetization suffers a reversal from positive value to negative value below the specific temperature (compensated temperature, T_{comp}) [12–14].

In order to reveal the internal logical relation between the negative magnetization and the positive EB, a series of isomorphic crystal complexes $[CH_3NH_2CH_3]_n[Cr^{III}{}_{1-x}Fe^{III}{}_xFe^{II}(HCO_2)_6]_n$ (x = 0.8 for **1**, x = 0.7 for **2**, x = 0.5 for **3**) were synthesized successfully. By adjusting the doping contents of Cr^{III} ions in this system, the magnetization reversals and EB could be controlled simultaneously. During the cooling process, the field-cooled magnetization of **1** and **2** did not increase monotonically with the temperature decreasing; instead, the magnetization of these two samples reduced to a negative value or a small positive value after arriving at the maximum, while for complex **3**, the magnetization had a dramatic increase in a similar temperature range. These results indicate that **1**, **2** and **3** belong to the *N*-, *P*-, and *Q*-type ferrimagnet, respectively, according to Néel's notation [15]. More striking, the directions of EB are highly dependent on thermo-magnetization. Complex **1** exhibits positive EB with negative magnetization, while for **2**, the negative EB behavior can be observed with positive magnetization. The origin of the H_i could be assigned to the magnetic interactions between the uncompensated spin of the host ferrimagnetic matrix and the guest antiferromagnets induced by anti-site defects of Fe^{III} and Fe^{II} ions [4]. Furthermore, we illustrate how the direction of EB is influenced by the sublattice spin configuration in the field cooling magnetization process of the host ferrimagnetic matrix.

2. Materials and Methods

All the chemical reagents for synthesis and analysis were obtained from commercial suppliers and used without further purification. The elemental analyses (C, H, N) were performed on a Perkin-Elmer 240 C elemental analyzer. The powder X-ray diffraction (PXRD) was recorded on a Rigaku D/Max-2500 diffractometer at 50 kV, 40 mA for a Cu-target tube, and a graphite monochromator. The molar ratios of Fe and Cr were determined by inductively coupled plasma (ICP, PerkinElmer Avio 500 ICP-OES). The simulation of the PXRD pattern was carried out by the single-crystal data and diffraction-crystal module of the Mercury (Hg) program available free of charge via the Internet at http://www.iucr.org. The magnetic data were collected by using the crushed crystals of the sample (**1–4**) on a Quantum Design MPMS-XL SQUID magnetometer equipped with a 5 T magnet. The data were corrected using Pascal's constants to calculate the diamagnetic susceptibility and an experimental correction for the sample holder was applied.

Synthesis of $[CH_3NH_2CH_3]_n[Cr^{III}{}_{1-x}Fe^{III}{}_xFe^{II}(HCO_2)_6]_n$

$[CH_3NH_2CH_3]_n[Cr^{III}{}_{0.2}Fe^{III}{}_{0.8}Fe^{II}(HCOO)_6]_n$ (**1**): 0.25 mmol $CrCl_3{\cdot}6H_2O$, 3 mmol $FeCl_3{\cdot}6H_2O$ and 2 mmol $FeCl_2{\cdot}4H_2O$ were dissolved in 10 mL mixture solution containing water, formate acid, and N,N-dimethylformamide (DMF) in a volume ratio of 0.5:1:1. The mixture was sealed in Teflon-lined stainless steel vessel, heated at 120 °C for 2 d under autogenous pressure, and then cooled to room temperature in 36 h. Black hexagonal prism-like crystals of **1** were harvested in about 20% yield based on Cr^{3+} ions. Elemental analysis (%) calcd. for **1**, $C_8H_{14}Cr_{0.2}Fe_{1.8}NO_{12}$ (427.12): C 22.50, H 3.30, N 3.28%; found for **1**: C 22.79, H 3.45, N 3.67%.

$[CH_3NH_2CH_3]_n[Cr^{III}{}_{0.3}Fe^{III}{}_{0.7}Fe^{II}(HCOO)_6]_n$ (**2**): The synthetic procedure of **2** is similar to that of **1**, but with 0.5 mmol $CrCl_3{\cdot}6H_2O$, 3 mmol $FeCl_3{\cdot}6H_2O$ and 2 mmol $FeCl_2{\cdot}4H_2O$. Black hexagonal prism-like crystals of **2** were harvested in about 20% yield based on Cr^{3+} ions. Elemental analysis (%) calcd. for **2**, $C_8H_{14}Cr_{0.3}Fe_{1.7}NO_{12}$ (426.73): C 22.52, H 3.31, N 3.28%; found for **2**: C 22.82, H 3.34, N 3.58%.

$[CH_3NH_2CH_3]_n[Cr^{III}_{0.5}Fe^{III}_{0.5}Fe^{II}(HCOO)_6]_n$ (**3**): The synthetic procedure of **3** is similar to that of **1**, but with 1 mmol $CrCl_3{\cdot}6H_2O$, 3 mmol $FeCl_3{\cdot}6H_2O$ and 2 mmol $FeCl_2{\cdot}4H_2O$. Black hexagonal prism-like crystals of **2** were harvested in about 20% yield based on Cr^{3+} ions. Elemental analysis (%) calcd. for **1**, $C_8H_{14}Cr_{0.5}Fe_{1.5}NO_{12}$ (426.35): C 22.56, H 3.31, N 3.29%; found for **3**: C 22.83, H 3.14, N 3.46%.

$[CH_3NH_2CH_3]_n[Cr^{III}Fe^{II}(HCOO)_6]_n$ (**4**): The block crystals of complex **4** were synthesized by solvothermal reaction. Then, 4 mmol $CrCl_3{\cdot}6H_2O$ and 2 mmol $FeCl_2{\cdot}4H_2O$ were dissolved in 10 mL mixture solution containing water, formate acid, and N,N-dimethylformamide in a volume ratio of about 0.5:1:1. The mixture was sealed in a Teflon-lined stainless steel vessel, heated at 140 °C for 2 d under autogenous pressure, and then cooled to room temperature in 36 h. Rosy color hexagonal prism-like crystals of **4** were harvested in about 20% yield based on Cr^{3+} ion. Elemental analysis (%) calcd. for **4**, $C_8H_{14}NCrFeO_{12}$ (424.04): C 22.66, H 3.33, N 3.30%; found for **4**: C 22.87, H 3.13, N 3.63%.

3. Results and Discussion

3.1. Syntheses and Crystal Structures

Complexes **1–4** were all fabricated by the solvothermal reaction for two days according to our previous report [12,16]. The detailed structure analyses of these complexes are illustrated in Figure 1. The lattice parameters of **1–3** are larger than that of **4** ($[CH_3NH_2CH_3]_n[Cr^{III}Fe^{II}(HCOO)_6]_n$) [17], while smaller than that of $[CH_3NH_2CH_3]_n[Fe^{III}Fe^{II}(HCOO)_6]_n$ [12]. Moreover, in complexes **1–3**, as the contents of the Fe^{III} increase, the *c* and *V* parameters increase slightly. All of this indicates that the doping of Cr^{III} into the lattice of $[CH_3NH_2CH_3]_n[Fe^{III}Fe^{II}(HCOO)_6]_n$ was achieved successfully in complexes **1–3**. As shown in Figure 1a, in the unit cell of these isomorphic complexes, there are two independent metal sites occupied by divalent and trivalent metal ions, respectively. These two kinds of metal centers are both coordinated in octahedral geometry defined by six carboxylate oxygen atoms from anti, anti formate and could be served as $(4^9{\cdot}6^6)$ (yellow polyhedron in Figure 1) and $(4^{12}{\cdot}6^3)$ (dark blue polyhedron in Figure 1) nodes, respectively. In **1–4**, the M-O bond lengths of $(4^9{\cdot}6^6)$ nodes are in the range of 2.1197 (17) Å–2.125 (2) Å, while the bond lengths of M-O in $(4^{12}{\cdot}6^3)$ nodes range from 1.9747 (15) (Å) to 2.005 (2) (Å), which suggests that the trivalent metal (Cr^{III} and Fe^{III}) were mainly located in $(4^{12}{\cdot}6^3)$ nodes and the Fe^{II} ions were in $(4^9{\cdot}6^6)$ nodes. In these networks, the $[CH_3NH_2CH_3]^+$ cations which contained the disorder N atoms were anchored into the cavities formed by metal ions and formates to balance the surplus negative charge of the network.

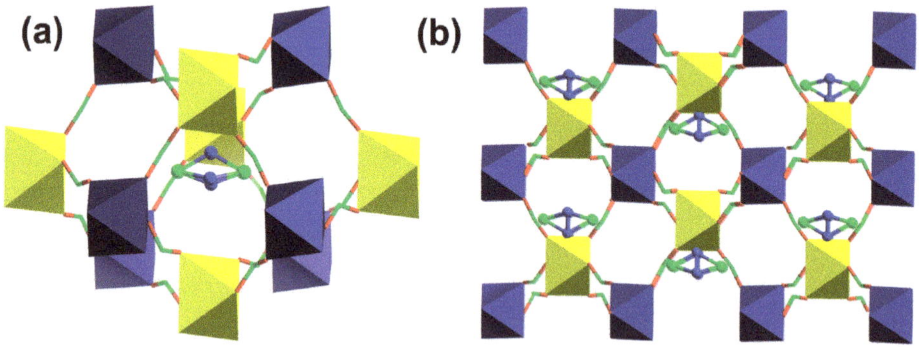

Figure 1. (**a**) The polyhedron view of the cavities in isomorphic complexes **1–4** constructed by M^{III}/Fe^{II} ions and formates; (**b**) The polyhedron view of the two sublattices of M^{III} (blue) and Fe^{II} ions (yellow).

3.2. Magnetic Properties

The phase purity, metal contents, and the valence states of metal ions in **1–4** were confirmed by PXRD, ICP, and XPS, respectively (Figures S1–S3). The magnetic performances were measured by using the crystalline sample of **1–4**. As shown in Figure 2, the magnetic susceptibilities of **1–4** were measured in the range from 300 to 2 K under an applied field of 1 kOe and depicted as the plots of $\chi_m T$ vs. T and χ_m vs. T (inset in Figure 2), in which χ_m is the molar magnetic susceptibility per pair of one high spin Fe^{II} ion and one M^{III} ion. During the cooling process, the value of the $\chi_m T$ of **1–3** decreased gradually until it reached the minimum at *ca.* 37 K, which reveals the overall antiferromagnetic interactions in **1–3** (Figure 2). For **4**, during the cooling process, the $\chi_m T$ curve increases until reaching the maximum of 95.23 cm^3 K mol^{-1} at *ca.* 6.0 K. The magnetic susceptibility data above 50 K of **1–4** are fitted by the modified Curie–Weiss law $\chi_m = \chi_0 + C/(T - \theta)$ giving $\chi_0 = -0.00135$ cm^3 K mol^{-1}, -0.00136 cm^3 K mol^{-1}, -0.00070 cm^3 K mol^{-1}, -0.00026 cm^3 K mol^{-1}, $C = 9.09$ cm^3 K mol^{-1}, 8.53 cm^3 K mol^{-1}, 7.41 cm^3 K mol^{-1} and 5.76 cm^3 K mol^{-1}, θ value = -53.93 K, -48.35 K, -28.14 K and 4.93 K for **1–4** (Figure S4) [18]. The above results illustrate that as the doping content of Cr^{III} increased, the Curie constants of **1–4** decreased and reached the value of 5.68 cm^3 K mol^{-1} (the value expected for magnetically isolated octahedral Cr^{III} and Fe^{II} ions with spin-orbital coupling) in **4**. Moreover, the negative Weiss constants in **1–3** and positive value in **4** suggest the antiferromagnetic coupling in **1–3** and the ferromagnetic interactions in **4**. From **4** to **1**, as the doping content of Cr^{III} decreased, the Weiss constants become more negative but could not reach the value of -54.75 K in $[CH_3NH_2CH_3]_n[Fe^{III}Fe^{II}(HCO_2)_6]_n$. The regular change of the Curie and Weiss constants from **1** to **4** further validates the successful doping of Cr^{III} ions.

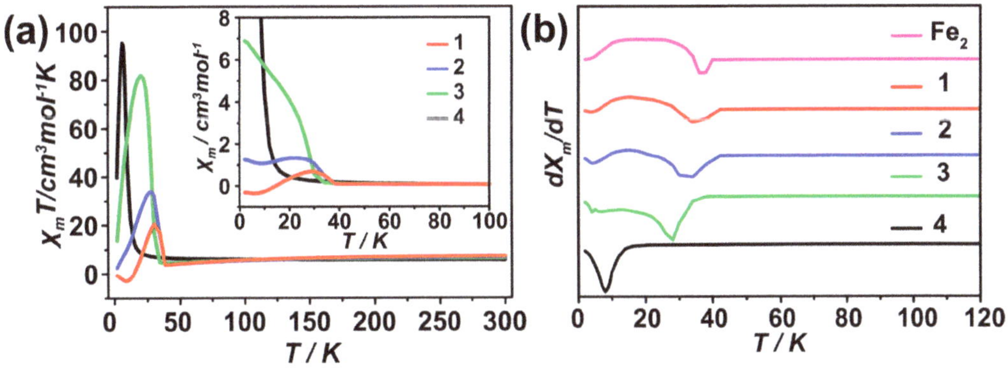

Figure 2. (**a**) The $\chi_m T$ vs. T plots of **1–4** per one high spin Fe^{II} ion and one M^{III} ion under applied field 1000 Oe (inset: the corresponding χ_m vs. T plots); (**b**) The derivative of χ_m vs. T plots of **1–4** and Fe_2 ($[CH_3NH_2CH_3]_n[Fe^{III}Fe^{II}(HCO_2)_6]_n$).

Below 50 K, the χ_m plots of **1** have a similar trend with the isomorphic iron complex [12,19–21]. During the cooling process, the value of χ_m increased rapidly to the maximum of 0.68 cm^3 mol^{-1} at about 30 K and then went down and became negative at the compensation temperature (15 K). The negative value of χ_m disappeared when the applied field was larger than 3 kOe (Figure S5a). The χ_m curves of **2** reached the maximum at 23 K and decreased to the minimum at about 9 K; after that, the value of χ_m increased upon further cooling (Figure 2 and Figure S5b), which is similar to that of **1** under a larger applied field exceeding 3 kOe. The χ_m value of **3** increases monotonically as the temperature decreases (Figure 2 and Figure S5). Thus, based on the temperature dependence of magnetization, **1**, **2**, and **3** belong to Néel's *N*-, *P*- and *Q*-type ferrimagnets, respectively [15]. Moreover, the zero field-cooled magnetization (ZFCM) and field-cooled magnetization (FCM) revealed the phase transition temperatures of **1–4** are 35 K, 33 K, 33 K, and 8 K, respectively (Figure S6). Due to the magnetic interaction between Fe^{III} and Fe^{II} being much stronger than that between Cr^{III} and Fe^{II}, as the doping contents of Cr^{III} increase, the phase-transition temperature of these complexes decreases gradually. Moreover, ZFCM also confirmed the characteristics of *N*-, *P*- and *Q*-type ferrimagnets in **1–3**. The ZFCM and FCM curves of **1** diverge at about 35 K and have a crossover at about 17 K, which is typical behavior of the *N*-type ferrimagnet. On heating, the ZFCM curve of **2** coincides with the curve of FCM at T_N. Compared to ZFCM, the FCMs of complexes **1–4** are quite different, in which the curves of **1** and **2** have a maximum value at 30 K and 23 K, respectively, suggesting the *N* and *P*-type ferrimagnetism again.

The magnetic nature and strength of magnetic interactions between Cr^{III}-Fe^{II} and Fe^{III}-Fe^{II} could be comprehended by the orbital orthogonal theory (Figure 3a) [16]. In **1–3**, the ($4^{12}{\cdot}6^3$) nodes consist of Cr^{III} and Fe^{III} ions, so ferromagnetic (between Cr^{III} and Fe^{II} ions) and antiferromagnetic interactions (between Fe^{III} and Fe^{II} ions) could be found in these complexes, leading to a ferro-ferrimagnetic state in **1–3** (Figure 3b) [22]. Thus, the spins of Fe^{III} ions in **1–3** should be antiparallel to those of ferromagnetic coupled spins of Cr^{III} and Fe^{II}. Due to the magnetic interactions between Fe^{II} and Fe^{III} being stronger than that of Fe^{II} and Cr^{III}, the total magnetization parallel to the Fe^{II} sublattice is weakened by Fe^{III} ions below T_N and enhanced by Cr^{III} ions upon further cooling. Thus, by adjusting the content of Cr^{III} ions (the values of x) in **1–3**, the thermal response of the magnetization could be regulated precisely.

The increase in the χ_m value with a maximum of **1** and **2** could be attributed to the fact that the spins of Fe^{II} ions are ordered firstly and parallel to the applied field at high temperatures above 35 K (T_N) [12]. After that, as the temperature decreased, the magnetization of Fe^{III} was enhanced and the total magnetization decreased upon cooling. As the temperature further decreased, the magnetization of Fe^{III} exceeded the sum of the magnetization of Cr^{III} and ordered Fe^{II}, finally leading to the negative magnetization of **1**. Compared to the Fe^{III} ions, due to the magnetization of ordered Cr^{III} ions being hard to compensate for the gap of Fe^{II} ions, **1** still exhibited negative magnetization in the range of 18 to 2 K. This indicates that restricted by the amount of the doping Cr^{III} ions, the total magnetization of Cr^{III} and Fe^{II} is still smaller than that of Fe^{III} below T_N (Néel temperature). Meanwhile, in **2**, owing to the doping content of Cr^{III} ions increased, the sum of magnetization of Cr^{III} and ordered Fe^{II} could exceed the magnetization of Fe^{III} ions below T_N; thus, no negative magnetization can be found. The contribution of different metal ions to magnetization of **1–4** can also be seen in the plots of the first derivative of χ_m (**1–4** and Fe2) vs. T (Figure 2b). In more detail, for **1** and **2**, the maximum of the derivative appears in 35–10 K, which suggests a decrease in χ_m value in the corresponding temperature range. In **3**, the χ_m plot exhibits a dramatically increasing trend in the range of 35 to 22 K, and after that, the increasing trend slows down until the temperature is below 7 K. Then, the χ_m value increases with a steep slope again during the further cooling process. The two-step increase trend of χ_m being dependent on the decreased temperature is more obvious in Figure 2b, which exhibits two valleys at about 6 K and 30 K. The first dramatic increase trend of χ_m on the cooling process could be attributed to the ferrimagnetic order of Fe^{III} and Fe^{II} spins. With further cooling below T_N, the sum of Fe^{II} magnetization in **3** exceeds that of Fe^{III} spins, and thus, the total magnetization increases constantly. In the range of 7 to 2 K, the order spins of Cr^{III} ions accelerate the raising of magnetization. In comparison, the derivative of χ_m of **4** has only one minimum at about 9 K.

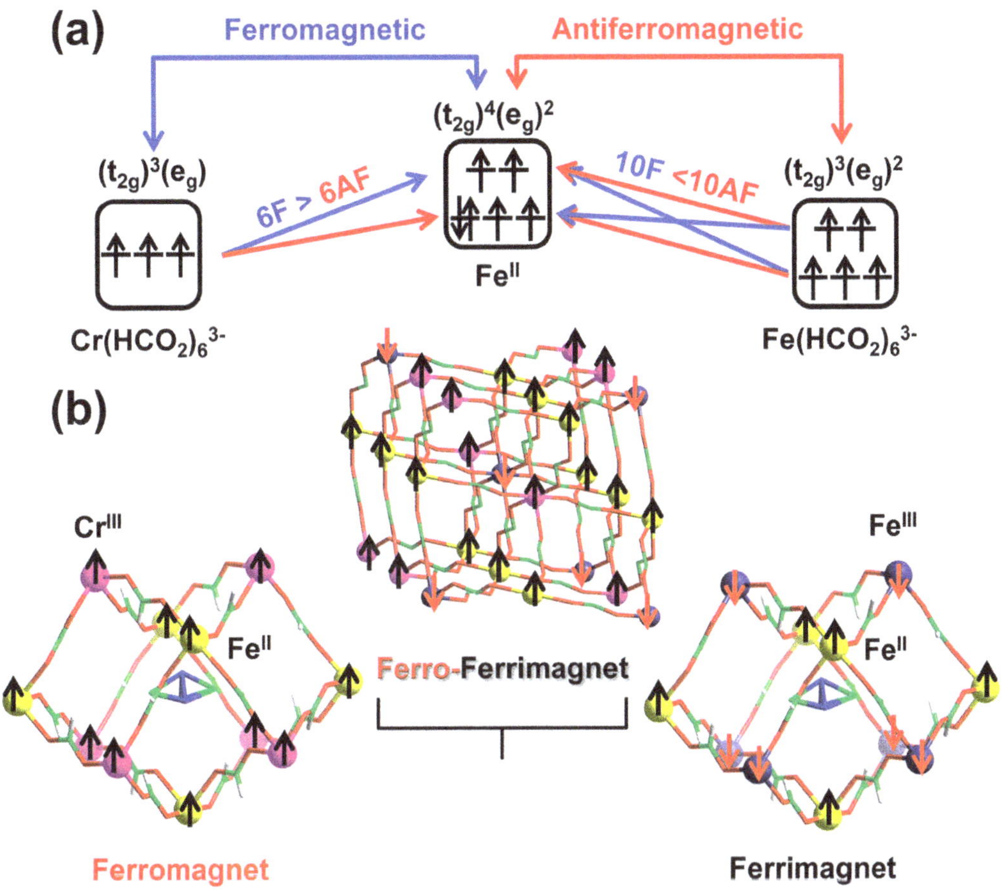

Figure 3. Schematic diagram illustrating the mixed ferro-ferrimagnetism. (**a**) Magnetic interactions between Cr^{III}-Fe^{II} and Fe^{III}-Fe^{II} based on orbital orthogonal theory; (**b**) The ferromagnetic, ferro-ferrimagnetic and ferri-magnetic state.

3.3. Exchange Bias

After that, the hysteresis loops of complexes **1**–**4** after the 1000 Oe field cooling process were collected in the range from −5 T to 5 T (Figure 4). Both **1** and **2** exhibit an asymmetric hysteresis loop between ±5 T. To quantify the shift of the hysteresis loop, we define the values of shifted and average coercive fields as $H_i = (H_{C+} + H_{c-})/2$ and $H_C = (H_{C+} - H_{c-})/2$, in which H_{C+} and H_{c-} are coercivity of right-left branches. The H_i and H_C of **1** is 946 Oe and 5771 Oe under 1 kOe field cooling. Unlike that of **1**, the center of the hysteresis loop of **2** shifts to the negative field with H_i −546 Oe and H_C 7038 Oe. The hysteresis loop of **3** also shifts to the negative field with H_i −98 Oe and H_C 6535 Oe. The hysteresis loop at 2 K after cooling in different fields of **1**–**3** is shown in Figure S7 and the corresponding H_i and H_C are summarized in Figure S8, in which the H_i and H_C tend to decrease slightly with the increase in the cooling field, except for the H_i of **1** under zero-field cooling. Distinct from the above three complexes, the hysteresis loop of **4** is symmetric with a coercivity of 284 Oe and saturated magnetization of about 5.9 $N\beta$ at 5 T.

Since the EB effect was first observed in nanoparticles composed of an FM cobalt core and an AFM cobalt oxide shell, it usually occurs at the interface of composite FM/AFM material heterostructures. In those heterostructures, the spin of the ferromagnetic-like component was pinned by the antiferromagnet through the field cooling process. Compared with the heterogeneous structures, the design and synthesis of single-phase bulk materials with EB effects is more challenging, due to the pinning phenomenon between different magnetic phases or multiple magnetic sublattices being hard to achieve [23]. For example, Ajaya Nayak and colleagues designed a ferrimagnetic Heusler alloy with giant tunable EB by putting ferromagnetic clusters induced by anti-site defects in the antiferromagnetic matrix [4]. In the alloy, the antiferromagnetic matrix acted as a highly anisotropic magnetic moment (μ_i) being responsible for the internal magnetic field (H_i).

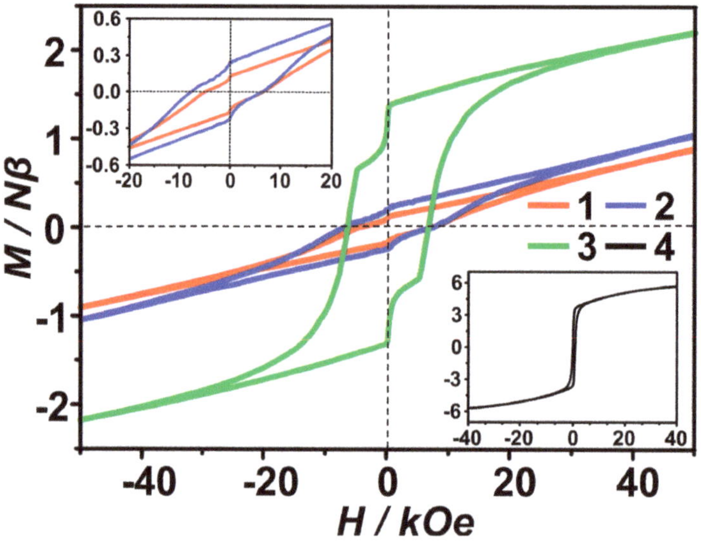

Figure 4. Plots of field-cooled hysteresis loops of **1–4** at 2 K under 1 kOe field cooling.

Distinct from the previous reports, in **1–3**, the anti-site defects induce local antiferromagnetic moments embedded in the ferrimagnetic matrix. On cooling, the spin of Fe^{II} ions became ordered first, which induced the arrangement of the spins of Fe^{III} ions. More important, the spins of the guest antiferromagnetic clusters are also arranged (Figure 5a,b). The spin of the cluster is unchanged regardless of the external magnetic field. Thus, the uncompensated spins suffered antiferromagnetic interactions from the spins of the guest at low field. The coupling between the net magnetization and the guest is equivalent to providing an additional positive magnetic field (H_i). For **1**, below T_{comp} with a low applied field, the uncompensated Fe^{III} lattice is antiparallel to the applied magnetic field. And the antiferromagnetic clusters act as an internal field (H_i) to transfer an antiferromagnetic coupling with Fe^{III} lattice (Figure 5b), like that of **2** and **3** (Figure 5a). However, when the external magnetic field increases, the overall magnetic moments reverse (Figure 5c), but the spins of the antiferromagnetic clusters are unchanged, and thus, the spin of the Fe^{III} lattice (uncompensated magnetic moment) suffers an additional antiferromagnetic effect, which is equivalent to adding an additional negative magnetic field to the spin of Fe^{III} lattice. Therefore, in the process of demagnetization from a positive field, a smaller magnetic field is required, and a larger magnetic field is required during the magnetization process from negative to positive fields (Figure S9 right). In **2** and **3**, the magnetization (Figure 5a) of Cr^{III} and Fe^{II} are larger than that of Fe^{III} ions, and there is no pole reversal like that of **1**. And the uncompensated spin suffers from antiferromagnetic coupling of the clusters, which is equivalent to adding an additional positive magnetic field to the spin of Fe^{II} lattice

(Figure S9 left). In short, the uncompensated spin in **1** and **2** experiences a force antiparallel or parallel to the applied, which provides a positive or negative H_i (Figure 5). Obvious negative exchange bias was also found in **2** under zero-field cooling, which may be caused by the pinning between the host and guest during the magnetization process (Figure S8).

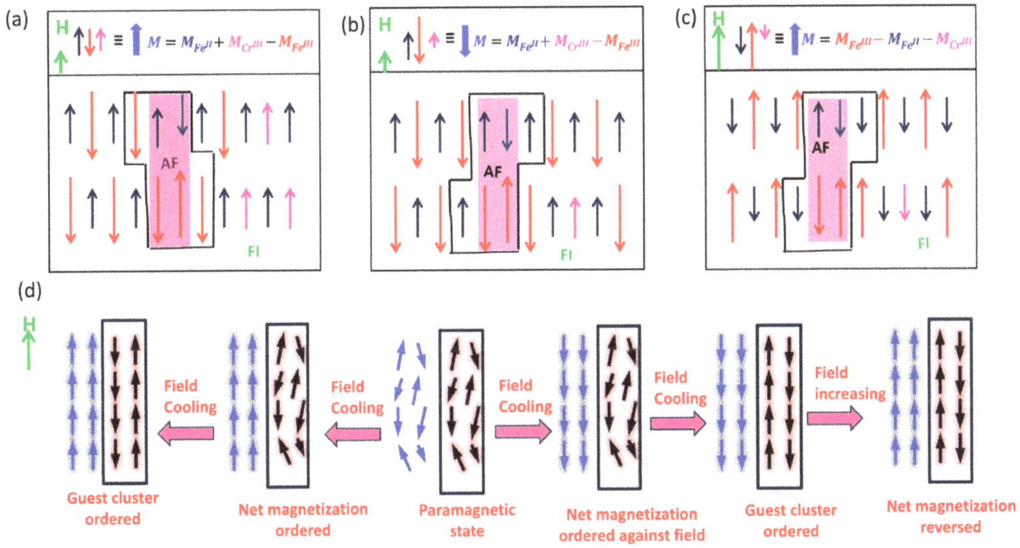

Figure 5. (**a**,**b**) Spin orientation of the host ferrimagnet and the guest antiferromagnetic cluster in *P*-/*Q*- and *N*-type ferrimagnets at low applied fields; (**c**) Spin orientation of that in the *N*-type ferrimagnets at high applied fields; (**d**) Predigested interface spin configuration of the host ferrimagnet (blue arrow) and the guest antiferromagnetic clusters (black arrow) in different conditions.

It is worth noting that the above discussion is only from a perspective of coupling; however, the magnetic anisotropy is also very important in determining the magnetic behavior of *N*-, *P*-type ferrimagnets and EB. In the EB materials, it is generally required that the ferromagnetic contribution part is ordered first. The ordered ferromagnetic induces the orientation of the antiferromagnetic part that does not flip with the external field. In this work, the anisotropy of the Fe^{II} makes it order earlier than that of the Fe^{III} during the cooling process. And as pointed out by Randy S. Fishman, the persistence of negative magnetization in small fields is caused by the spin-orbit energy cost for flipping L of Fe^{II} once it is aligned with the magnetic field [24]. Therefore, the single ion anisotropy of Fe^{II} is an important condition for non-monotonic-increased magnetization cooling and EB. And the coercivity of **2** and **3** is larger than that of **1**, which indicates that the doping of Cr^{III} improved the anisotropy of the system. And it is found that there is a step jump in the hysteresis loop at zero field in **1**–**3**, which is more pronounced in **3**. In a similar complex $[CH_3NH_2CH_3]_n[Fe^{II}(HCOO)_3]_n$, a step jump in the hysteresis loop at zero field was also found [25]. In that complex, the quantum tunneling of magnetization of the single Fe^{II} ions is responsible for this phenomenon, which is due to the magnetic phase separation caused by the competition of the super-exchange interactions conducted with the formate group and the hydrogen bonding. Meanwhile, neutron diffraction studies show that, in $[CH_3NH_2CH_3]_n[Fe^{III}Fe^{II}(HCOO)_6]_n$, in order phase, the spins of Fe^{II} ions are not completely parallel, but have a rotation angle with the *c*-axis, which indicates that the spins of Fe^{II} have canted parallel arrangements. In **1**–**3**, spin arrangements of Fe^{II} ions may also be similar to $[CH_3NH_2CH_3]_n[Fe^{III}Fe^{II}(HCOO)_6]_n$. And thus, at high fields, the canted angles of the spins of Fe^{II} will be suppressed or enhanced by the magnetic field, depending on whether the spins are parallel or antiparallel to the external magnetic field. When demagnetizing to zero field, this angle should return to the situation without an external field, which would lead to a sharp change in the magnetic moment and a jump in the hysteresis loop at zero field.

4. Conclusions

In summary, we established an approach to design single-phase magnetic materials with controllable positive or negative EB by altering spin configuration in the field cooling magnetization process of the host ferrimagnetic matrixes. The exchange anisotropy could be attributed to the coupling effect between the guest intrinsic anti-site defects antiferromagnetic clusters and the host ferrimagnetic matrix. By controlling the content of the Cr^{III} ions, a series of ferro-ferrimagnets was realized to give *N*-, *P*- and *Q*-type ferrimagnets. Both the magnetizations of the *N*- and *P*-type ferrimagnets do not increase monotonically with temperature decreasing and exhibit the maximum at low temperatures. However, the insight mechanism of these two ferrimagnets is fundamentally different. In the *N*-type, the sublattice with smaller magnetization order is ordered quickly during the cooling process, which leads to a spin reversal at a high applied field; thus, the material exhibits the positive EB. In the *P*-type ferrimagnet, the sublattice with similar magnetization order is ordered slowly upon cooling, and thus, no spin reversals could be observed, leading to the negative EB. All these results reveal the internal logic of the negative magnetization and positive EB. Furthermore, it provided a potential way to achieve positive or negative EB magnetic materials by controlling the spin configurations of the host ferrimagnetic matrix. This work not only provides a feasible method for the design and synthesis of EB materials but also shows that the simple and direct structure design nature of the molecular-based materials can bring new possibilities for assembling specific magnetic materials and understanding their mechanism.

Supplementary Materials: The following supporting information can be downloaded at: https://www.mdpi.com/article/10.3390/magnetochemistry11020010/s1, Table S1: Crystal data and structure refinements for **1–4**; Table S2: Selected bond lengths for complexes **1–4**; Figure S1: The powder X-ray diffraction (PXRD) patterns of **1–4**; Figure S2: the high-resolution XPS spectra of the Cr 2p region for **1–3**; Figure S3: the high-resolution XPS spectra of the Fe 2p region for **1–3**; title; Figure S4: The Curie–Weiss plots of **1–4**; Figures S5–S8: The magnetic measurements; Figure S9: Schematic diagram of the interface host uncompensated spin and the guest antiferromagnetic spin during the magnetization reversal with external field. See refs [26–29].

Author Contributions: Conceptualization, J.Z. and X.B.; methodology, Y.Z. and J.Z.; software, Y.Z., Z.Y. and J.Z.; validation, Z.Y., J.Z. and F.L.; formal analysis, Y.Z. and J.Z.; investigation, Y.Z., N.L. and J.Z.; resources, J.Z. and X.B.; data curation, Z.Y., N.L. and J.Z.; writing—original draft preparation, Y.Z. and J.Z.; writing—review and editing, Z.Y., N.L. and J.Z.; visualization, Z.Y.; supervision, X.B.; project administration, J.Z. and X.B.; funding acquisition, J.Z. and X.B. All authors have read and agreed to the published version of the manuscript.

Funding: This research was supported by the NSFC of China (22271217 and 22035003).

Data Availability Statement: The original contributions presented in this study are included in the article and Supplementary Materials. Further inquiries can be directed to the corresponding authors.

Conflicts of Interest: The authors declare no conflicts of interest.

Abbreviations

The following abbreviations are used in this manuscript:

EB	Exchange Bias
FM	Ferromagnetic
AFM	Anti-ferromagnetic
TM	Transition metal
T_{comp}	Compensated temperature
PXRD	Powder X-ray diffraction
ICP	Inductively coupled plasma
XPS	X-ray photoelectron spectroscopy
ZFCM	Zero field cooled magnetization
FCM	Field cooled magnetization

References

1. Nogués, J.; Schuller, I.K. Exchange bias. *J. Magn. Magn. Mater.* **1999**, *192*, 203–232. [CrossRef]
2. Wu, S.M.; Cybart, S.A.; Yu, P.; Rossell, M.D.; Zhang, J.X.; Ramesh, R.; Dynes, R.C. Reversible electric control of exchange bias in a multiferroic field-effect device. *Nat. Mater.* **2010**, *9*, 756–761. [CrossRef] [PubMed]
3. Meiklejohn, W.H.; Bean, C.P. A new magnetic anisotropy. *IEEE Trans. Magn.* **2001**, *37*, 3866–3876. [CrossRef]
4. Nayak, A.K.; Nicklas, M.; Chadov, S.; Khuntia, P.; Shekhar, C.; Kalache, A.; Baenitz, M.; Skourski, Y.; Guduru, V.K.; Puri, A.; et al. Design of compensated ferrimagnetic Heusler alloys for giant tunable exchange bias. *Nat. Mater.* **2015**, *14*, 679–684. [CrossRef] [PubMed]
5. Kuch, W.; Chelaru, L.I.; Offi, F.; Wang, J.; Kotsugi, M.; Kirschner, J. Tuning the magnetic coupling across ultrathin antiferromagnetic films by controlling atomic-scale roughness. *Nat. Mater.* **2006**, *5*, 128–133. [CrossRef]
6. Stamps, R.L. Mechanisms for exchange bias. *J. Phys. D Appl. Phys.* **2000**, *33*, R247–R268. [CrossRef]
7. Nogués, J.; Lederman, D.; Moran, T.J.; Schuller, I.K. Positive exchange bias in FeF_2-Fe bilayers. *Phys. Rev. Lett.* **1996**, *76*, 4624–4627. [CrossRef] [PubMed]
8. Gilbert, D.A.; Olamit, J.; Dumas, R.K.; Kirby, B.J.; Grutter, A.J.; Maranville, B.B.; Arenholz, E.; Borchers, J.A.; Liu, K. Controllable positive exchange bias via redox-driven oxygen migration. *Nat. Commun.* **2016**, *7*, 11050. [CrossRef]
9. Koon, N.C. Calculations of Exchange Bias in Thin Films with Ferromagnetic/Antiferromagnetic Interfaces. *Phys. Rev. Lett.* **1997**, *78*, 4865–4868. [CrossRef]
10. Hong, T.M. Simple mechanism for a positive exchange bias. *Phys. Rev. B* **1998**, *58*, 97–100. [CrossRef]
11. Kiwi, M. Exchange bias theory. *J. Magn. Magn. Mater.* **2001**, *234*, 584–595. [CrossRef]
12. Zhao, J.P.; Hu, B.W.; Lloret, F.; Tao, J.; Yang, Q.; Zhang, X.F.; Bu, X.H. Magnetic Behavior Control in Niccolite Structural Metal Formate Frameworks $[NH_2(CH_3)_2][Fe^{III}M^{II}(HCOO)_6]$ (M = Fe, Mn, and Co) by Varying the Divalent Metal Ions. *Inorg. Chem.* **2010**, *49*, 10390–10399. [CrossRef] [PubMed]
13. Yoshii, K. Positive exchange bias from magnetization reversal in $La_{1-x}Pr_xCrO_3$ ($x \sim 0.7$–0.85). *Appl. Phys. Lett.* **2011**, *99*, 142501. [CrossRef]
14. Bora, T.; Ravi, S. Sign reversal of magnetization and tunable exchange bias field in $NdCr_{1-x}Fe_xO_3$ (x = 0.05–0.2). *J. Magn. Magn. Mater.* **2015**, *386*, 85–91. [CrossRef]
15. Néel, L. Propriétés magnétiques des ferrites; ferrimagnétisme et antiferromagnétisme. *Ann. Phys.* **1948**, *12*, 137–198. [CrossRef]
16. Zhao, J.-P.; Han, S.-D.; Liu, F.-C. Tunable Ferromagnetic Strength in Niccolite Structural Heterometallic Formate Framework Based on Orthogonal Magnetic Orbital Interactions. *Inorg. Chem.* **2019**, *58*, 1184–1190. [CrossRef] [PubMed]
17. Yang, L.; Li, J.; Pu, T.C.; Kong, M.; Zhang, J.; Song, Y. Study of the relationship between magnetic field and dielectric properties in two ferromagnetic complexes. *RSC. Adv.* **2017**, *7*, 47913–47919. [CrossRef]
18. Pramanik, P.; Ghosh, S.; Yanda, P.; Joshi, D.C.; Pittala, S.; Sundaresan, A.; Mishra, P.K.; Thota, S.; Seehra, M.S. Magnetic ground state, field-induced transitions, electronic structure, and optical band gap of the frustrated antiferromagnet $GeCo_2O_4$. *Phys. Rev. B* **2019**, *99*, 134422. [CrossRef]
19. Hagen, K.S.; Naik, S.G.; Huynh, B.H.; Masello, A.; Christou, G. Intensely Colored Mixed-Valence Iron(II) Iron(III) Formate Analogue of Prussian Blue Exhibits Néel N-Type Ferrimagnetism. *J. Am. Chem. Soc.* **2009**, *131*, 7516–7517. [CrossRef]
20. Tian, Y.-Q.; Yao, S.-Y.; Guo, Y.-J.; Song, Y.; Zhang, G. Mixed Valence Iron Formate $\{[NH_4][Fe^{II}Fe^{III}(CHOO)_6]\}_\infty$: A Molecule-Based Ferrimagnet Showing Temperature Compensation Behavior. *Chin. J. Inorg. Chem.* **2010**, *26*, 385–390.
21. Mączka, M.; Kucharska, E.; Gągor, A.; Pikul, A.; Hanuza, J. Synthesis, magnetic and vibrational properties of two novel mixed-valence iron(II)-iron(III) formate frameworks. *J. Solid State Chem.* **2018**, *258*, 163–169. [CrossRef]

22. Ohkoshi, S.; Iyoda, T.; Fujishima, A.; Hashimoto, K. Magnetic properties of mixed ferro-ferrimagnets composed of Prussian blue analogs. *Phys. Rev. B* **1997**, *56*, 11642–11652. [CrossRef]
23. Sun, Y.; Cong, J.Z.; Chai, Y.S.; Yan, L.Q.; Zhao, Y.L.; Wang, S.G.; Ning, W.; Zhang, Y.H. Giant exchange bias in a single-phase magnet with two magnetic sublattices. *Appl. Phys. Lett.* **2013**, *102*, 172406. [CrossRef]
24. Fishman, R.S.; Reboredo, F.A. Magnetic anisotropy in the Fe(II)Fe(III) bimetallic oxalates. *Phys. Rev. B* **2008**, *77*, 144421. [CrossRef]
25. Tian, Y.; Wang, W.; Chai, Y.; Cong, J.; Shen, S.; Yan, L.; Wang, S.; Han, X.; Sun, Y. Quantum Tunneling of Magnetization in a Metal-Organic Framew. *Phys. Rev. Lett.* **2014**, *112*, 017202. [CrossRef]
26. CrysAlisPro 1.171.39.7e, Empirical Absorption Correction Using Spherical Harmonics, Implemented in SCALE3 ABSPACK Scaling Algorithm; Rigaku Oxford Diffraction, 2015.
27. Sheldrick, G.M. SHELXT-Integrated space-group and crystal-structure determination. *Acta Crystallogr. Sect. A: Found. Adv.* **2015**, *71*, 3–8. [CrossRef] [PubMed]
28. Sheldrick, G. Crystal structure refinement with SHELXL. *Acta Crystallogr. Sect. C Struct. Chem.* **2015**, *71*, 3–8. [CrossRef] [PubMed]
29. Dolomanov, O.V.; Bourhis, L.J.; Gildea, R.J.; Howard, J.A.K.; Puschmann, H. OLEX2: A complete structure solution, refinement and analysis program. *J. Appl. Crystallogr.* **2009**, *42*, 339–341. [CrossRef]

magnetochemistry

MDPI

Article

Magnetic Anisotropy and Slow Magnetic Relaxation in Two Mononuclear Octahedral Cobalt(II) Complexes

Hui-Hui Cui, Dou-Zun Wang, Shixiang Li, Leixin Wang, Xinrui Yu, Xiancong Liu, Jin Wang *, Miao Wang * and Yanfeng Tang *

School of Chemistry and Chemical Engineering, Nantong University, Nantong 226019, China; cuihuihui@ntu.edu.cn (H.-H.C.); wangdz360@163.com (D.-Z.W.); 18537737153@163.com (S.L.); 13939987081@163.com (L.W.); 18382082524@163.com (X.Y.); 16637336671@163.com (X.L.)

* Correspondence: wangjin110@ntu.edu.cn (J.W.); wangmiao@ntu.edu.cn (M.W.); tangyf@ntu.edu.cn (Y.T.)

Abstract: Two mononuclear octahedral Co(II) complexes, $[Co(L)X_2]$ (L = 1-(prop-2-en-1-yl)-1*H*-imidazole, X = NCS^- (**1**) and $NCSe^-$ (**2**)), have been synthesized and characterized. The central Co(II) ions in two complexes adopt an octahedral geometry, coordinated by four N atoms from the ligand and two N atoms from the anion. Direct-current magnetic data revealed large easy-plane magnetic anisotropy in both **1** and **2**. Dynamic magnetic measurements demonstrated that **1** and **2** display field-induced slow magnetic relaxation. For **1** and **2**, the Raman mechanism is found to the dominant process in the whole temperature range. Compared to **1**, the magnetic relaxation of **2** is faster, likely due to the presence of the hydrogen bonding system in **2**.

Keywords: cobalt(II) complexes; magnetic anisotropy; slow magnetic relaxation

1. Introduction

Single-molecule magnets (SMMs) are a fascinating class of materials which can retain magnetization after removing the external magnetic field due to the existence of an energy barrier. SMMs derive their magnetic properties from individual metal ions. This unique characteristic makes them particularly interesting for applications in high-density data storage [1], quantum computing [2] and molecular electronics [3]. Single-ion magnets (SIMs) are a unique subclass of SMMs [4,5]. The simplicity of SIMs, which have only one metal center, facilitates the study of their magneto-structural relationships. Additionally, their design flexibility allows for precise optimization of magnetic properties by adjusting ligands and coordination environments. The molecular magnetism in SIMs originates from the energy barrier (U) that prevents the reversal of magnetization between the 'spin up' (M_S = +S) and 'spin down' (M_S = −S) states. This barrier depends on the ground spin state (S) and magnetic anisotropy (D) with $U = |D|S^2$ for integer S and $U = |D|(S^2 - 1/4)$ for non-integer S [1].

Since Ishikawa et al. reported the first SIM based on the lanthanide ion [6], numerous mononuclear compounds containing lanthanide ions have been documented. Additionally, several SIMs including 3d-transition metals such as Mn(III) [7], Fe(II) [8], Co(II) [9–14] and Ni(I) [15] have been found to exhibit slow magnetic relaxation. Experimental and theoretical studies suggest that Co(II) ions with high spin are excellent candidates for constructing SIMs. This system can exhibit significant magnetic anisotropy with a flexible zero-field splitting (ZFS) parameter by modulating the polyhedron geometry [16–19].

For hexa-coordinate high spin Co(II) complexes with SIM properties, both negative and positive magnetic anisotropies can be observed [20–26]. Of these, Co(II) complexes with trigonal prismatic geometry usually show negative axial magnetic anisotropy. By employing a macrocyclic ligand, Novikov et al. reported easy-axis magnetic anisotropy for a cobalt(II) cage [22]. In contrast, most Co(II) complexes with distorted octahedral geometry typically exhibit high easy-plane magnetic anisotropy due to the unquenched orbital contribution of the t_{2g} level and low-lying excited states [16]. For example, the first octahedral Co(II)-SIM, the *cis*-[Co(dmphen)$_2$(NCS)$_2$] complex (dmphen = 2,9-dimethyl-1,10-phenanthroline) reported by Vallejo et al., exhibits positive magnetic anisotropy with $D = +98\ cm^{-1}$, which results in field-induced slow magnetic relaxation [20]. Indeed, the mechanism of slow relaxation in positive D systems remains incompletely understood, despite attempts to explain it [21].

The above results clearly indicate that magnetic anisotropy can be modulated by changing the coordination environment of the Co(II) complexes. For most reported hexa-coordinate Co(II)-SIMs, bidentate or multidentate ligands are utilized to introduce distortions due to the chelate effect. Conversely, certain monodentate ligands containing N atoms can provide a weak ligand field, resulting in significant magnetic anisotropy for Co(II)-SIMs [26]. To explore this further, we synthesized two hexa-coordinate Co(II) complexes in this study using a monodentate nitrogen-containing ligand, 1-(prop-2-en-1-yl)-1*H*-imidazole. Both complexes exhibit distorted octahedral geometry. For **1**, no significant intermolecular interactions are observed, while for **2**, weak C–H···Se hydrogen bonding interactions link the molecules, forming a supermolecular chain. Magnetic measurements revealed significant easy-plane magnetic anisotropy, with D values of +70.4(7) cm^{-1} for **1** and +64.0(2) cm^{-1} for **2**, respectively. Additionally, both complexes display field-induced SIM behavior, with the Raman mechanism identified as the dominant process in both cases.

2. Materials and Methods

2.1. Materials

All chemicals were purchased from commercial sources (J&K Scientific, Beijing, China) and used as received without additional purification. Powder X-ray diffraction (PXRD) data for the polycrystalline samples were recorded at room temperature on a Bruker D8 ADVANCE diffractometer (Bruker, Berlin, Germany) in a 2θ range of 5.0–50.0°. The PXRD patterns verified the purity of the samples used in the magnetic measurements, consistent with the simulated data derived from single-crystal X-ray diffraction, as presented in Figures S1 and S2 of the Supplementary Materials. Elemental analyses were carried out on an Elementar Vario ELIII elemental analyzer.

2.2. Synthesis of [Co(L)(NCS)$_2$] (1)

$CoCl_2$ (0.065 g, 0.5 mmol) and KSCN (0.097 g, 1 mmol) were dissolved in 10 mL of ethanol and stirred for 30 min. Then, another 5 mL of dichloromethane solution containing L (0.18 g, 1 mmol) was added. The solution was stirred for 1 h and filtered. The diffusion of diethyl ether into the above solution over one week gave red crystals of **1** at 60 %. Elemental analysis (%) calcd. for $C_{26}H_{32}CoN_{10}S_2$: C, 51.39; H, 5.31; N, 23.05. Found: C, 51.37; H, 5.33; N, 23.06.

2.3. Synthesis of [Co(L)(NCSe)$_2$] (2)

Red crystals of complex **2** were synthesized using a procedure similar to that of complex **1**, except that an equivalent amount of KSeCN was used in place of KSCN. Yield: 63%. Elemental analysis (%) calcd. for $C_{26}H_{32}CoN_{10}Se_2$: C, 44.52; H, 4.60; N, 19.97. Found: C, 44.51; H, 4.63; N, 19.94.

2.4. X-Ray Single-Crystal Structure Determination

Single-crystal X-ray diffraction data for **1** and **2** were obtained at 296 K using a Bruker APEX II diffractometer (Bruker, Germany) with a CCD area detector and Mo Kα radiation (λ = 0.71073 Å) [27]. The APEXII program was utilized to determine unit cell parameters and collect the data. Integration of the data was performed using the SAINT program [28], applying corrections for Lorentz factor and polarization. Absorption corrections were processed with SADABS [29]. The structures were solved via the direct method, and refinements were conducted with full-matrix least-squares on F^2 using the SHELXTL software package [30]. Non-hydrogen atoms were refined anistropically. The hydrogen atoms were positioned based on calculations and refined isotropically, with their vibration parameters linked to their associated non-hydrogen atoms.

2.5. Magnetic Measurements

Magnetic susceptibility measurements for complexes **1** and **2** were carried out using a Quantum Design SQUID VSM magnetometer (Quantum Design, San Diego, CA, USA) on finely ground powders obtained from single crystals. Direct-current (dc) susceptibility data were recorded over a temperature range of 2–300 K under dc fields up to 70 kOe. Alternating current (ac) susceptibility measurements were performed under different external fields with frequencies ranging from 1 to 1000 Hz. The magnetic susceptibility data were corrected for the diamagnetic contributions of the sample holder and the samples themselves by using Pascal's constants [31].

3. Results and Discussion

3.1. Structural Description

The molecular structures of complexes **1** and **2** were determined through single-crystal X-ray diffraction analysis. The crystallographic and structure refinement data are listed in Table 1. The selected bond lengths and angles are presented in Table 2.

Table 1. Summary of crystallographic data for complexes **1** and **2**.

Compound	1	2
Molecular formular	$C_{26}H_{32}CoN_{10}S_2$	$C_{26}H_{32}CoN_{10}Se_2$
M (g mol^{-1})	607.66	701.46
Temperature (K)	296	296
Crystal system	Triclinic	Triclinic
Space group	*P*-1	*P*-1
a/Å	8.875 (4)	8.9168 (16)
b/Å	9.620 (4)	9.7118 (17)
c/Å	10.632 (4)	10.6394 (19)
α (°)	69.923 (7)	107.091 (3)
β (°)	65.435 (6)	113.791 (3)
γ (°)	86.636 (6)	94.205 (3)
V/Å^3	771.6 (6)	786.2 (2)
Z	1	1
ρ_{calc}, g/cm^3	1.308	1.482
μ/mm^{-1}	0.724	2.893
F (000)	317.0	353.0
θ range [°]	4.5/55.176	6.47/55.372
Reflns collected	4743	4896
R_{int}	0.0523	0.1411

Table 1. *Cont.*

Compound	1	2
Indep. reflns	3388	3508
Data/restr./paras	3388/0/186	3508/0/178
Goodness-of-fit on F^2	1.216	1.057
R_1, wR_2 [$I > 2\sigma(I)$] [a]	0.0410/0.1155	0.0676/0.1698
R_1, wR_2 [all data] [a]	0.0459/0.1277	0.0776/0.1794

[a] $R_1 = \Sigma||F_o| - |F_c||/\Sigma|F_o|$; $wR_2 = [\Sigma[w(F_o^2 - F_c^2)^2]/\Sigma[w(F_o^2)^2]]^{1/2}$.

Table 2. Selected bond lengths (Å) and angles (°) for complexes **1** and **2**.

1		2	
Co1–N2^1	2.1656 (18)	Co1–N2^1	2.163 (3)
Co1–N2	2.1656 (18)	Co1–N2	2.163 (3)
Co1–N4	2.140 (2)	Co1–N4	2.131 (3)
Co1–N4^1	2.1401 (19)	Co1–N4^1	2.131 (3)
Co1–N3	2.163 (2)	Co1–N3	2.149 (3)
Co1–N3^1	2.163 (2)	Co1–N3^1	2.149 (3)
N2^1–Co1–N2	180.0	N2^1–Co1–N2	180.00 (12)
N4–Co1–N2	87.23 (7)	N4–Co1–N2	92.70 (11)
N4^1–Co1–N2	92.77 (7)	N4^1–Co1–N2	87.30 (11)
N2–Co1–N3	91.29 (8)	N2–Co1–N3	88.81 (12)
N4–Co1–N4^1	180.0	N4–Co1–N4^1	180.0
N3^1–Co1–N3	180.0	N3^1–Co1–N3	180.0

Both compounds are crystalized in the triclinic system with the *P*-1 space group. The structures of both complexes are highly similar. As shown in Figure 1, the asymmetric unit of complexes **1** and **2** each contain one crystallographic independent Co^{2+} ion, four L ligands and two pseudohalide coligands. The cobalt center in both complexes adopts a hexa-coordinate geometry featuring a [CoN_6] coordination mode. The average Co–N bond length is about 2.16 Å for complex **1** and 2.15 Å for complex **2**. The Co–N_{SCN} bond length in **1** (~2.16 Å) is slightly longer than the Co–N_{SeCN} bond length in **2** (~2.14 Å). Both the average Co-N bond length and the Co–N_X (X = SCN or SeCN) bond lengths in **1** and **2** are slightly longer than those reported for other mononuclear octahedral Co(II) complexes with imidazole ligands (Table S1). In **1** and **2**, the angles formed by two diagonal atoms and the central cobalt ion are equal to 180°, indicating that the geometry of both complexes closely approaches an ideal octahedral polyhedron. In both **1** and **2**, the CoN_6 unit is slightly compressed along the N4–Co1–N4^1 bond, which will be referred to as the *Z*-axis. Additionally, a geometric analysis was conducted using the SHAPE program to evaluate deviations from ideal symmetry [32,33]. The deviation values from a perfect octahedron for most octahedral Co(II) complexes with imidazole ligands, as summarized in Table S1, are all relatively small. For complexes **1** and **2**, the values are 0.053 and 0.050 (Table S2), respectively, indicating a slight distortion from the ideal geometry.

It is well established that intermolecular interactions have a substantial impact on the magnetic behavior of mononuclear complexes. A detailed analysis of the crystal packing in **1** and **2** shows a considerable difference between them. As illustrated in Figures S3 and S4, no significant intermolecular interactions are observed in **1**, while in **2**, the molecules form supermolecular chains along the *b*-axis through weak C–H···Se hydrogen bonding interactions. The shortest distances between neighboring Co(II) ions in complexes **1** and **2** are 8.875 Å and 8.917 Å, respectively (Figures S5 and S6).

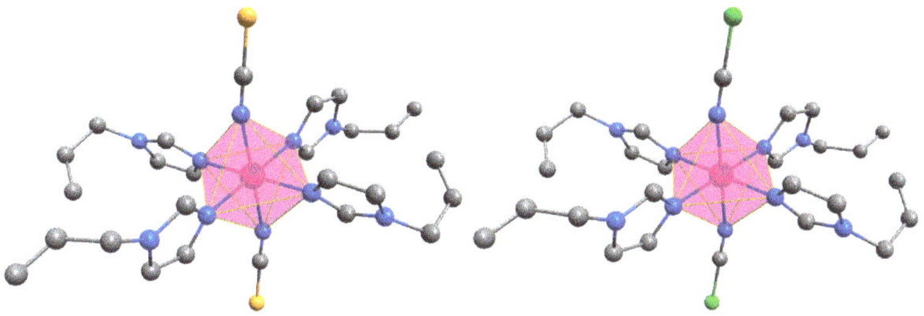

Figure 1. Coordination geometry of **1** (**left**) and **2** (**right**). Pink, blue, yellow, green and gray spheres represent Co, N, S, Se and C atoms, respectively.

The IR spectra confirm the proposed structures of the complexes (Figure S7). The IR spectra revealed a strong absorption band at 1520 cm^{-1} in both **1** and **2**, which were assigned to the ν (C=N) of the imidazole ring [34]. The vibration of $\nu_{C\equiv N}$ was observed around 2085 cm^{-1} in both complexes, confirming the presence of the pseudohalide ion [35].

3.2. Static Magnetic Properties

Variable-temperature dc magnetic susceptibility measurements were performed on polycrystalline samples of **1** and **2**. As shown in Figures 2 and S8, at room temperature, the $\chi_M T$ values for **1** and **2** are 3.09 and 3.17 cm^3 K mol^{-1}, respectively. These values exceed the theoretical spin-only value of 1.875 cm^3 K mol^{-1} for a mononuclear high-spin Co(II) center with $S = 3/2$ and $g = 2.0$. Comparable $\chi_M T$ values have been observed in previously studied Co(II) complexes featuring an octahedral coordination geometry [36–39]. For complexes **1** and **2**, the $\chi_M T$ values show a smooth decrease as the temperature decreases, remaining stable until around 100 K. Below this point, they drop sharply, reaching final values of 1.81 cm^3 K mol^{-1} at 2 K and 1.96 cm^3 K mol^{-1} at 2.5 K, respectively. This reduction is most likely due to the spin–orbit coupling effects. Moreover, the field-dependent magnetizations of both complexes were measured from 0 to 7 T (Figures 2 and S8).

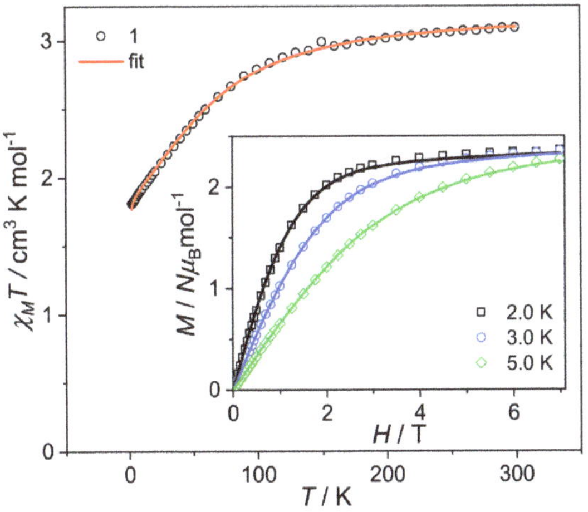

Figure 2. The dc susceptibility data of **1**. Solid lines indicate the best fits with the PHI program [40].

To investigate the magnitude and sign of the anisotropy parameter, the magnetic susceptibilities and field-dependent magnetization data were analyzed simultaneously using the PHI program [40] according to the following spin Hamiltonian [Equation (1)]:

$$H = D(\hat{S}_z^2 - S(S+1)/3) + E(\hat{S}_x^2 - \hat{S}_y^2) + \mu_B g \hat{S} \cdot H \quad (1)$$

where D and E are the axial and rhombic ZFS parameters, respectively. The optimal parameters are summarized in Table 3. Although the derived magnitudes of the ZFS parameters are not highly precise, the large positive D values (+70.4(7) cm^{-1} for **1**; +64.0(2) cm^{-1} for **2**) have been definitively determined, indicating the easy-plane anisotropy for both complexes.

Table 3. The fitting results of the magnetization data by the PHI program [40] for **1** and **2**.

	Fittings of the dc Magnetic Data with Equation (1)				
	D (cm^{-1})	$\lvert E \rvert$ (cm^{-1})	$g_{x,y}$	g_z	zj (cm^{-3})
1	+70.4(7)	0 (1)	2.53 (1)	2.74 (1)	−0.013 (2)
2	+64.0(2)	0 (8)	2.60 (1)	2.66 (1)	/

3.3. *Dynamic Magnetic Properties*

AC susceptibility measurements were conducted on powdered samples of **1** and **2**. Figure S9 illustrates the frequency-dependent behavior of the out-of-phase component of the AC susceptibility at 1.8 K under different external dc fields. No slow magnetic relaxation was observed under a zero dc field for both **1** and **2**. Under applied magnetic fields, a peak in χ'' initially shifted to lower frequencies until reaching 1000 Oe for **1** and 2000 Oe for **2**, after which it began to shift to higher frequencies. To determine the optimal applied dc field, the plot of ln(τ) vs. H was derived from the variable-frequency AC susceptibility data collected under different dc fields (Figure S10). It can be observed that when the applied dc field is 1000 Oe for **1** and 2000 Oe for **2**, the ln(τ) reaches its maximum value. Therefore, 1000 Oe and 2000 Oe were identified as the optimal fields for observing slow magnetic relaxation in **1** and **2**, respectively. As shown in Figures 3 and S11–S14, both complexes exhibit frequency and temperature dependence in their χ'' and χ' signals.

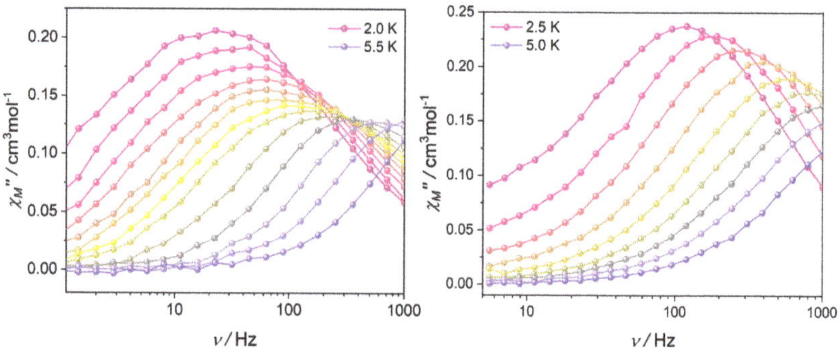

Figure 3. Frequency dependence of the out-of-phase ac susceptibility under an applied dc field of 1000 Oe for **1** (**left**) and 2000 Oe for **2** (**right**). The solid lines are guides for the eye.

Assuming that the Raman and Orbach processes are considered a constant $k(T)$ due to their weak field-dependence, the ln(τ) vs. H curves of **1** and **2** were fitted using Equation (2):

$$\tau^{-1} = AH^4T + \frac{B_1}{1 + B_2H^2} + k(T) \quad (2)$$

where the first term represents the direct process, the second term corresponds to the quantum tunneling of magnetization (QTM), and the last term accounts for the combination of Raman and Orbach processes. The fitting curve agrees well with the experimental data, giving reasonable parameters, which are listed in Table 4. As illustrated in Figure 4a,b, the QTM process dominates the magnetic relaxation for both **1** and **2** at low fields. However, at the optimal field (1000 Oe for **1** and 2000 Oe for **2**), the temperature-dependent combination process ($k(T)$) becomes the prevailing mechanism. It is worth noting that the direct process does not dominate across the entire field range studied.

Table 4. The parameters fit by Equation (2) for **1** and **2**.

	A (s^{-1} T^{-4} K^{-1})	B_1 (s^{-1})	B_2 (T^{-2})	k (T) (s^{-1})
1	6016.3	340.6	1145.8	70.7
2	21,346.5	2167.2	707.4	554.6

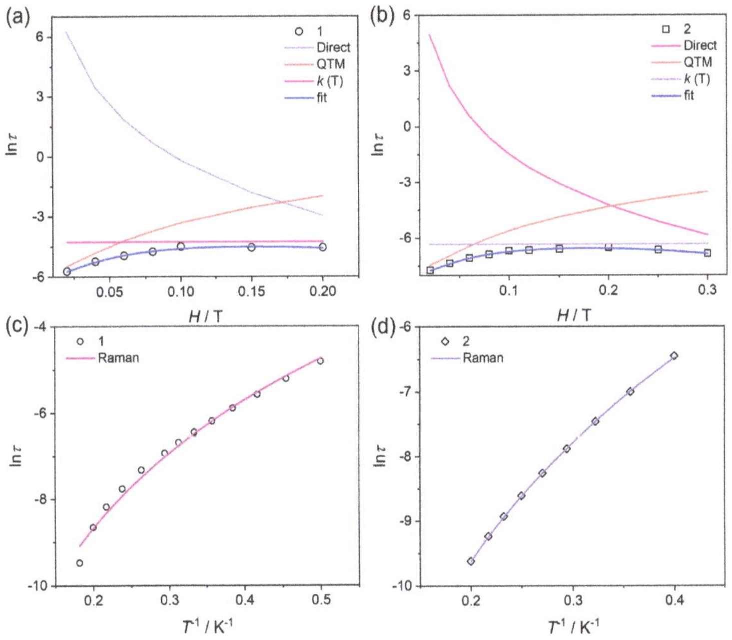

Figure 4. Field dependence of the magnetization relaxation rates for **1** (**a**) and **2** (**b**). The blue line represents the best fit using Equation (2). The other lines represent the contribution of direct, QTM and other processes, respectively. Relaxation time of the magnetization ln(τ) vs. T^{-1} plots for **1** (**c**) and **2** (**d**).

At fixed temperatures for **1** (2.0–5.5 K) and **2** (2.5–5.0 K), semicircular Cole–Cole plots were generated and fitted using the generalized Debye model [41] (Figure S15), assuming a single relaxation pathway to determine the magnetization relaxation time (τ) and the α parameter. The α parameter ranged from 0.44 to 0.08 for complex **1** and from 0.24 to 0.11 for complex **2**, indicating a broad distribution of relaxation times at lower temperatures that narrows as the temperature increases. The relaxation times extracted from the Debye fits were utilized to construct the ln(τ) vs. T^{-1} plots for **1** and **2**. It is noteworthy that a Raman process plays an important role in most reported Co(II)-based SIMs with octahedral geometry, while a direct one-phonon contribution cannot be ignored at low temperatures [26]. Therefore, a model including both direct and Raman relaxation mechanisms was initially employed. However, the results, which are consistent with the findings obtained from fitting the ln(τ) vs. H curve, indicate that the direct mechanism is not the main contributor to the relaxation process. Consequently, the Raman process alone, described as $\tau^{-1} = CT^n$, was used to fit the ln(τ) vs. T^{-1} data (Figure 4c,d). The best-fit parameters were $C = 5.7\ s^{-1}\ K^{-n}$ and $n = 4.3$ for **1**, and $C = 10.1\ s^{-1}\ K^{-n}$ and $n = 4.5$ for **2**.

The cobalt(II) ions in complexes **1** and **2** show the same donor set and exhibit very similar bond lengths and angles, with comparable degrees of distortion. However, the AC magnetic properties demonstrate that the peaks in out-of-phase susceptibilities for **2** appear at higher frequencies than those for **1**, suggesting a faster relaxation process in **2**. Since the experimental magnetic data show that the magnetic anisotropies of **1** and **2** are very similar, the differences in their dynamic susceptibility behavior are attributed to the distinct pseudohalide ions. The substitution of SCN^- in **1** with $SeCN^-$ in **2** leads to a significant change in the crystal packing. In **1**, there are no significant intermolecular interactions, while in **2**, weak C–H···Se hydrogen bonding interactions link the molecules to form supermolecular chains along the *b*-axis. As reported by Boča et al., differences in hydrogen bonding may contribute to the difference in dynamic magnetic properties [42,43]. For instance, the hydrogen boding systems can enhance coupling of the dimeric units within the supramolecular chains, leading to the absence of the out-of-phase component and ultimately preventing the system from behaving as single molecule magnets [42]. Moreover, Song et al. have reported that the structural rigidity significantly impacts the slow magnetic relaxation process in the SIM system [44]. Increased structural rigidity can enhance spin–phonon coupling, which will accelerate the magnetic relaxation. Consequently, the hydrogen boding systems in **2**, which provide additional stabilization to the crystal structure, may be responsible for its faster relaxation process compared to **1**.

4. Conclusions

We synthesized two new mononuclear compounds, $[Co(L)(NCS)_2]$ (**1**) and $[Co(L)(NCSe)_2]$ (**2**), where L represents 1-(prop-2-en-1-yl)-1*H*-imidazole. Both complexes feature a similar octahedral Co(II) center, coordinated by four N atoms from the ligand L and two N atoms from the anion. Analysis of the dc magnetic data indicates the presence of easy-plane magnetic anisotropy with significant positive D values (+70.4 cm^{-1} for **1** and +64.0 cm^{-1} for **2**). Ac magnetic susceptibility measurements confirm that both complexes exhibit field-induced slow magnetic relaxation behavior. Moreover, the Raman mechanism is found to be the dominant relaxation process across the entire temperature range for both complexes.

Supplementary Materials: The following supporting information can be downloaded at https://www.mdpi.com/article/10.3390/magnetochemistry11020011/s1. Figures S1 and S2: XRD patterns for complexes 1 and 2; Table S1: Selected previously obtained mononuclear octahedral Co(II) complexes supported by imidazole ligands; Table S2: The results of the continuous shape measure (CSM) analyses of 1 and 2 by SHAPE software; Figure S3: View of the crystal packing along the a-axis in 1; Figure S4: View of the crystal packing along the a-axis in 2; Figures S5 and S6: The intermolecular Co–Co distances in 1 and 2; Figure S7: The IR spectra of complexes 1 and 2; Figure S8: Variable-temperature dc susceptibility data of 2; Table S3: The fitting results of the magnetization data with Equation (3) for 1 and 2; Figure S9: Frequency dependence of out-of-phase ($\chi M''$) ac susceptibility under the different applied static fields for 1 and 2; Figure S10: Field dependence of the magnetization relaxation rates for 1 and 2; Figures S11 and S12: Frequency dependence of in-phase ($\chi M'$) ac susceptibility under applied dc field of 1000 Oe for 1 and 2000 Oe for 2; Figures S13 and S14: Temperature dependence of in-phase ($\chi M'$) and out-of-phase ($\chi M''$) ac susceptibility at different ac frequencies for 1 and 2; Figure S15: Cole–Cole plot obtained from the ac susceptibility data under a different temperature range for **1** and **2**.

Author Contributions: The manuscript was prepared with the contributions of all authors. They specifically contributed as follows: H.-H.C.—Investigation, writing—original draft; D.-Z.W.—Investigation; S.L.—Formal analysis; L.W.—Formal analysis; X.Y.—Formal analysis; X.L.—Formal analysis; J.W.—Project administration, supervision; M.W.—Review and editing; Y.T.—Supervision. All authors have read and agreed to the published version of the manuscript.

Funding: This work was supported by the National Natural Science Foundation of China (22101144), the Natural Science Foundation of Jiangsu Province (BK20210835) and the Natural Science Research Project of Jiangsu Colleges and Universities (21KJB150003).

Institutional Review Board Statement: Not applicable.

Informed Consent Statement: Not applicable.

Data Availability Statement: CCDC 2414366 and 2414367 contain the supplementary crystallographic data for compounds 1 and 2, respectively. These data can be obtained free of charge via https://www.ccdc.cam.ac.uk/structures/ (accessed on 5 February 2025) or from Cambridge Crystallographic Data Centre, 12 Union Road, Cambridge CB2 1EZ, UK; fax: (+44) 1223-336-033; or e-mail: deposit@ccdc.cam.ac.uk. The original contributions presented in this study are included in the article and Supplementary Materials; further inquiries can be directed to the corresponding authors.

Acknowledgments: We are very grateful to the Nantong University Analytical Testing Center for its support for testing.

Conflicts of Interest: The authors declare no conflicts of interest.

References

1. Gatteschi, D.; Sessoli, R.; Villain, J. *Molecular Nanomagnets*; Oxford University Press, Inc.: New York, NY, USA, 2006.
2. Clemente-Juan, J.M.; Coronado, E.; Gaita-Arino, A. A magnetic polyoxometalates: From molecular magnetism to molecular spintronics and quantum computing. *Chem. Soc. Rev.* **2012**, *41*, 7464–7478. [CrossRef] [PubMed]
3. Sanvito, S. Molecular spintronics. *Chem. Soc. Rev.* **2011**, *40*, 3336–3355. [CrossRef] [PubMed]
4. Craig, G.A.; Murrie, M. 3d single-ion magnets. *Chem. Soc. Rev.* **2015**, *44*, 2135–2147. [CrossRef] [PubMed]
5. Feng, M.; Tong, M.-L. Single ion magnets from 3d to 5f: Developments and strategies. *Chem. Eur. J.* **2018**, *24*, 7574–7594. [CrossRef]
6. Naoto, I.; Sugita, M.; Ishikawa, T.; Koshihara, A.S.; Kaizu, Y. Lanthanide double-decker complexes functioning as magnets at the single-molecular level. *J. Am. Chem. Soc.* **2003**, *125*, 8694–8695.
7. Vallejo, J.; Pascual-Álvarez, A.; Cano, J.; Castro, I.; Julve, M.; Lloret, F.; Krzystek, J.; De Munno, G.; Armentano, D.; Wernsdorfer, W.; et al. Field-induced hysteresis and quantum tunneling of the magnetization in a mononuclear manganese(III) complex. *Angew. Chem. Int. Ed.* **2013**, *52*, 14075–14079. [CrossRef]
8. Zadrozny, J.M.; Atanasov, M.; Bryan, A.M.; Lin, C.-Y.; Rekken, B.D.; Power, P.P.; Neese, F.; Long, J.R. Slow magnetization dynamics in a series of two-coordinate iron(ii) complexes. *Chem. Sci.* **2013**, *4*, 125–130. [CrossRef]
9. Bunting, P.C.; Atanasov, M.; Damgaard-Moller, E.; Perfetti, M.; Crassee, I.; Orlita, M.; Overgaard, J.; van Slageren, J.; Neese, F.; Long, J.R. A linear cobalt(II) complex with maximal orbital angular momentum from a non-Aufbau ground state. *Science* **2018**, *362*, eaat7319. [CrossRef]
10. Rechkemmer, Y.; Breitgoff, F.D.; van der Meer, M.; Atanasov, M.; Hakl, M.; Orlita, M.; Neugebauer, P.; Neese, F.; Sarkar, B.; van Slageren, J. A fourcoordinate cobalt(II) single-ion magnet with coercivity and a very high energy barrier. *Nat. Commun.* **2016**, *7*, 10467. [CrossRef]
11. Cui, H.-H.; Lu, F.; Chen, X.-T.; Zhang, Y.-Q.; Tong, W.; Xue, Z.-L. Zero-field slow magnetic relaxation and hysteresis loop in four-coordinate Co^{II} single-ion magnets with strong easy-axis anisotropy. *Inorg. Chem.* **2019**, *58*, 12555–12564. [CrossRef]
12. Gupta, S.K.; Rao, S.V.; Demeshko, S.; Dechert, S.; Bill, E.; Atanasov, M.; Neese, F.; Meyer, F. Air-stable four-coordinate cobalt(II) single-ion magnets: Experimental and ab initio ligand field analyses of correlations between dihedral angles and magnetic anisotropy. *Chem. Sci.* **2023**, *14*, 6355–6374. [CrossRef] [PubMed]
13. Ferreira, P.S.; Malta, J.F.; Bandeira, N.A.G.; Allgaier, A.; van Slageren, J.; Paixão, J.A.; Almeida, M.; Pereira, L.C.J.; Gomes, P.T. Enhancing SIM behaviour in a mononuclear tetrahedral [Co(*N,N'*-2-iminopyrrolyl)$_2$] complex. *Chem. Commun.* **2022**, *58*, 9682–9685. [CrossRef]
14. Cui, H.-H.; Xu, H.; Zhang, T.; Chen, Q.; Luo, S.; Wang, M.; Wang, J.; Chen, L.; Zhang, M.; Tang, Y. Magnetic anisotropy and relaxation in four-coordinate cobalt(II) single-ion magnets with a [$Co^{II}O_4$] Core. *Inorg. Chem.* **2024**, *63*, 9050–9057. [CrossRef]

15. Poulten, R.C.; Page, M.J.; Algarra, A.G.; Le Roy, J.J.; López, I.; Carter, E.; Llobet, A.; Macgregor, S.A.; Mahon, M.F.; Murphy, D.M.; et al. Synthesis, electronic structure, and magnetism of $[Ni(6\text{-}Mes)_2]^+$: A two-coordinate nickel(I) complex stabilized by bulky N-heterocyclic carbenes. *J. Am. Chem. Soc.* **2013**, *135*, 13640–13643. [CrossRef] [PubMed]
16. Sahu, P.K.; Kharel, R.; Shome, S.; Goswami, S.; Konar, S. Understanding the unceasing evolution of Co(II) based single-ion magnets. *Coord. Chem. Rev.* **2023**, *475*, 214871.
17. Sarkar, A.; Dey, S.; Rajaraman, G. Role of coordination number and geometry in controlling the magnetic anisotropy in Fe^{II}, Co^{II}, and Ni^{II} single-ion magnets. *Chem. Eur. J.* **2020**, *26*, 14036–14058. [CrossRef]
18. Tripathi, S.; Dey, A.; Shanmugam, M.; Narayanan, R.S.; Chandrasekhar, V. *Cobalt(II) Complexes as Single-Ion Magnets in Organometallic Magnets;* Chandrasekhar, V., Pointillart, F., Eds.; Springer International Publishing: Cham, Switzerland, 2019; pp. 35–75. [CrossRef]
19. Juráková, J.; Šalitroš, I. Co(II) single-ion magnets: Synthesis, structure, and magnetic properties. *Monatsh. Chem.* **2022**, *153*, 1001–1036. [CrossRef]
20. Vallejo, J.; Castro, I.; Ruiz García, R.; Cano, J.; Julve, M.; Lloret, F.; De Munno, G.; Wernsdorfer, W.; Pardo, E. Field-induced slow magnetic relaxation in a six-coordinate mononuclear cobalt(II) complex with a positive anisotropy. *J. Am. Chem. Soc.* **2012**, *134*, 15704–15707. [CrossRef]
21. Gómez-Coca, S.; Urtizberea, A.; Cremades, E.; Alonso, P.J.; Camón, A.; Ruiz, E.; Luis, E. Origin of slow magnetic relaxation in kramers ions with non-uniaxial anisotropy. *Nat. Commun.* **2014**, *5*, 4300. [CrossRef]
22. Novikov, V.V.; Pavlov, A.A.; Nelyubina, Y.V.; Boulon, M.E.; Varzatskii, O.A.; Voloshin, Y.Z.; Winpenny, R.E.P. A trigonal prismatic mononuclear cobalt(II) complex showing single-molecule magnet behavior. *J. Am. Chem. Soc.* **2015**, *137*, 9792–9795. [CrossRef]
23. Zhu, Y.-Y.; Zhu, M.-S.; Yin, T.-T.; Meng, Y.-S.; Wu, Z.-Q.; Zhang, Y.-Q.; Gao, S. Cobalt(II) coordination polymer exhibiting single-ion-magnet-type field-induced slow relaxation behavior. *Inorg. Chem.* **2015**, *54*, 3716–3718. [CrossRef] [PubMed]
24. Zhang, J.; Li, J.; Yang, L.; Zhang, Y.-Q.; Song, Y. Magnetic anisotropy from trigonal prismatic to trigonal antiprismatic Co(II) complexes: Experimental observation and theoretical prediction. *Inorg. Chem.* **2018**, *57*, 3903–3912. [CrossRef] [PubMed]
25. Zhang, Y.-Z.; Gómez-Coca, S.; Brown, A.J.; Saber, M.R.; Zhang, X.; Dunbar, K.R. Trigonal antiprismatic Co(II) single molecule magnets with large uniaxial anisotropies: Importance of Raman and tunneling mechanisms. *Chem. Sci.* **2016**, *7*, 6519–6527. [CrossRef] [PubMed]
26. Zhou, J.; Song, J.; Yuan, A.; Wang, Z.; Chen, L.; Ouyang, Z.-W. Slow magnetic relaxation in two octahedral cobalt(II) complexes with positive axial anisotropy. *Inorg. Chim. Acta.* **2018**, *479*, 113–119. [CrossRef]
27. *SMART & SAINT Software Reference Manuals*, version 6.45; Bruker Analytical X-ray Systems, Inc.: Madison, WI, USA, 2003.
28. *SAINT, Program for Data Extraction and Reduction*; Siemens Analytical X-ray Instruments: Madison, WI, USA, 1996.
29. Sheldrick, G.M. *SADABS: Software for Empirical Absorption. Correction*, version 2.05; University of Göttingen: Göttingen, Germany, 2002.
30. Sheldrick, G.M. *SHELXL14: Program for Crystal Structure Refinement*; University of Göttingen: Göttingen, Germany, 2014.
31. Bain, G.A.; Berry, J.F. Diamagnetic corrections and Pascal's constants. *J. Chem. Educ.* **2008**, *85*, 532–536. [CrossRef]
32. Llunell, M.; Casanova, D.; Cirera, J.; Alemany, P.; Alvarez, S. *SHAPE*, version 2.1; Universitat de Barcelona: Barcelona, Spain, 2013.
33. Alvarez, S.; Alemany, P.; Casanova, D.; Cirera, J.; Llunell, M.; Avnir, D. Shape maps and polyhedral interconversion paths in transition metal chemistry. *Coord. Chem. Rev.* **2005**, *249*, 1693–1708. [CrossRef]
34. Kurdziel, K.; Głowiak, T. X-ray and spectroscopic characterization of octahedral cobalt(II) and nickel(II) complexes with 1-(prop-2-en-1-yl)-1*H*-imidazole in the solid state and electron-donor properties of the latter in aqueous solution. *Polyhedron* **2000**, *19*, 2183–2188. [CrossRef]
35. Maslejova, A.; Uhrinova, S.; Mroziriski, J.; Zurowska, B.; Munoz, M.C.; Julve, M. Study of the mutual influence of ligands in cobalt(II) complexes containing thiocyanate and imidazole derivatives. *Inorg. Chimi. Acta* **1997**, *255*, 343–349. [CrossRef]
36. Osipov, N.G.; Faraonov, M.A.; Yakushev, I.A.; Denisov, N.N.; Otsuka, A.; Kitagawa, H.; Konarev, D.V. Slow magnetic relaxation in a complex of photochromic spiropyran in a merocyanine form and cobalt(II) hexafluoroacetylacetonate. *Dalton Trans.* **2024**, *53*, 3159–3166. [CrossRef]
37. Hu, Z.-B.; Jing, Z.-Y.; Li, M.-M.; Yin, L.; Gao, Y.-D.; Yu, F.; Hu, T.-P.; Wang, Z.; Song, Y. Important role of intermolecular interaction in cobalt(II) single-ion magnet from single slow relaxation to double slow relaxation. *Inorg. Chem.* **2018**, *57*, 10761–10767. [CrossRef]
38. Chen, S.-Y.; Cui, H.-H.; Zhang, Y.-Q.; Wang, Z.; Ouyang, Z.-W.; Chen, L.; Yan, H.; Xue, Z.-L. Magnetic anisotropy and relaxation behavior of six-coordinate tris(pivalato)-Co(II) and -Ni(II) complexes. *Dalton Trans.* **2018**, *47*, 10162–10171. [CrossRef] [PubMed]
39. Walsh, J.P.S.; Bowling, G.; Ariciu, A.-M.; Jailani, N.F.M.; Chilton, N.F.; Waddell, P.G.; Collison, D.; Tuna, F.; Higham, L.J. Evidence of slow magnetic relaxation in $Co(AcO)_2(py)_2(H_2O)_2$. *Magnetochemistry* **2016**, *2*, 23. [CrossRef]
40. Chilton, N.F.; Anderson, R.P.; Turner, L.D.; Soncini, A.; Murray, K.S. PHI: A powerful new program for the analysis of anisotropic monomeric and exchange-coupled polynuclear d- and f-block complexes. *J. Comput. Chem.* **2013**, *34*, 1164–1175. [CrossRef] [PubMed]

41. Cole, K.S.; Cole, R.H. Dispersion and absorption in dielectrics I. Alternating current characteristics. *J. Chem. Phys.* **1941**, *9*, 341–351. [CrossRef]
42. Smolko, L.; Černák, J.; Dušek, M.; Titiš, J.; Boča, R. Tetracoordinate Co(II) complexes containing bathocuproine and single molecule magnetism. *New J. Chem.* **2016**, *40*, 6593–6598. [CrossRef]
43. Smolko, L.; Černák, J.; Kuchár, J.; Rajnák, C.; Titiš, J.; Boča, R. Field-induced slow magnetic relaxation in mononuclear tetracoordinate cobalt(II) complexes containing a neocuproine ligand. *Eur. J. Inorg. Chem.* **2017**, *2017*, 3080–3086. [CrossRef]
44. Hu, Z.-B.; Gui, L.-A.; Li, L.-H.; Xiao, T.-T.; Hand, A.-T.; Tin, P.; Ozerov, M.; Peng, Y.; Ouyang, Z.; Wang, Z.; et al. Co^{II} single-ion magnet and its multi-dimensional aggregations: Influence of the structural rigidity on magnetic relaxation process. *Chinese Chem. Lett.* **2025**, *36*, 109600. [CrossRef]

magnetochemistry

MDPI

Article

An Organic–Inorganic Hybrid Semiconducting Quantum Spin Liquid Candidate: $(BEDT\text{-}TTF)_3[Cu_2(\mu\text{-}C_2O_4)_3 \cdot CH_3CH_2OH \cdot 1.2H_2O]$

Bin Zhang [1,*,†], Yan Zhang [2,*,†], Dongwei Wang [3], Zheming Wang [4,*], Guangcai Chang [5], Zengqiang Gao [5], Yanjun Guo [3], Fen Liu [6], Zhijuan Zhao [6], Xiaoyu Zhang [6], Baolong Qu [6], Peng Xu [3], Jiaou Wang [5], Fenliang Dong [3], Tongling Liang [6], Yang Sun [6], Deliang Yang [6], Qiaolian Li [6], Xiaofei Luo [6], Rongjuan Feng [6], Mei Liu [6] and Xueying Zhang [6]

1 Organic Solid Laboratory, BNLMS, CMS & Institute of Chemistry, Chinese Academy of Sciences, Beijing 100190, China
2 Institute of Condensed Matter and Material Physics, School of Physics, Peking University, Beijing 100871, China
3 CAS Key Laboratory of Standardization and Measurement for Nanotechnology, National Center for Nanoscience and Technology, Beijing 100190, China
4 State Key Laboratory of Rare Earth Materials Chemistry and Applications, BNLMS, College of Chemistry and Molecular Engineering, Peking University, Beijing 100871, China
5 BSRF, Institute of High Energy Physics, Chinese Academy of Sciences, Beijing 100047, China
6 Institute of Chemistry, Chinese Academy of Sciences, Beijing 100190, China
* Correspondence: zhangbin@iccas.ac.cn (B.Z.); zhang_yan@pku.edu.cn (Y.Z.); zmw@pku.edu.cn (Z.W.)
† These authors contributed equally to this work.

Abstract: The organic–inorganic hybrid $(BEDT\text{-}TTF)_3[Cu_2(\mu\text{-}C_2O_4)_3 \cdot CH_3CH_2OH \cdot 1.2H_2O]$ (I) was obtained using the electrocrystallization method. It comprises a θ^{21}-phase organic donor layer and a two-dimensional inorganic antiferromagnetic honeycomb lattice. Cu(II) is octahedrally coordinated by three bisbidenetate oxalates, exhibiting Jahn–Teller distortion. CH_3CH_2OH and H_2O molecules are located within the cavities of the honeycomb lattice. The total formal charge of the three donor molecules was assigned to be +2 based on the bond lengths in the TTF core, which corresponded to the Raman spectra. It is a semiconductor with σ_{rt} = 0.04 S/cm and E_α = 40 meV. No long-range ordering was observed above 2 K from zero-field cooling/field cooling magnetization, as confirmed by specific heat measurements. The spin frustration with f > 10 from the antiferromagnetic copper-oxalate-framework was observed. It is a candidate quantum spin liquid.

Keywords: Jahn–Teller effect; Cu(II); oxalate; single crystal; conductivity; spin frustration; quantum spin liquid candidate

1. Introduction

In 1973, P. W. Anderson proposed that in a $S = 1/2$ antiferromagnetic interaction triangular lattice, magnetic ordering or freezing is not observed due to spin frustration, resulting in a quantum spin liquid [1]. In 1987, he suggested that the insulating antiferromagnet La_2CuO_4 acts as the parent compound of the cuprate superconductor after hole doping [2,3]. Quantum spin liquids with two-dimensional lattices, including triangular lattices, Kagome lattices and honeycomb lattices, have been proposed [4,5]. Quantum spin liquids have emerged as a central focus in condensed matter physics due to their strong connection to magnetism and conductivity. Numerous molecular superconductors have been identified within hybrid organic–inorganic charge–transfer complexes [6]. These complexes also provide strong support for research into quantum spin liquid candidates. The behavior of quantum spin liquids was identified in organic–inorganic hybrid charge–transfer complexes, such as κ-(BEDT-TTF)$_2$[Cu$_2$(CN)$_3$] (BEDT-TTF = bis(ethylenedithio)tetrathiafulvalene), κ-(BEDT-TTF)$_2$[Ag$_2$(CN)$_3$], EtMe$_3$Sb[Pd(dmit)$_2$]$_2$ (dmit = 1,3-dithiol-2-thione-4,5-dithiolate) and κ-H$_3$(cat-EDT-TTF)$_2$. κ-(BEDT-TTF)$_2$[Cu$_2$(CN)$_3$] features a triangular lattice, while (EDT-TTF-CONH$_2$)$_6$[Re$_6$Se$_8$(CN)$_6$] exhibits a Kagome lattice [7–11].

Oxalate ($C_2O_4^{2-}$), one of the most commonly used short connectors, plays a crucial role in molecular superconductors, molecular conductors and molecular magnets. The exchange coupling in oxalate-bridged binuclear Cu(II) has been extensively studied [12]. Long-range ordering has been reported in metal–oxalate frameworks, ranging from one-dimensional chains such as K$_2$[Fe(μ-C$_2$O$_4$)(C$_2$O$_4$)], A[Fe(μ-C$_2$O$_4$)Cl$_2$] (A = alkyl ammonium), K$_2$[Co(μ-C$_2$O$_4$)(C$_2$O$_4$)] and [Co(μ-C$_2$O$_4$)(μ-HOC$_3$H$_6$OH)] to two-dimensional layers like [(C$_4$H$_9$)$_4$N][MCr(μ-C$_2$O$_4$)$_3$] (M = Mn^{2+}, Fe^{2+}, Co^{2+}, Ni^{2+}, Cu^{2+}, Zn^{2+}), A[$M^{II}Fe^{III}$(μ-C$_2$O$_4$)$_3$] (A = ammonium; M = Mn, Fe), [Fe(μ-C$_2$O$_4$)(μ-CH$_3$OH)] and [Fe$_2$(μ-C$_2$O$_4$)$_3$]·4H$_2$O, and extending to three-dimensional metal–oxalate networks such as, [Zn^{II}(bpy)$_3$][$M^{II}Cr^{III}$(μ-C$_2$O$_4$)$_3$][ClO$_4$] (Z = Mn, Fe, Co, Ni; M = Mn, Fe, Co, Ni, Cu, Zn), [Co(bpy)$_3$][Co$_2$(μ-C$_2$O$_4$)$_3$]ClO$_4$, [(Me$_4$N)$_6$[Mn$_3$Cr$_4$(μ-C$_2$O$_4$)$_{12}$]·6H$_2$O and [Mn(μ-C$_2$O$_4$)]·0.25H$_2$O. Additionally, a single-chain magnet, [C$_{12}$H$_{24}$O$_6$K]$_{0.5}$[(C$_{12}$H$_{24}$O$_6$)(FC$_6$H$_4$NH$_3$)]$_{0.5}$[Co(H$_2$O)$_2$Cr(μ-C$_2$O$_4$)$_2$(C$_2$O$_4$)], has been reported [13–26].

The magnetism and conductivity of $M(C_2O_4)_2^{2-}$, characterized by a square anion, have been investigated. A molecular conductor with a room-temperature conductivity of 320 S/cm was observed in Mg$_{0.33}$[Pt(C$_2$O$_4$)$_2$]·5.3H$_2$O. It exhibits a metal-to-insulator transition alongside a spin–Peierls transition [27,28]. When the metal cation was replaced with an organic donor, the paramagnetic conductors (BEDT-TTF)$_2$[Pt(C$_2$O$_4$)$_2$] and (BEDT-TTF)$_2$[Cu(C$_2$O$_4$)$_2$] were obtained, and a metal–insulator transition took place at 65 K [29–31].

The magnetism and conductivity of organic–inorganic hybrid charge–transfer complexes composed of an organic donor and $M(C_2O_4)_3^{3-}$, featuring an octahedral anion, were also examined. The first paramagnetic superconductor containing a paramagnetic metal ion, β″-(BEDT-TTF)$_3$[Fe(C$_2$O$_4$)$_3$(H$_3$O)·C$_6$H$_5$CN], along with the semiconductor paramagnets β″-(BEDT-TTF)$_3$[KFe(C$_2$O$_4$)$_3$·C$_6$H$_5$CN] and β″-(BEDT-TTF)$_3$[(NH$_4$)Fe(C$_2$O$_4$)$_3$·C$_6$H$_5$CN], were synthesized [32,33]. The insulating TTF$_2$[Fe(μ-C$_2$O$_4$)Cl$_2$], (BEDT-TTF)[Fe(C$_2$O$_4$)Cl$_2$]·CH$_2$Cl$_2$, semiconductive [BEDT-TTF]$_2$[Fe(μ-C$_2$O$_4$)Cl$_2$] and metallic β-BETS$_2$[Fe(μ-C$_2$O$_4$)Cl$_2$] (BETS = bis(ethylenedithio)tetraselenafulvalene) antiferromagnet were derived from a one-dimensional [Fe(μ-C$_2$O$_4$)Cl$_2$$^-$]$_n$ zigzag chain. A π–d interaction between the donor and the anion was observed [14,34–37]. The metallic ferromagnets β-(BEDT-TTF)$_3$[CrMn(μ-C$_2$O$_4$)$_3$] and β-BETS$_3$[CrMn(μ-C$_2$O$_4$)$_3$]·CH$_2$Cl$_2$ were derived from an oxalate-bridged two-dimensional [CrMn(μ-C$_2$O$_4$)$_3$] anion, which formed a honeycomb lattice. Their magnetic behavior is identical to that of [(C$_4$H$_9$)$_4$N][CrMn(μ-C$_2$O$_4$)$_3$] [18,38–40].

The Jahn–Teller effect on the charge–transfer complexes within copper-oxalate frameworks has previously been studied [41]. The ferromagnetic insulator [(CH$_3$)$_3$NH]$_2$[Cu(μ-C$_2$O$_4$)(C$_2$O$_4$)]·2.5H$_2$O and weak ferromagnetic insulator [(C$_2$H$_5$)$_3$NH]$_2$[Cu(μ-C$_2$O$_4$)(C$_2$O$_4$)]·H$_2$O, derived from the one-dimensional zigzag chain [Cu(μ-C$_2$O$_4$)(C$_2$O$_4$)$^{2-}$], were obtained and characterized [42]. Spin frustration with [Cu$_2$(μ-C$_2$O$_4$)$_3$$^{2-}$], which has a honeycomb lattice, as the quantum spin liquid candidate was discovered in organic–inorganic hybrid charge–transfer complexes, including the insulator [(C$_3$H$_7$)$_3$NH]$_2$[Cu$_2$(μ-C$_2$O$_4$)$_3$·2.2H$_2$O], the semiconductor θ^{21}-(BEDT-TTF)$_3$[Cu$_2$(μ-C$_2$O$_4$)$_3$·2CH$_3$OH] and the conductor θ^{21}-BETS$_3$[Cu$_2$(μ-C$_2$O$_4$)$_3$·2CH$_3$OH] [43–45]. A quantum spin liquid candidate from a three-dimensional copper-oxalate-framework with a hyperhoneycomb lattice was studied [46,47]. Further research on the role of the Jahn–Teller effect in organic–inorganic hybrids containing a copper–oxalate framework will help identify molecular conductors/superconductors and quantum spin liquid candidates. A new hybrid organic–

inorganic change–transfer complex, θ^{21}-(BEDT-TTF)$_3$[Cu_2(μ-C_2O_4)$_3$·C_2H_5OH·1.2H_2O] (**I**), was obtained, and related work is presented here.

2. Experiment and Discussion

BEDT-TTF was acquired from the TCI company (Tokyo, Japan). [$(C_2H_5)_3NH$]$_2Cu_2$(μ-C_2O_4)$_3$ was synthesized from $Cu(NO_3)_2$·6H_2O, and $H_2C_2O_4$·2H_2O and $(C_2H_5)_3N$ were synthesized from CH_3OH, which was confirmed through elemental analysis, IR spectra and powder X-ray diffraction [46]. A total of 5.0 mg of BEDT-TTF and 30.0 mg of [$(C_2H_5)_3NH$]$_2Cu_2$(μ-C_2O_4)$_3$ were dissolved in a mixture of 25.0 mL of distilled C_6H_5Cl and 5.0 mL of distilled C_2H_5OH, and the mixture was then placed in an electrocrystallization cell. The cell was subjected to a constant source of 0.20 mA at room temperature. Shiny black, thin-plate crystals of compound (**I**) were obtained on the cathode after two weeks. The single crystals, with fresh crystal surface cleavage carried out using scotch tape, were utilized for EDS and XPS experiments [48]. The element analysis carried out using EDS revealed a composition ratio of S:Cu = 12:1, indicating a donor-to-Cu ratio of 3:2. The characteristic photoelectron peaks of Cu2p, S2p, O1S and C1S were detected in the XPS spectra.

A piece of the single crystal was selected for use in single-crystal X-ray diffraction experiments at the Beijing Radiation Synchrony Facility using λ = 0.70 Å at 290 K, a Rigaku SuperNova diffractometer with Mo radiation (λ = 0.71073 Å) at 290 K, 180 K and 120 K and a Rigaku XtalAB synergy R, HyPix diffractometer with Cu (λ = 1.54184 Å) at 100 K. The crystal structure was solved using the direct method and refined with the full-matrix least-square of F^2 using the SHELXL program [49]. The non-hydrogen atoms on BEDT-TTF, Cu, oxalate were refined anisotropically. The hydrogen atoms on BEDT-TTF were located through calculation, and refined isotropically. The solvent molecules present in the cavity were defined by the solvent MASK program. The crystals from different cells were subjected to a single-crystal X-ray diffraction experiment. The crystal structure remained the same. The compound remained stable from 290 K to 100 K. The crystallographic data from the experiments carried out at 290 K, 180 K and 120 K are listed in Table S1. The following discussion refers to the data from the experiment carried out at 100 K.

Compound **I** crystallizes in a triclinic system with the following cell parameters: a = 8.6724(2) Å; b = 16.7949(6) Å; c = 19.5607 Å; α = 74.096(3)°; β = 89.296(2)°; γ = 88.086(2)°; V = 2738.5(1) Å^3; and Z = 2, $P\bar{1}$ at 100 K. There are 2 full and 2 half BEDT-TTF molecules, 2 Cu, 3 oxalates, 2 half ethanol molecules, 0.7 H_2O and 0.5 H_2O in an independent unit (Figure 1). This is different from θ^{21}-(BEDT-TTF)$_3$[Cu_2(μ-C_2O_4)$_3$·2CH_3OH] and θ^{21}-(BETS-TTF)$_3$[Cu_2(μ-C_2O_4)$_3$·2CH_3OH], which have three donor molecules, two Cu, three oxalates and two methanol molecules which coexist in an independent unit.

The donor molecules are stacked face-to-face along the a axis to create a donor column. Donor columns are arranged side-by-side along the b axis to form a donor layer in the form of a θ^{21}-phase, which is observed in θ^{21}-(BEDT-TTF)$_3$[Cu_2(μ-C_2O_4)$_3$·2CH_3OH] and θ^{21}-(BETS-TTF)$_3$[Cu_2(μ-C_2O_4)$_3$·2CH_3OH] (Figure 2) [44,45]. One of the ethylene groups on BEDT-TTF is disordered at 290 K and ordered below 180 K. This is similar to [$(C_2H_5)_3NH$]$_2$[Cu_2(μ-C_2O_4)$_3$], which exhibits a disorder-to-order transition of its ethylene group at 170 K [46]. There are also S···S contacts between donor molecules.

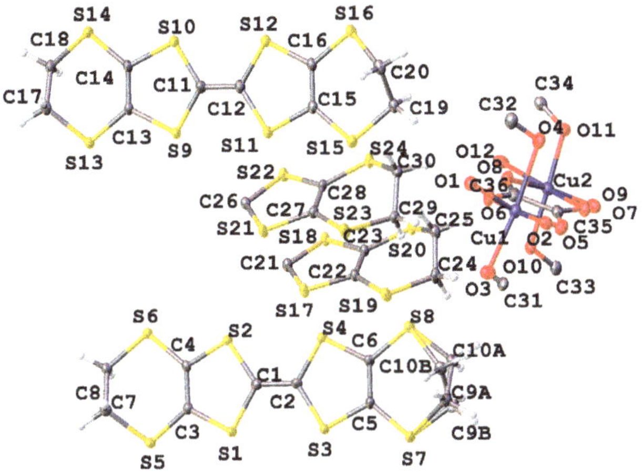

Figure 1. Atoms in an independent unit of **I**.

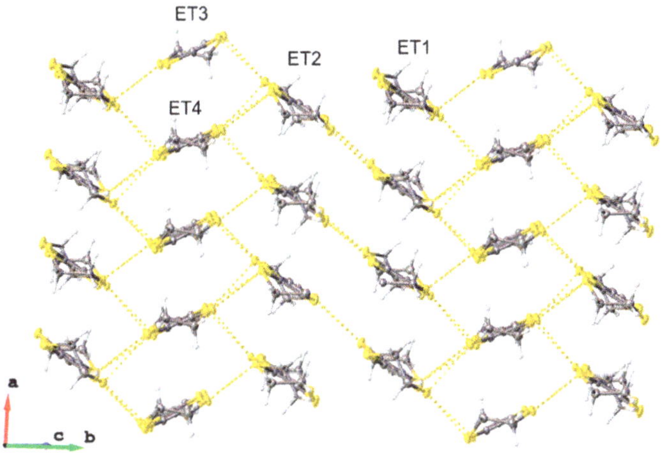

Figure 2. Donor arrangement of **I** on the ab plane. Dashed yellow lines are S···S contacts. ET is an abbreviation for BEDT-TTF.

Cu^{2+} is octahedrally coordinated by six O atoms from three bisbidentate oxalate ligands. The Cu1–O distances are 1.960(4)~2.045(4) Å in the equatorial plane and 2.230(4) Å and 2.362(6) Å from the apex. The Cu2–O distances are 1.969(5)~1.999(5) Å in the equatorial plane and 2.231(4) Å and 2.327(6) Å from the apex. The elongated bonds in the Jahn–Teller distorted octahedron around Cu are highlighted with blue solid lines (Figure 3). Cu1 and Cu2 are adjacent to each other. A honeycomb lattice is formed in the *ab* plane as observed in $[(C_3H_7)_3NH]_2[Cu_2(\mu\text{-}C_2O_4)_3\cdot 2.2H_2O]$, θ^{21}-(BEDT-TTF)$_3[Cu_2(\mu\text{-}C_2O_4)_3\cdot 2CH_3OH]$ and θ^{21}-(BETS-TTF)$_3[Cu_2(\mu\text{-}C_2O_4)_3\cdot 2CH_3OH]$ (Figure 3) [43–45]. Solvent molecules occupy the cavities of the honeycomb lattice. The two donor layers are separated by an anion sheet along the *c* axis (Figure 4). Hydrogen bonds exist between the donor and anion, as well as between the anion and solvent molecules (Figure 4).

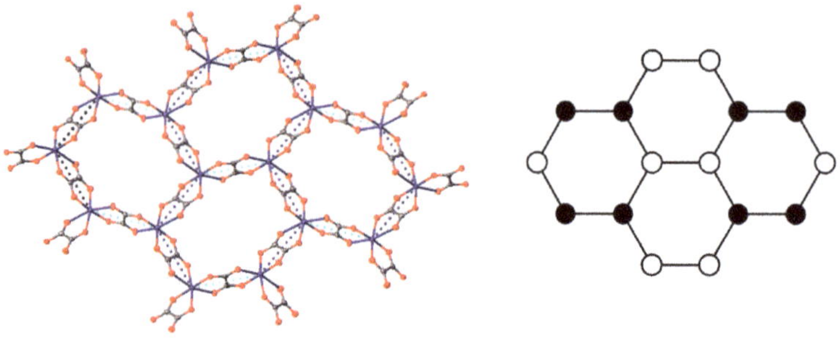

Figure 3. Anion layer of **I** viewed along the *c* axis.

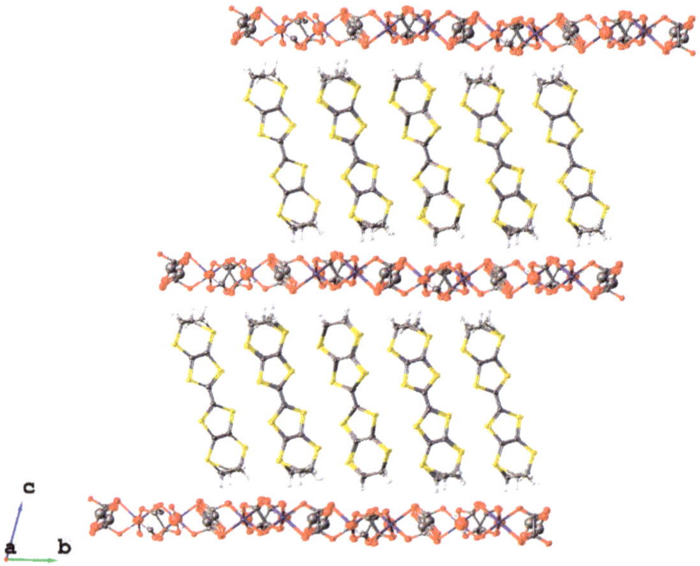

Figure 4. Packing diagram of **I** viewed along the a axis. Color code: Cu, blue; O red: S, yellow; C gray. Hydrogen bond, dashed blue line.

Considering the bond length of a TTF core with a standard deviation of 0.1 of δ, the oxidation state of ET1 is +2/3, while ET2 and ET3 have oxidation states greater than +2/3, and ET4 has an oxidation state less than +1/2. The oxidation states of ET2 and ET3 decreased as the temperature increased. At 290 K, the oxidation states are +0.70 for ET1, +2/3 for ET2, +2/3 for ET3 and +0.43 for ET4. The total oxidation state in an independent unit is +1.98 (Table 1) [50].

The C=C stretching frequency of the charge–transfer complexes of BEDT-TTF serves as a reliable metric for determining the oxidation state of BEDT-TTF [51]. In the Raman spectra (λ = 514.5 nm), a broad band was observed at 1490 cm^{-1} (Figure 5). This corresponds to charge–transfer complexes with BEDT-$TTF^{+2/3}$ and is similar to the 1493 cm^{-1} band observed for θ^{21}-(BEDT-TTF)$_3$[Cu_2(μ-C_2O_4)$_3$·2CH_3OH] [44]. The weak band at 1427 cm^{-1} is attributed to ET2, corresponding to an oxidation state ranging from +1/4 to +3/4. It aligns with the formal charge assigned based on the bond lengths in the TTF core.

Table 1. Bond lengths of the TTF core in BEDT-TTF of **I** at 100 K.

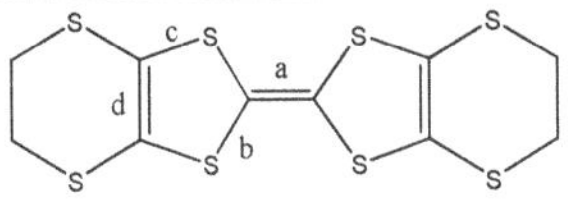

$$\delta = (b + c) - (a + d)$$

$$Q = 6.347 - 7.463\delta$$

	a	*b*	*c*	*d*	*δ*	Q
	1.370	1.741	1.748	1.363		
		1.740	1.756	1.359		
ET1		1.735	1.750		0.763	0.65
		1.746	1.758			
	~1.370 *	~1.741	~1.753	~1.361		
	1.379	1.737	1.748	1.362		
		1.734	1.746	1.355		
ET2		1.733	1.752		0.744	0.80
		1.734	1.741			
	~1.370	~1.735	~1.747	~1.359		
	1.382	1.722	1.744	1.352		
ET3		1.728	1.746		0.736	0.85
	~1.382	~1.725	~1.745	~1.352		
	1.354	1.753	1.756	1.337		
ET4		1.756	1.765		0.824	0.20
	~1.354	~1.755	~1.761	~1.337		
Total						1.975 **

* The average value of bond lengths. ** Total charge of two and two half donors in an independent unit: $\Sigma Q = Q_{ET1} + Q_{ET2} + 1/2Q_{ET3} + 1/2_{ET4}$.

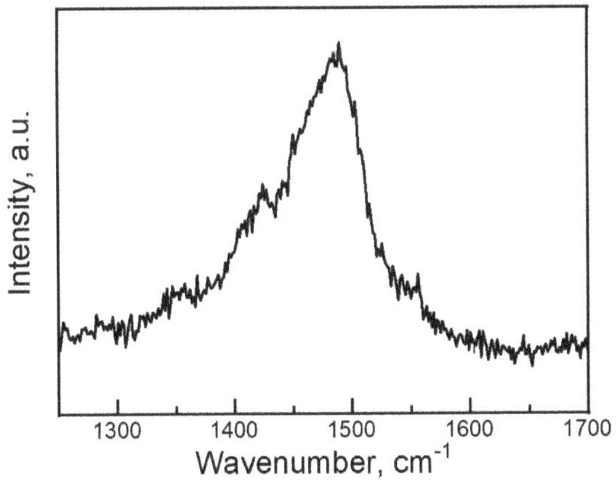

Figure 5. Raman spectrum of **I** with λ = 514.5 nm.

The conductivity measurement was conducted using a four-probe method in a Quantum Design PPMS 9 system. Gold wire was affixed to the best-developed surface of a single crystal with gold paste. The room-temperature conductivity was σ_{rt} = 0.02 S/cm. This value is lower than σ_{rt} = 4 S/cm for θ^{21}-(BEDT-TTF)$_3$[Cu_2(μ-C_2O_4)$_3$·2CH_3OH], and significantly lower than σ_{rt} = 140 S/cm for θ^{21}-(BETS-TTF)$_3$[Cu_2(μ-C_2O_4)$_3$·2CH_3OH] [44,45]. The conductivity decreased as the temperature dropped, with E_α = 40 meV indicating semiconductor behavior (Figure 6).

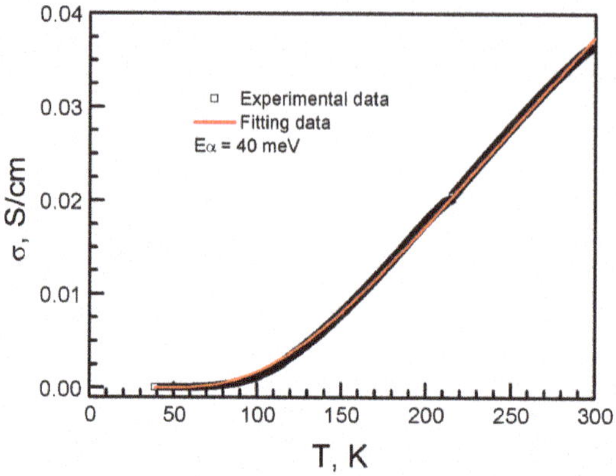

Figure 6. Temperature-dependent conductivity of **I**.

The unpaired electron of Cu(II) is located in the magnetic orbital $d_{x^2-y^2}$ within the equatorial plane. The orbital corresponding to the elongated octahedral Cu(II) is the d_{z^2} orbital, as indicated by the blue solid line highlighted in Figure 3. The magnetic configuration is a stripy antiferromagnetic state [52]. Since the antiferromagnetic interaction is stronger than the ferromagnetic interaction between oxalate-bridged Cu(II) atoms, an antiferromagnetic behavior is expected [12].

The magnetic susceptibility measurement was carried out on Quantum Design MPMS 7XL system. The susceptibility data were corrected by the Pascal constant of the sample and the sample holder [53].

At 300 K, the χT value was 0.389 cm^3 K mol^{-1} and g was 2.04. This is consistent with an isolated, spin only Cu(II) ion with $S = 1/2$ and $g = 2.00$ [54]. As the temperature decreased, the χ increased smoothly, exhibiting a bend around 230 K, followed by an upturn at 22 K. The data fitted to the Curie–Weiss law yielded C = 0.503(2) cm^3 mol^{-1}, $\theta = -87(2)$ K and $R = 9.6 \times 10^{-7}$ above 230 K, while C = 0.405(1) cm^3 mol^{-1}, $\theta = -25.2(7)$ K and $R = 1.3 \times 10^{-5}$ from 100 K to 230 K (Figure 7). No bifurcation is observed from zero-field-cooling magnetization and field-cooling magnetization (ZFCM/FCM) above 2 K under 100 G (Figure 8). No long-range ordering was detected above 2 K. As such, it is a spin-frustrated complex with $f > 12$ [54].

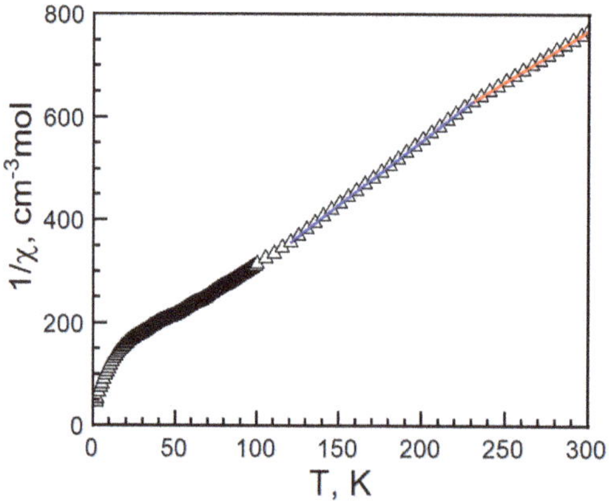

Figure 7. $1/\chi$ vs. T (empty black triangle) of **I**. The red and solid lines are fit to the Curie–Weiss law.

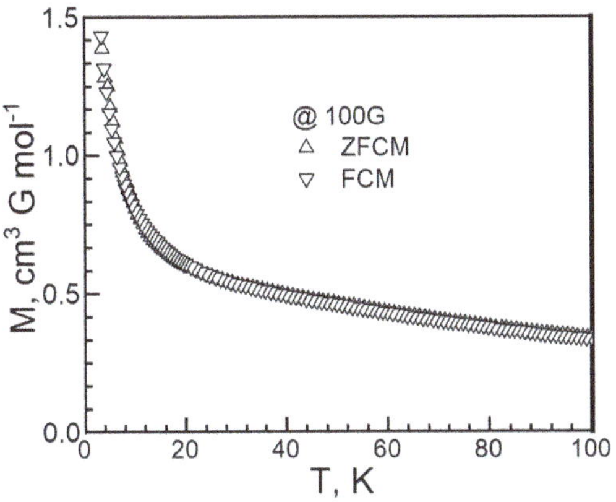

Figure 8. ZFCM/FCM of polycrystal under 100G.

Specific heat measurements were performed from 2 to 120 K under 0 T and 8 T, utilizing a Quantum Design PPMS9 system (Figure 9). No λ-peak was detected in the range of 2 to 120 K. When combined with the X-ray diffraction experiment from 280 K to 120 K, there was no long-range order observed above 2 K.

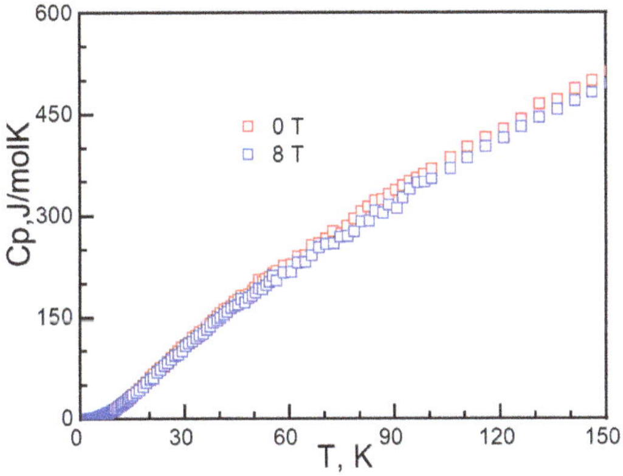

Figure 9. Specific heat of **I** from 2 K to 120 K.

Isothermal magnetization at 2 K increased with the magnetic field, reaching 0.162 N_β at 65 kG (Figure 10). This value surpasses the 0.077 N_β for θ^{21}-(BEDT-TTF)$_3$[Cu_2(μ-C_2O_4)$_3$·2CH_3OH], 0.076 N_β of θ^{21}-(BETS-TTF)$_3$[Cu_2(μ-C_2O_4)$_3$·2CH_3OH] and 0.040 N_β for [(C_3H_7)$_3$NH]$_2$[Cu_2(μ-C_2O_4)$_3$·2.2H_2O] at 2 K and 65 kG [42–44].

Compared with θ^{21}-(BEDT-TTF)$_3$[Cu_2(μ-C_2O_4)$_3$·2CH_3OH] and θ^{21}-(BETS-TTF)$_3$ [Cu_2(μ-C_2O_4)$_3$·2CH_3OH], the conductivity is contributed by the organic donor. The solvent molecule influences the magnetic properties of compounds.

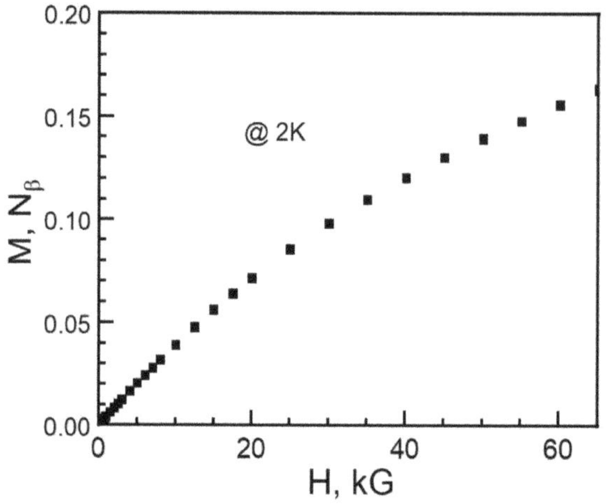

Figure 10. Isothermal magnetization of **I** at 2 K.

3. Conclusions

In this study, we show an organic–inorganic hybrid charge–transfer complex composed of a θ^{21}-phase organic donor layer and an inorganic sheet featuring a honeycomb lattice. The Jahn–Teller effect on the transport and magnetism of organic-inorganic hybrid charge–transfer complexes, influenced by solvent molecules hosted in the honeycomb Cu-oxalate-framework, was studied. It exhibits a semiconducting behavior with E_α = 40 meV. An antiferromagnetic interaction with the spin frustration f > 10 occurs, with no long-range order observed above 2 K. Thus, it is a candidate semiconducting quantum spin liquid.

Supplementary Materials: The following supporting information can be downloaded at: https://www.mdpi.com/article/10.3390/magnetochemistry11020012/s1, Table S1. The crystallographic data of **I** at 290 K, 180 K, 120 K. Table S2. Bond lengths of Cu–O in **I** at 290 K, 180 K and 120 K. Table S3. Bond lengths of the TTF core in BEDT-TTF of **I** at 290 K, 180 K and 120 K. X-ray crystallographic data supplied by CCDC 2417702/2412001/2417459/2417460.

Author Contributions: The manuscript was written with the contributions of all authors. Funding acquisition B.Z.; B.Z. and Y.Z. synthesized the sample.; Y.Z. Performed the transport and magnetic measurements; G.C., Z.G., Z.W., T.L. and Y.S. performed the X-ray experiments and data analysis: D.W. performed specific heat experiment and data analysis; Y.G. performed Raman experiment and data validation; F.L., Z.Z., X.Z. (Xiaoyu Zhang), B.Q., P.X. and J.W. performed the XPS experiments and data validation, F.D. performed thin film of crystal, D.Y. performed the EDS experiment, Q.L. and X.L., performed element anlysis and R.F. performed IR spectra measuremetn. M.L. and X.Z. (Xueying Zhang) performed data valiation, B.Z., Y.Z. and Z.W. analyzed all of the data. B.Z., Y.Z. and Z.W. wrote the main manuscript text. All authors have read and agreed to the published version of the manuscript.

Funding: This research was funded by the National Natural Science Foundation of China: 22273109, 22073106, 21573242 and 21173230.

Institutional Review Board Statement: Not applicable.

Informed Consent Statement: Not applicable.

Data Availability Statement: CCDC 2417702/2412001/2417459/2417460 contain the supplementary crystallographic data for compounds **I** at 100 K, 120 K, 180 K and 290 K, respectively. These data can be obtained free of charge via https://www.ccdc.cam.ac.uk/structures/ (accessed on 21 January

2025) or from Cambridge Crystallographic Data Centre, 12 Union Road, Cambridge CB2 1EZ, UK; fax: (+44) 1223-336-033; or e-mail: deposit@ccdc.cam.ac.uk. The original contributions presented in this study are included in the article/Supplementary Material. Further inquiries can be directed to the corresponding authors.

Conflicts of Interest: The authors declare no conflicts of interest.

References

1. Anderson, P.W. Resonating Valence Bonds: A New Kind of Insulator. *Mater. Res. Bull.* **1973**, *8*, 153–160. [CrossRef]
2. Bednorz, J.G.; Muller, K.A. Possible High T_c Superconductivity in the Ba−La−Cu−O System. *Z. Phys. B. Condensed Matter* **1986**, *64*, 189–193. [CrossRef]
3. Anderson, P.W. The Resonating Valence Bond State in La_2CuO_4 and Superconductivity. *Science* **1987**, *235*, 1197–1198. [CrossRef] [PubMed]
4. Balents, L. Spin liquids in frustrated magnets. *Nature* **2010**, *464*, 199–208. [CrossRef] [PubMed]
5. Broholm, C.; Cava, R.J.; Kivelson, S.A.; Nocera, D.G.; Norman, M.R.; Senthil, T. Quantum Spin Liquids. *Science* **2020**, *367*, eaay0668. [CrossRef] [PubMed]
6. Williams, J.M.; Schultz, A.J.; Geiser, U.; Carlson, K.D.; Kini, A.M.; Wang, H.H.; Kwok, W.; Whangbo, M. Organic Superconductors—New Benchmarks. *Science* **1991**, *252*, 1501–1508. [CrossRef] [PubMed]
7. Pratt, F.L.; Baker, P.J.; Blundell, S.J.; Lancaster, T.; Ohira-Kawamura, S.; Baines, C.; Shimizu, Y.; Kanoda, K.; Watanabe, I.; Saito, G. Magnetic and non-magnetic phase of quantum spin liquid. *Nature* **2011**, *471*, 612–616. [CrossRef]
8. Shimizu, Y.; Hiramatsu, T.; Maesato, M.; Otsuka, A.; Yamochi, H.; Ono, A.; Itoh, M.; Yoshida, M.; Takigawa, M.; Yoshida, Y.; et al. Pressure-Tuned Exchange Coupling of q Quan Spin Liquid in the Molecular Triangular Lattice κ-$(ET)_2Ag_2(CN)_3$. *Phys. Rev. Lett.* **2016**, *117*, 107203. [CrossRef] [PubMed]
9. Yamashita, M.; Nakata, N.; Senshu, Y.; Nagata, M.; Yamamoto, H.M.; Kato, R.; Shibauchi, T.; Matsuda, Y. Highly Mobile Gapless Excitations in a Two-Dimensional Candidate Quantum Spin Liquid. *Science* **2010**, *328*, 1246–1268. [CrossRef]
10. Isono, T.; Kamo, H.; Ueda, A.; Takahashi, K.; Kimata, M.; Tajima, H.; Tsuchiya, S.; Terashima, T.; Uji, S.; Mori, H. Gapless Quantum Spin Liquid in an Organic Spin-1/2 Triangular-Lattice κ-H_3(cat-EDT-TTF)$_2$. *Phys. Rev. Lett.* **2014**, *112*, 177201. [CrossRef] [PubMed]
11. Baudron, S.A.; Batail, P.; Coulon, C.; Clérac, R.; Canadell, E.; Laukhin, V.; Melzi, R.; Wzietek, P.; Jérome, D.; Auban-Senzier, P.; et al. (EDT-TTF-CONH$_2$)$_6$[$Re_6Se_8(CN)_6$], a Metallic Kagome-Type Organic−Inorganic Hybrid Compound: Electronic Instability, Molecular Motion, and Charge Localization. *J. Am. Chem. Soc.* **2005**, *127*, 11785–11797. [CrossRef]
12. Cano, J.; Alemany, P.; Alvarez, S.; Verdaguer, M.; Ruiz, E. Exchange Coupling in Oxalato-Bridged Copper(II) Binuclear Compounds: A Density Functional Study. *Chem. Eur. J.* **1998**, *4*, 476–484. [CrossRef]
13. Hursthouse, M.B.; Light, M.E.; Price, D.J. One-dimensional Magnetism in Anhydrous Iron and Cobalt Ternary Oxalates with Rare Trigonal-Prismatic Metal Coordination Environment. *Angew. Chem. Int. Ed.* **2004**, *43*, 472–475. [CrossRef]
14. Xu, H.; Wang, Z.; Liu, T.; Gao, S. Synthesis, Structure, and Magnetic Properties of (A)[Fe^{III}(oxalate)Cl_2] (A = Alkyl Ammonium Cations) with Anionic 1D [Fe^{III}(oxalate)Cl_2]$^-$ Chains. *Inorg. Chem.* **2007**, *46*, 3089–3096. [CrossRef] [PubMed]
15. Duan, Z.; Zhang, Y.; Zhang, B.; Zhu, D. $Co(C_2O_4)(HO(CH_2)_3OH)$: An Antiferromagnetic Neutral Zigzag Chain Compound Showing Long-Range Ordering of Spin Canting. *Inorg. Chem.* **2008**, *47*, 9152–9154. [CrossRef] [PubMed]
16. Zhang, B.; Zhang, Y.; Chang, G.; Wang, Z.; Zhu, D. Crystal-to-Crystal Transformation from $K_2[Co(C_2O_4)_2(H_2O)_2]\cdot 4H_2O$ to $K_2[Co(\mu\text{-}C_2O_4)(C_2O_4)]$. *Magnetochemistry* **2021**, *7*, 77. [CrossRef]
17. Mathoniere, C.; Nuttall, C.J.; Carling, G.; Day, P. Ferrimagnetic Mixed-Valency and Mixed-Metal Tris(oxalato)iron(III) Compounds: Synthesis, Structure, and Magnetism. *Inorg. Chem.* **1996**, *35*, 1201–1206. [CrossRef]
18. Tamaki, H.; Zhong, Z.J.; Matsumoto, N.; Kida, S.; Koikawa, M.; Achiwa, N.; Hashimoto, Y.; Okawa, H. Design of Metal-Complex Magnets. Syntheses and Magnetic Properties of Mixed-Metal Assemblies {$NBu_4[MCr(ox)_3]$} (NBu_4^+ = Tetra(n-buty1)ammonium Ion; ox^{2-} = Oxalate Ion; M = Mn^{2+}, Fe^{2+}, Co^{2+}, Ni^{2+}, Cu^{2+}, Zn^{2+}). *J. Am. Chem. Soc.* **1992**, *114*, 6974–6979. [CrossRef]
19. Zhang, B.; Zhang, Y.; Zhang, J.; Li, J.; Zhu, D. A neutral molecular-based layered magnet $[Fe(C_2O_4)(CH_3OH)]_n$ exhibiting magnetic ordering at $T_N \approx 23$ K. *Dalton Trans.* **2008**, 5037–5040. [CrossRef]
20. Rousse, G.; Rodriguez-Carvajal, J. Oxalate-mediated long-range antiferromagnetism order in $Fe_2(C_2O_4)_3\cdot 4H_2O$. *Dalton Trans.* **2016**, *45*, 14311–14319. [CrossRef] [PubMed]
21. Decurtins, S.; Schmalle, H.W.; Schneuwly, P.; Ensling, J.; Gutlich, P. A Concept for the Synthesis of 3-Dimensional Homo- and Bimetallic Oxalate-Bridged Networks $[M_2(ox)_3]_n$. Structural, Mössbauer, and Magnetic Studies in the Field of Molecular-Based Magnets. *J. Am. Chem. Soc.* **1994**, *116*, 9521–9528. [CrossRef]
22. Hernandez-Molina, M.; Lloret, F.; Ruiz-Perez, C.; Julve, M. Weak Ferromagnetism in Chiral 3-Dimensional Oxalato-Bridged Cobalt(II) Compounds. Crystal Structure of $Co(bpy)_3][Co_2(ox)_3]ClO_4$. *Inorg. Chem.* **1998**, *37*, 4131–4145. [CrossRef]

23. Coronado, E.; Galan-Mascaros, J.R.; Gomez-Garzia, C.J.; Martinez-Agudo, J.M. Molecule-Based Magnets Formed by Bimetallic Three-Dimensional Oxalate Networks and Chiral Tris(bipyridyl) Complex Cations. The Series Tris(bipyridyl) Complex Cations. The Series $[Z^{II}(bpy)_3][ClO_4][M^{II}Cr^{III}(ox)_3]$ (Z^{II} = Ru, Fe, Co, and Ni; M^{II} = Mn, Fe, Co, Ni, Cu, and Zn; ox = Oxalate Dianion). *Inorg. Chem.* **2001**, *40*, 113–120.
24. Mon, M.; Grancha, T.; Verdaguer, M.; Train, C.; Armentano, D.; Pardo, E. Solvent-Dependent Self-Assembly of an Oxalato-Based ThreeDimensional Magnet Exhibiting a Novel Architecture. *Inorg. Chem.* **2016**, *55*, 6845–6847. [CrossRef] [PubMed]
25. Zhang, B.; Zhang, Y.; Zhang, J.; Xiang, H.; Zhu, D. $Mn(C_2O_4)(H_2O)_{0.25}$: An antiferromagnetic oxalato-based cage compound. *Dalton Trans.* **2011**, *40*, 5430–5432. [CrossRef] [PubMed]
26. Coronado, E.; Galan-Mascaros, J.R.; Marti-Gastaldo, C. Single chain magnets based on the oxalate ligand. *J. Am. Chem. Soc.* **2008**, *130*, 14987–14989. [CrossRef]
27. Thomas, T.W.; Che-Hsiung, H.; Labes, M.M.; Gomm, P.S.; Underhill, A.E.; Watkins, D.M. Temperature Dependence of Conductivity of Some Partly-Oxidised Platinum Complexes. *J. Chem. Soc. Dalton* **1972**, 2050–2053. [CrossRef]
28. Braude, A.; Carneiro, K.; Jacobsen, C.S.; Mortensen, K.; Turner, D.J.; Underhill, A.E. Solid-state properties of one-dimensional metals based on bis(oxalate)platinate anions with divalent cations. *Phys. Rev. B* **1987**, *36*, 7835–7846. [CrossRef] [PubMed]
29. Gartner, S.; Heinen, I.; Schveitzer, D.; Nuber, B.; Keller, H.J. BEDT-TTF salts with square platinates (II) as counterions: $(BEDT\text{-}TTF)_4[Pt(C_2O_4)_2]$. A new organic metal. *Synth. Met.* **1989**, *31*, 199–213. [CrossRef]
30. Wang, X.; Ge, C.; Xing, X.; Wang, P.; Zhang, D.; Wu, P.; Zhu, D. Preparation and properties of a new organic conductor $(BEDT\text{-}TTF)_4Cu(C_2O_4)_2$. *Synth. Met.* **1991**, *39*, 355–358. [CrossRef]
31. Wang, P.; Bandow, S.; Maruyama, Y.; Wang, X.; Zhu, D. Physical and structure properties of a new organic conductor $(BEDT\text{-}TTF)_4Cu(C_2O_4)_2$. *Synth. Met.* **1991**, *44*, 147–157. [CrossRef]
32. Kurmoo, M.; Graham, A.W.; Day, D.; Simon, J.C.; Hursthouse, M.B.; Caufield, J.L.; Singleton, J.; Pratt, F.L.; Hayes, W.; Ducasse, L.; et al. Superconducting and Semiconducting Magnetic Charge Transfer Salts: $(BEDT\text{-}TTF)_4AFe(C_2O_4)_3\cdot C_6H_5CN$. *J. Am. Chem. Soc.* **1995**, *117*, 12209–12217. [CrossRef]
33. Cassoux, P. A Molecular Paramagnetic Superconductor. *Science* **1996**, *272*, 1277–1278. [CrossRef]
34. Zhang, B.; Wang, Z.; Fujiwara, H.; Kobayashi, H.; Kurmoo, M.; Inoue, K.; Mori, T.; Gao, S.; Zhang, Y.; Zhu, D. Tetrathiafulvalene $[Fe^{III}(C_2O_4)Cl_2]$: An organic-inorganic hybrid exhibiting canted antiferromagnetism. *Adv. Mater.* **2005**, *17*, 1988–1991. [CrossRef]
35. Zhang, B.; Wang, Z.; Zhang, Y.; Takahashi, K.; Okano, Y.; Cui, H.; Kobayashi, H.; Inoue, K.; Kurmoo, M.; Pratt, F.; et al. Hybrid organic-inorganic conductor with a magnetic chain anion: κ-$BETS_2[Fe^{III}(C_2O_4)Cl_2]$ [BETS = bis(ethylenedithio)tetraselenafulvalene]. *Inorg. Chem.* **2006**, *45*, 3275–3280. [CrossRef]
36. Zhang, B.; Zhang, Y.; Gao, Z.; Chang, G.; Su, S.; Wang, D.; Guo, Y.; Zhu, D. Synthesis, Crystal Structure, and Characterization of the Charge-transfer Salt: $(BEDT\text{-}TTF)[Fe(C_2O_4)Cl_2](CH_2Cl_2)$, (BEDT-TTF = bis(ethylenedithio)tetrathiafulvalene). *Eur. J. Inorg. Chem.* **2014**, *2014*, 4028–4032. [CrossRef]
37. Zhang, B.; Zhang, Y.; Wang, Z.; Yang, D.; Gao, Z.; Wang, D.; Guo, Y.; Zhu, D.; Mori, T. $[BEDT\text{-}TTF][Fe(C_2O_4)Cl_2]$: An organic–inorganic hybrid semiconductive antiferromagnet. *Dalton Trans.* **2016**, *45*, 16561–16565. [CrossRef] [PubMed]
38. Palacio, F.; Miller, J.S. A dual-function material. *Nature* **2000**, *408*, 421–422. [CrossRef] [PubMed]
39. Coronado, E.; Galan-Mascaros, J.R.; Gomez-Garcia, C.J.; Laukin, V. Coexistence of ferromagnetism and metallic conductivity in a molecular-based layered compound. *Nature* **2000**, *408*, 447–449. [CrossRef] [PubMed]
40. Alberola, A.; Coronado, E.; Galan-Mascaros, J.R.; Gimenez-Saiz, C.; Gimene-Garcia, C.J. A Molecular Metal Ferromagnet from the Organic Donor Bis(ethylenedithio)tetraselenafulvalene and Bimetallic Oxalate Complexes. *J. Am. Chem. Soc.* **2003**, *125*, 10774–10775. [CrossRef] [PubMed]
41. Jahn, H.A.; Teller, E. Stability of polyatomic molecules in degenerate electronic states-I—Orbital degeneracy. *Proc. R. Soc. Lond. Ser. A Math. Phys. Sci.* **1937**, *161*, 220–235.
42. Zhang, B.; Zhang, Y.; Wang, Z.; Sun, Y.; Liang, T.; Liu, M.; Zhu, D. Two One-dimensional Copper-Oxalate Frameworks with the Jahn–Teller Effect: $[(CH_3)_3NH]_2[Cu(\mu\text{-}C_2O_4)(C_2O_4]\cdot 2.5H_2O$ (I) and $[(C_2H_5)_3NH]_2[Cu(\mu\text{-}C_2O_4)(C_2O_4)]\cdot H_2O$. *Magnetochemistry* **2023**, *9*, 120. [CrossRef]
43. Zhang, B.; Zhang, Y.; Wang, Z.; Wang, D.; Baker, P.; Pratt, F.; Zhu, D. Candidate Quantum Spin Liquid due to Dimensional Reduction of a Two-Dimensional Honeycomb Lattice. *Sci. Rep.* **2014**, *4*, 6451. [CrossRef] [PubMed]
44. Zhang, B.; Zhang, Y.; Zhu, D. $(BEDT\text{-}TTF)_3[Cu_2(C_2O_4)_3](CH_3OH)_2$: An organic-inorganic hybrid antiferromagnetic semiconductor. *Chem. Commun.* **2012**, *48*, 197–199. [CrossRef] [PubMed]
45. Zhang, B.; Zhang, Y.; Wang, Z.; Gao, S.; Guo, Y.; Liu, F.; Zhu, D. $BETS_3[Cu_2(C_2O_4)_3](CH_3OH)_2$: An organic–inorganic hybrid antiferromagnetic metal (BETS = bisethylene(tetraselenfulvalene)). *CrystEngComm* **2013**, *15*, 3529–3535. [CrossRef]
46. Zhang, B.; Zhang, Y.; Zhu, D. $[(C_2H_5)_3NH]_2Cu_2(C_2O_4)_3$: A three-dimensional metal-oxalato framework showing structurally related dielectric and magnetic transitions at around 165 K. *Dalton Trans.* **2012**, *14*, 8509–8511. [CrossRef] [PubMed]
47. Zhang, B.; Baker, P.; Zhang, Y.; Wang, D.; Wang, Z.; Su, S.; Zhu, D. Quantum Spin Liquid from a Three-Dimensional Copper-Oxalate Framework. *J. Am. Chem. Soc.* **2018**, *140*, 122–125. [CrossRef]

48. Geim, A.K.; Novoselov, K.S. The rise of graphene. *Nat. Mater.* **2007**, *6*, 183–191. [CrossRef] [PubMed]
49. Sheldrick, G.M. *SHELXL-2012/7*, University of Göttingen: Göttingen, Germany, 2014.
50. Guionneau, P.; Kepert, C.J.; Bravic, G.; Chasseau, D.; Truter, M.R.; Kurmoo, M.; Day, P. Determing the charge distribution in BEDT-TTD salts. *Synth. Met.* **1997**, *86*, 1973–1974. [CrossRef]
51. Wang, H.H.; Ferraro, J.R.; Williams, J.M.; Geiser, U.; Schlueter, J.A. Rapid Raman Spectroscopic Determination of the Stoichiometry of Microscopic Quantities of BEDT-TTF-based Organic Conductors and Superconductors. *J. Chem. Soc. Chem. Commun.* **1994**, 1893–1894. [CrossRef]
52. Price, C.C.; Perkins, N.B. Critical Properties of the Kitaev-Heisenberg Model. *Phy. Rev. Lett.* **2012**, *109*, 187201. [CrossRef]
53. Kahn, O. *Molecular Magnetism*; John Wiley & Sons Inc.: New York, NY, USA, 1993.
54. Ramirez, A.P. Strongly geometrically frustrated magnets. *Annu. Rev. Mater. Sci.* **1994**, *24*, 453–480. [CrossRef]

magnetochemistry

MDPI

Article

Magnetic Relaxation in a Heterolanthanide Binuclear Complex Involving a Nitronyl Nitroxide Biradical †

Yan Zhou, Junfang Xie, Chaoyi Jin, Yue Ma and Licun Li *

Department of Chemistry, Key Laboratory of Advanced Energy Materials Chemistry, College of Chemistry, Nankai University, Tianjin 300071, China
* Correspondence: llicun@nankai.edu.cn
† Dedicated to Prof. Dai-Zheng Liao on the occasion of his 85th birthday.

Abstract: The reaction of nitronyl nitroxide biradical NITPhMeImbis(5-(2-methylimidazole)-1,3-bis(1′-oxyl-3′-oxido-4′,4′,5′,5′-tetramethyl-4,5-hydro-1*H*-imidazol-2-yl)-benzene) with $Dy(hfac)_3$ and $La(hfac)_3$ (hfac = hexafluoroacetylacetonate) afforded a heterolanthanide complex $[Dy_{0.56}La_{1.44}(hfac)_7(NITPhMeImbisH)]$ (**1**). In this complex, the biradical NITPhMeImbis ligand chelates one Ln(III) ion via its two neighboring NO units and simultaneously binds the La(III) ion through another NO group to form a dinuclear structure. Direct current (dc) magnetic measurement shows the dominant ferromagnetic couplings in Complex **1**. Spin dynamics studies exhibit visible frequency-dependent peaks of χ'' signals under a dc field evidenced by field-induced magnetic relaxation behavior, which is a combination of Orbach and QTM processes, giving the U_{eff}, τ_0 and τ_{QTM} values of 15.14 K, 3.04×10^{-7} s and 3.61×10^{-4} s, respectively.

Keywords: nitronyl nitroxide biradical; heterolanthanide; crystal structure; magnetic relaxation

1. Introduction

Due to the high magnetic moments and strong magnetic anisotropies, lanthanides are of great interest for designing single molecule magnets (SMMs) [1–4], which have potential applications in high-density memory storage [5], the quantum sciences [6–8] and molecular spintronics [9,10]. High performance SMMs have been achieved by the use of lanthanides, which show blocking temperatures of the magnetization (T_B) near liquid nitrogen temperatures [11,12]. Despite these great advances, it is still a challenge to enhance magnet behaviors of SMMs toward practical application. To develop performant SMMs, one of the promising strategies is to design Ln-radical architectures in which strong magnetic exchanges between the Ln centers and the radicals could be achieved, leading to effectively suppress the quantum tunneling of magnetization (QTM) of Ln ions and thus toward improving the performance of SMMs [13]. For instance, a series of Ln-radical species exhibiting remarkably high T_B and large coercive fields has been obtained using N-based radicals [14–17].

On the other hand, recently, another approach, with the aim of enhancing lanthanide SMM properties, is to design Ln-Ln′ heterolanthanide complexes that combine two kinds of magnetic anisotropy in one molecule, offering a promising pathway to SMMs with improved performances [18–22]. However, the synthesis of heterolanthanide molecular complexes is difficult due to the similar chemistries of lanthanides and still remains challenging. So far, different synthetic approaches have been developed to obtain Ln-Ln′ heterolanthanide complexes in which the distinct lanthanide ions stay in selected sites, e.g., by step-by-step approaches [23–26] or connecting preformed lanthanide containing species [27–29].

Using these synthetic procedures, a few heterolanthanide SMMs have been reported [20,30]. However, the heterolanthanide SMMs involving nitronyl nitroxide radical remain unexplored. The nitronyl nitroxide–heterolanthanide complexes will provide a new way for improving the magnetic performances of the magnetic system in which three different spins behave in a synergistic way. Therefore, in this study, we use nitronyl nitroxide biradical NITPhMeImbis (5-(2-methylimidazole)-1,3-bis(1′-oxyl-3′-oxido-4′,4′,5′,5′-tetramethyl-4,5-hydro-1*H*-imidazol-2-yl)-benzene, Scheme 1) to construct a heterolanthanide system. The NITPhMeImbis nitronyl nitroxide biradical could construct Ln-Ln′ complexes as a step-by-step approach since it could firstly chelate a Ln ion by two neighboring NIT moieties and then the remaining NO units could be ligated to another different Ln′ ion. In this study, Dy(III) and La(III) ions were chosen to construct heterolanthanide-radical complex because the Dy(III) ion is a Kramer ion with strong magnetic anisotropy and the diamagnetic La(III) ion could decrease undesirable intermolecular dipole interactions. Accordingly, one DyLa heterolanthanide binuclear complex [$Dy_{0.56}La_{1.44}(hfac)_7$(NITPhMeImbisH)] (NITPhMeImbisH = 5-(2-methyl-3-hydroimidazolium)-1,3-bis(1′-oxyl-3′-oxido-4′,4′,5′,5′-tetramethyl-4,5-hydro-1*H*-imidazol-2-yl)-benzene) was achieved, which exhibits magnetic relaxation behavior.

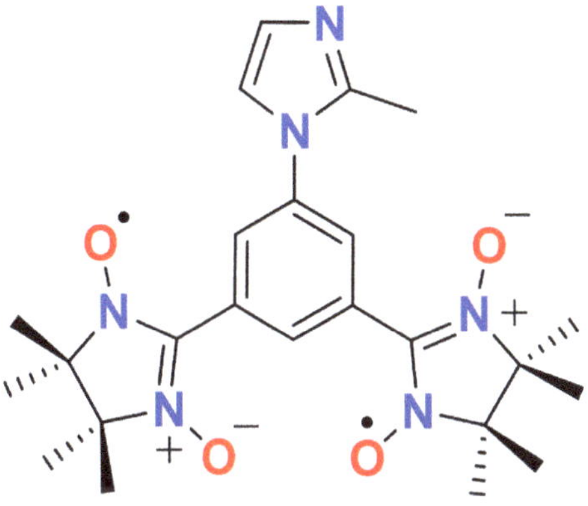

Scheme 1. NITPhMeImbis ligand.

2. Results and Discussion

2.1. Synthesis and Crystal Structure

The nitronyl nitroxide biradical ligand NITPhMeImbis was used to prepare a DyLa heterolanthanide complex by step-by-step strategy. The NITPhMeImbis biradical was firstly coordinated with $Dy(hfac)_3$ via two adjacent NO units of the biradical to a form mononuclear Dy complex and then the resulting mononuclear precursor was ligated to $La(hfac)_3$ using uncoordinated NO groups. As a result, a heterolanthanide binuclear DyLa complex [$Dy_{0.56}La_{1.44}(hfac)_7$(NITPhMeImbisH)] with a statistical distribution was obtained.

Complex **1** crystallizes in the monoclinic $P2_1/c$ space group. As shown in Figure 1, the NITPhMeIm ligand chelates one $Ln(hfac)_3$ unit through its two adjacent NO units and meanwhile coordinates with one $La(hfac)_4$ unit via another NO group to lead to a dinuclear structure. It should be noted that one Ln atom is octa-coordinated with three hfac coligands while the other is in nona-coordinated with four hfac coligands. Thus, to charge balance, the N atom of imidazole ring in the biradical ligand was protonated.

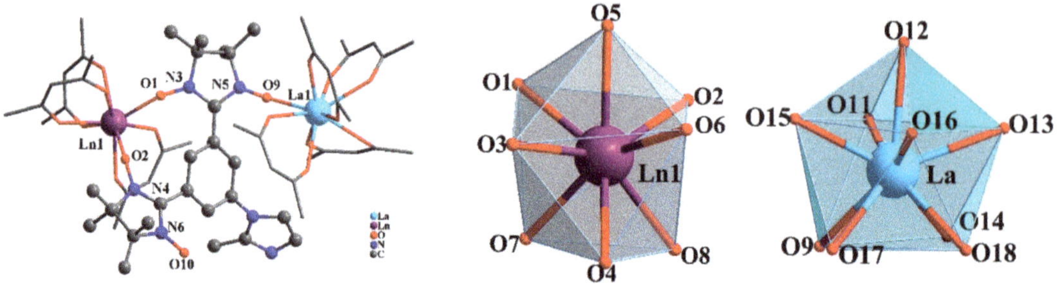

Figure 1. The binuclear structure of Complex **1** (**left**). Coordination polyhedra of Ln1 and La in Complex **1** (**right**).

Continuous shape measure analysis (CShM) [31] was performed to elucidate the geometry of the Ln1(III) and La(III) ion. The octa-coordinated Ln1(III) ion exhibits a capped trigonal prism(C_{2v}) geometry while the nona-coordinated La(III) ion exhibits a spherical capped square antiprism(C_{4v}) geometry (Table S3). The octa-coordinated Ln1 is bonded by two O atoms from two NO groups of the biradical and six O atoms from three hfac coligands. Two Ln1-O_{rad} bond lengths are 2.443(5) and 2.471(7) Å, respectively, and the Ln–O_{hfac} bond distances vary from 2.378(8) to 2.462(6) Å. These bond lengths are obviously longer than those which exist in similar nitronyl nitroxide biradical-Dy^{III} complexes [32–34]. As for this, it thus seems reasonable to consider this Ln coordination site to be occupied by either the Dy^{III} and La^{III} ions. The La atom is nine-coordinated by one O atom of the NO group (La–O_{rad}: 2.512(5) Å) and eight O atoms of the four hfac ligands (La–O_{hfac}: 2.415(6)–2.598(6)Å). The average La–O_{hfac} bond distance is 2.498(6) Å, which is close to those in reported La^{III} complexes with hfac coligands [35–37]. Inductively coupled plasma (ICP) measurements were therefore conducted to investigate the composition of Complex **1**. The result reveals a ratio of 2.59:1 for La^{III}: Dy^{III} and the Ln1 site shows a statistical distribution of the Dy and La ions.

The bond angles of Ln1–O1–N3 and Ln1–O2–N4 are 145.0(5)° and 148.6(6)°, respectively, and the torsion angles of Ln1–O1–N3–C1 and Ln1–O2–N4–C8 are $|-72.06(1)°|$ and 62.45(2)°, respectively. The N5–O9–La bond angle is 142.0(5)° and the La–O9–N5–C1 torsion angle is $|-85.65°|$. For Complex **1**, the dihedral angles between the benzene ring and the two NIT planes are 26.68(1)° and 29.48(4)°, respectively, while the dihedral angle formed by the imidazole and phenyl rings is 60.44(3)°. In the dinuclear molecule, the distance between two Ln^{III} ions is 8.736(7) Å while the nearest intermolecular Ln---Ln(Ln1$\cdots$La) distance is 12.653(7) Å. The stacking diagram is depicted in Figure S3, and the closest distance for non-coordinating NO groups is 9.450(2) Å.

2.2. Magnetic Properties

Direct current (dc) magnetic data of Complex **1** were measured under 1000 Oe between 2 K and 300 K. As seen in Figure 2a, at 300 K, the found $\chi_M T$ value was 8.59 cm^3 Kmol^{-1} at room temperature, which is lower than that expected for one Dy(III) in the eight-coordinated Ln site and one La(III) in the nine-coordinated site (Dy(III): $^6H_{15/2}$, $S = 5/2$, $L = 5$, $g = 4/3$, $C = 14.17$ cm^3 K mol^{-1}; La(III): diamagnetic) plus two radicals ($S = 1/2$); however, this value is close to the theoretical values (8.68 cm^3 K mol^{-1}) for mixed La(III) and Dy(III) ions with a ratio of 2.59:1 plus two radicals. This suggests that the La(III) ion is also found in the eight-coordinated Ln site. As the temperature decreases, the $\chi_M T$ value gradually increases and reaches the maximum of 11.80 cm^3 K mol^{-1} at 3 K and then slightly decreases to 11.80 cm^3 K mol^{-1} at 2 K, indicating the presence of ferromagnetic coupling in the system. In Complex **1**, the magnetic exchange between the Dy(III) ion and the coordinated NO unit is usually ferromagnetic [38,39]. Magnetic communication between two NIT units via the Ln(III) ion is expected to be antiferromagnetic while exchange coupling between two NIT moieties through m-phenyl ring is expected to be ferromagnetic based on spin polarization [40]. The isothermal M vs. H plot for Complex **1** has been determined to be in the range of 0–7 T at 2 K (Figure 2b). The value of M increases with the magnetic field and reaches a value of 5.16 Nβ at 70 kOe, which is lower than the anticipated saturation value, implying the existence of magnetic anisotropy of Complex **1**.

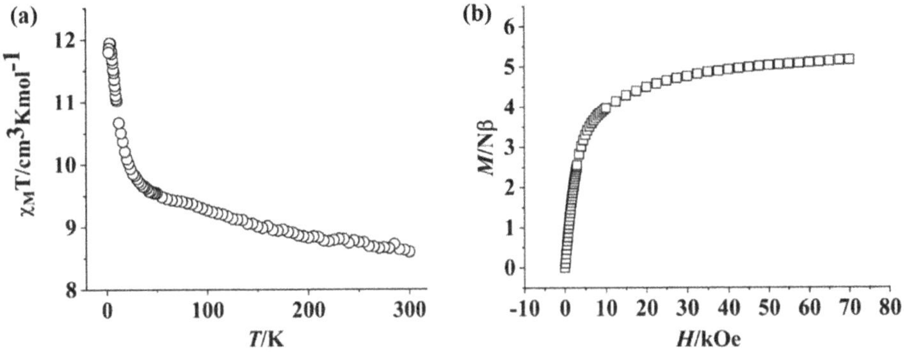

Figure 2. (**a**) $\chi_M T$ vs. T curve for Complex **1**. (**b**) M vs. H plot for Complex **1** at 2 K.

To explore the potential magnetic relaxation behavior of Complex **1**, ac susceptibility measurements without an applied dc field were performed on Complex **1**. Complex **1** displays pronounced frequency-dependent χ'' signals between 2 K and 10 K in the absence of a dc magnetic field, demonstrating the existence of magnetic relaxation behavior (Figure S4). However, there were no peaks to be found, suggesting significant quantum tunneling (QTM) of the magnetization in Complex **1** [13]. Consequently, the frequency-dependent ac magnetic susceptibilities of Complex **1** were measured at 2 K within a dc field range of 400–2400 Oe for suppressing QTM (Figure S5). The τ values obtained by fitting the ac susceptibility data through the Debye mode [41] were plotted vs. H (Figure S6) and then the optimized field was determined to be 800 Oe. Under the optimum dc field of 800 Oe, the χ'' signals exhibit distinct peaks in the range of 2.0 to 3.6 K (Figure 3). The relaxation time (τ) and the corresponding distribution parameter (α) were extracted by fitting the frequency-dependent ac data through the generalized Debye model [41]. The resulting lnτ versus T plot exhibits a linear temperature dependence at high temperature, but the derivation is observed at low temperature (Figure 4a). Consequently, the best fitting was achieved by considering the Orbach and QTM processes, i.e., $\tau^{-1} = \tau_{QTM}{}^{-1} + \tau_0{}^{-1}\exp(-U_{\text{eff}}/k_B T)$, leading to the anisotropic energy barrier U_{eff}/k_B = 15.14 K with $\tau_0 = 3.04 \times 10^{-7}$ s and $\tau_{QTM} = 3.61 \times 10^{-4}$ s. The resulting τ_0 value of 3.04×10^{-7} falls in the characteristic range for SMMs (10^{-5}–10^{-12} s) [2,42–44]. The Cole–Cole plot displays an almost semicircular shape (Figure 4a) and the obtained α is within the limit of 0.32–0.52 indicating the middle distribution of relaxation time. It is appealing to compare Complex **1** with the related Dy-nitronyl nitroxide biradical complexes in the literature. The related magnetic and structural parameters are summarized in Table 1. As seen, the magnetic relaxation behaviors of Dy-biradical complexes are very sensitive to the ligand field around the Dy atom as well as the ligand substituents. The energy barrier of Complex **1** remains superior, which could be ascribed to weak intermolecular dipole interactions arising from diamagnetic La(III) ions.

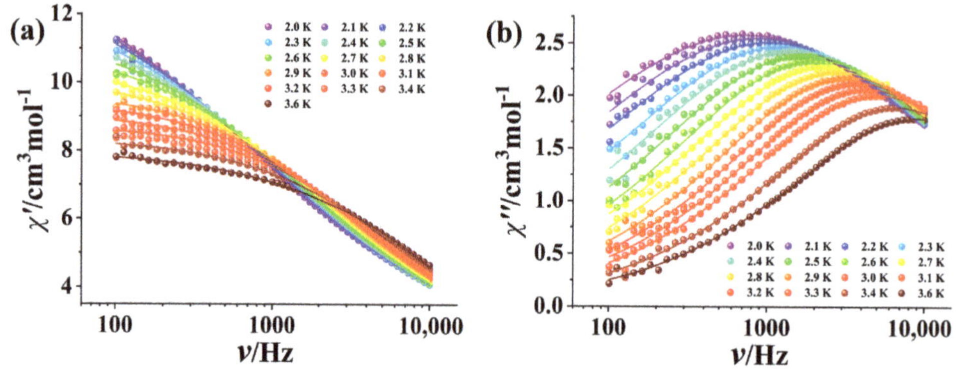

Figure 3. Frequency-dependent χ' (**a**) and χ'' (**b**) curves for Complex **1** under the 800 Oe dc field. (The solid lines are the fitting results of the Debye model.)

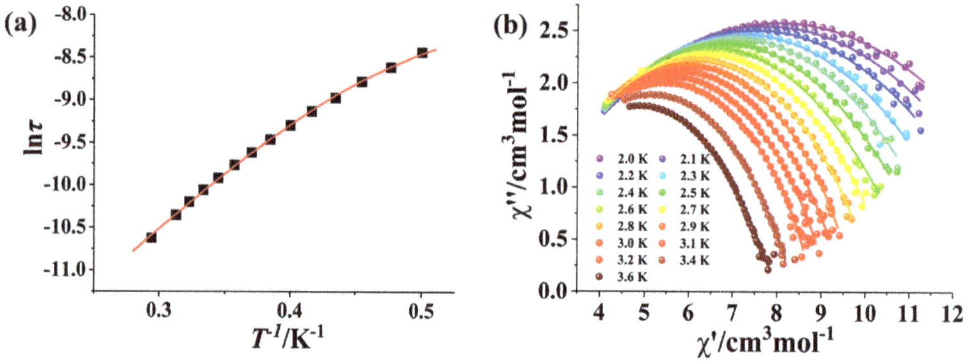

Figure 4. (**a**) The ln τ versus 1/T plot for Complex **1** (the red line is the fitting result of the combined Orbach and QTM processes). (**b**) Cole–Cole curves for Complex **1** under the 800 Oe dc field (the solid lines are the fitting results of the Debye model).

Table 1. Structural and magnetic parameters for related nitronyl nitroxide biradical-Dy complexes.

Compound	Configuration	U_{eff}/k_B (K)	τ_0 (s)	H_{dc} (Oe)	Ref.
$[Dy_2Cu_3(hfac)_{12}(NITPhMeImbis)_2]$	C_{2v}	13	5.80×10^{-8}	3000	[32]
$[Dy(hfac)_3(NITmbis)]_2$	C_{2v}	11.6	2.3×10^{-8}	0	[33]
$[DyMn(hfac)_5(NITPh\text{-}PyPzbis)]$	D_{2d}	/	/	/	[45]
$[DyNi(hfac)_5(NITPh\text{-}PyPzbis)]$	C_{2V}	23.30	6.33×10^{-7}	2000	[45]
$[DyCo(hfac)_5(NITPh\text{-}PyPzbis)]$	C_{2V}	14.91	6.80×10^{-8}	2500	[46]
$[Dy(hfac)_3Cu(hfac)_2(bisNITPhPyrim)]$	$C_{2v} \sim D_{2d}$	8.13	1.07×10^{-6}	500	[47]
$[Dy(hfac)_3(NITPymbis)_2]_2 \cdot C_7H_{16}$	C_{2v}	none	none	0	[48]
$[Dy(hfac)_3(NITFumbis)]_2$	C_{2v}	15	1.25×10^{-6}	3000	[49]
$[Dy_{0.56}La_{1.44}(hfac)_7(NITPhMeImbisH)]$	C_{2v}(Ln1), C_{4v}(La)	15.14	3.04×10^{-7}	800	This study.

NITPhMeImbis = (5-(2-methylimidazole)-1,3-bis(1′-oxyl-3′-oxido-4′,4′,5′,5′-tetramethyl-4,5-hydro-1H-imidazol-2-yl)-benzene; NITmbis = 1,3-bis-(1′-oxyl-3′-oxido-4′,4′,5′,5′-tetramethyl-4,5-hydro-1H-imidazol-2-yl)benzene; NITPh-PyPzbis = 5-(3-(2-pyridinyl)-1H-pyrazol-1-yl)-1,3-bis(1′-oxyl-3′-oxido- 4′,4′,5′,5′-tetramethyl-4,5-hydro-1H-imidazol-2-yl)benzene; bisNITPhPyrim = [5-(5-pyrimidyl)-1,3-bis(1′-oxyl-3′- oxido-4′,4′,5′,5′-tetramethyl-4,5-hydro-1H-imidazol-2-yl)]benzene; NITPymbis = 1,3-bis-(1′-oxyl-3′-oxido-4′,4′,5′,5′-tetramethyl-4,5-hydro-1H-imidazol-2-yl) pyridine; NITFumbis = 2,5-bis-(1′-oxyl-3′-oxido-4′,4′,5′,5′-tetramethyl-4,5-hydro-1H-imidazol-2-yl)furan.

3. Experimental Section

3.1. Materials and Characterizations

All reactions were carried out using commercial grade solvents without further purification. The specific synthesis process of nitronyl nitroxide biradical NITPhMeImbis [32] and Ln(hfac)$_3$·2H$_2$O [50] were prepared following previously reported methods. Elemental analyses were carried out by Perkin-Elmer 240 elemental analyzer. IR spectra were recorded on a Bruker Tensor 27 spectrometer with the samples prepared as KBr discs in the 4000–400 cm^{-1} range. Powder X-ray diffraction (PXRD) patterns were recorded with a Rigaku Ultima IV diffractometer. Inductively coupled plasma–optical emission spectrometry (ICP-OES) was performed on SPECTRO-BLUE to verify the composition of the complex. Magnetization measurements were conducted using a Quantum Design model MPMS3 SQUID and PPMS-9 magnetometers. DC magnetic susceptibility data were corrected using Pascal's constants [51].

3.2. Synthesis of [Dy$_{0.56}$La$_{1.44}$(hfac)$_7$(NITPhMeImbisH)]

A total of 0.01 mmol of Dy(hfac)$_3$·2H$_2$O (8.2 mg) in dry *n*-heptane (20 mL) was refluxed for 3 h. Then 8 mL of CH$_2$Cl$_2$ containing 4.7 mg (0.01 mmol) of NITPhMeImbis was added and reacted for 15 min. Subsequently, 0.02 mmol of La(hfac)$_3$·2H$_2$O(16.0 mg) in 8.0 mL of *n*-heptane, in which the hydration water molecules had been azeotropically removed, was dropped in the above solution. The resulting solution was then allowed to cool to room temperature and filtered, and the dark purple crystals suitable for crystallographic characterization were obtained after five days. Yield 56%; C$_{59}$H$_{40}$F$_{42}$Dy$_{0.56}$La$_{1.44}$N$_6$O$_{18}$ (2210.00 g mol^{-1}); Elem. Anal. (%) found (calcd.): C, 32.24 (32.06); N, 3.70 (3.80); H, 1.69 (1.81).

[Dy$_{0.56}$La$_{1.44}$(hfac)$_7$(NITPhMeImbisH)] (1): Yield 56%; C$_{59}$H$_{39}$F$_{42}$DyLaN$_6$O$_{18}$ (2219.37 g mol^{-1}); Elem. Anal. (%) found (calcd.): C, 32.24 (31.93); N, 3.70 (3.79); H, 1.69 (1.77). FT-IR (cm^{-1}): 1648 (s), 1605 (w), 1554 (w), 1527 (w), 1492 (m), 1369 (m), 1249 (s), 1195 (s), 1129 (s), 796 (s), 740 (m), 659 (s), 580 (s) and 547 (m).

For FT-IR spectra (Figure S1), the peaks at 1648 cm^{-1} and 796 cm^{-1} could be assigned to the stretching and bending vibrations of C=O from hfac- ligands, respectively. The C-F stretching vibrations of the CF$_3$ group appear at 1249, 1195 and 1129 cm^{-1} while the peak at 659 cm^{-1} is associated with the bending vibration of C–F from the hfac- ligand. The peaks at 1605, 1527 and 1492 cm^{-1} result from the stretching vibration of C=C (benzene ring) while those at 1554 and 1369 cm^{-1} could be assigned to the stretching vibration of the C=N and N–O groups, respectively [50]. The recorded experimental PXRD pattern of Complex **1** is in good agreement with the simulated pattern from single-crystal X-ray diffraction (Figure S2).

3.3. X-Ray Crystallography

Diffraction intensity data collection of Complex **1** was made on a Rigaku HyPix diffractometer with graphite monochromated Cu-Kα radiation (λ = 1.54184 Å) at a temperature of 113 K. Direct methods were utilized to solve crystal structures of Complex **1** using SHELXS-2018 and SHELXL-2018 [52,53]. All of the non-H atoms were refined anisotropically, and the H atom positions were given geometrically. To rationalize some disordered F and C atoms, commands of ISOR, SIMU and DELU were used. Crystallographic data and selected bond distances and angles are given in Table S1 and Table S2, respectively.

CCDC-2427073 contains the supplementary crystallographic data for Complex **1**. The data can be obtained free of charge from the Cambridge Structural Database through www.ccdc.cam.ac.uk/structures (accessed on 28 March 2025)

4. Conclusions

We have successfully synthesized one DyLa heterolanthanide binuclear complex [$Dy_{0.56}La_{1.44}(hfac)_7$(NITPhMeImbisH)] using a nitronyl nitroxide biradical ligand. In this heterolanthanide complex, one Ln(III) ion is chelated by two neighboring NO units of nitronyl nitroxide biradical while the La(III) ion is coordinated by another NO group. Interestingly, DyLa heterolanthanide binuclear complex exhibits field-induced magnetic behavior. To the best of our knowledge, this DyLa radical-heterolanthanide represents the first heterolanthanide complex involving nitronyl nitroxide radical. This work demonstrates that the radical-heterolanthanide approach is a promising way for designing new radical-Ln magnetic relaxation systems.

Supplementary Materials: The following supporting information can be downloaded at https://www.mdpi.com/article/10.3390/magnetochemistry11040026/s1. Table S1: The crystallographic data and refinement parameters for 1; Table S2: Selected bond lengths [Å] and bond angles [°] for Complex 1; Table S3: SHAPE analysis for the Ln coordination spheres for 1; Figure S1: The IR spectra of Complex 1; Figure S2: Powder X-ray diffraction (PXRD) patterns for Complex 1 at room temperature; Figure S3: Crystal packing diagram of Complex 1; Figure S4: Frequency dependences of the χ' (left) and χ'' (right) components of the ac susceptibility for 1 at 2 K in the absence of an applied dc field; Figure S5: Frequency dependences of the χ' (left) and χ'' (right) components of the ac susceptibility for Complex 1 at 2 K under different dc fields (The solid lines are the fitting results of the Debye model); Figure S6: The τ vs. H plot for Complex **1** at 2.0 K under the applied dc field.

Author Contributions: Synthesis and characterization of the materials: Y.Z., J.X. and C.J.; measurement and analysis of dc and ac magnetic data: J.X.; writing: Y.Z. and J.X.; supervision and writing–review: Y.M. and L.L. All authors have read and agreed to the published version of the manuscript.

Funding: This work was supported by the National Natural Science Foundation of China (No. 22371138 and 21773122).

Institutional Review Board Statement: Not applicable.

Informed Consent Statement: Not applicable.

Data Availability Statement: The original contributions presented in this study are included in the article/Supplementary Materials. Further inquiries can be directed to the corresponding authors.

Conflicts of Interest: The authors declare no conflicts of interest.

References

1. Ishikawa, N.; Sugita, M.; Ishikawa, T.; Koshihara, S.-Y.; Kaizu, Y. Lanthanide Double-Decker Complexes Functioning as Magnets at the Single-Molecular Level. *J. Am. Chem. Soc.* **2003**, *125*, 8694–8695. [PubMed]
2. Woodruff, D.N.; Winpenny, R.E.P.; Layfield, R.A. Lanthanide Single-Molecule Magnets. *Chem. Rev.* **2013**, *113*, 5110–5148. [PubMed]
3. Zabala-Lekuona, A.; Seco, J.M.; Colacio, E. Single-Molecule Magnets: From Mn12-ac to dysprosium metallocenes, a travel in time. *Coord. Chem. Rev.* **2021**, *441*, 213984.
4. Rinehart, J.D.; Long, J.R. Exploiting single-ion anisotropy in the design of f-element single-molecule magnets. *Chem. Sci.* **2011**, *2*, 2078–2085. [CrossRef]
5. Affronte, M. Molecular nanomagnets for information technologies. *J. Mater. Chem.* **2009**, *19*, 1731–1737. [CrossRef]
6. Leuenberger, M.N.; Loss, D. Quantum computing in molecular magnets. *Nature* **2001**, *410*, 789–793. [CrossRef]
7. Troiani, F.; Affronte, M. Molecular spins for quantum information technologies. *Chem. Soc. Rev.* **2011**, *40*, 3119–3129. [CrossRef]
8. Aromí, G.; Aguilà, D.; Gamez, P.; Luis, F.; Roubeau, O. Design of magnetic coordination complexes for quantum computing. *Chem. Soc. Rev.* **2012**, *41*, 537–546.
9. Bogani, L.; Wernsdorfer, W. Molecular spintronics using single-molecule magnets. *Nat. Mater.* **2008**, *7*, 179–186.
10. Clemente-Juan, J.M.; Coronado, E.; Gaita-Ariño, A. Magnetic polyoxometalates: From molecular magnetism to molecular spintronics and quantum computing. *Chem. Soc. Rev.* **2012**, *41*, 7464–7478.
11. Guo, F.-S.; Day, B.M.; Chen, Y.-C.; Tong, M.-L.; Mansikkamäki, A.; Layfield, R.A. Magnetic hysteresis up to 80 kelvin in a dysprosium metallocene single-molecule magnet. *Science* **2018**, *362*, 1400–1403. [PubMed]

12. Randall McClain, K.; Gould, C.A.; Chakarawet, K.; Teat, S.J.; Groshens, T.J.; Long, J.R.; Harvey, B.G. High-temperature magnetic blocking and magneto-structural correlations in a series of dysprosium(III) metallocenium single-molecule magnets. *Chem. Sci.* **2018**, *9*, 8492–8503. [PubMed]
13. Demir, S.; Jeon, I.-R.; Long, J.R.; Harris, T.D. Radical ligand-containing single-molecule magnets. *Coord. Chem. Rev.* **2015**, *289–290*, 149–176.
14. Demir, S.; Gonzalez, M.I.; Darago, L.E.; Evans, W.J.; Long, J.R. Giant coercivity and high magnetic blocking temperatures for N_2^{3-} radical-bridged dilanthanide complexes upon ligand dissociation. *Nat. Commun.* **2017**, *8*, 2144.
15. Mavragani, N.; Errulat, D.; Gálico, D.A.; Kitos, A.A.; Mansikkamäki, A.; Murugesu, M. Radical-Bridged Ln_4 Metallocene Complexes with Strong Magnetic Coupling and a Large Coercive Field. *Angew. Chem. Int. Ed.* **2021**, *60*, 24206–24213.
16. Zhang, P.; Luo, Q.-C.; Zhu, Z.; He, W.; Song, N.; Lv, J.; Wang, X.; Zhai, Q.-G.; Zheng, Y.-Z.; Tang, J. Radical-Bridged Heterometallic Single-Molecule Magnets Incorporating Four Lanthanoceniums. *Angew. Chem. Int. Ed.* **2023**, *62*, e202218540.
17. Bajaj, N.; Mavragani, N.; Kitos, A.A.; Chartrand, D.; Maris, T.; Mansikkamäki, A.; Murugesu, M. Hard single-molecule magnet behavior and strong magnetic coupling in pyrazinyl radical-bridged lanthanide metallocenes. *Chem* **2024**, *10*, 2484–2499.
18. Sato, R.; Suzuki, K.; Sugawa, M.; Mizuno, N. Heterodinuclear Lanthanoid-Containing Polyoxometalates: Stepwise Synthesis and Single-Molecule Magnet Behavior. *Chem. Eur. J* **2013**, *19*, 12982–12990.
19. Leng, J.-D.; Liu, J.-L.; Zheng, Y.-Z.; Ungur, L.; Chibotaru, L.F.; Guo, F.-S.; Tong, M.-L. Relaxations in heterolanthanide dinuclear single-molecule magnets. *Chem. Commun.* **2013**, *49*, 158–160.
20. Aguilà, D.; Velasco, V.; Barrios, L.A.; González-Fabra, J.; Bo, C.; Teat, S.J.; Roubeau, O.; Aromí, G. Selective Lanthanide Distribution within a Comprehensive Series of Heterometallic [LnPr] Complexes. *Inorg. Chem.* **2018**, *57*, 8429–8439. [CrossRef]
21. Li, Q.-W.; Liu, J.-L.; Jia, J.-H.; Leng, J.-D.; Lin, W.-Q.; Chen, Y.-C.; Tong, M.-L. Fluorescent single-ion magnets: Molecular hybrid $(HNEt_3)[Dy_xYb_{1-x}(bpyda)_2]$ (x = 0.135−1). *Dalton Trans.* **2013**, *42*, 11262–11270. [CrossRef] [PubMed]
22. Bartolomé, E.; Arauzo, A.; Fuertes, S.; Navarro-Spreafica, L.; Sevilla, P.; Fernández Cortés, H.; Settineri, N.; Teat, S.J.; Sañudo, E.C. Luminescent and magnetic [TbEu] 2D metal–organic frameworks. *Dalton Trans.* **2023**, *52*, 7258–7270. [CrossRef] [PubMed]
23. Costes, J.-P.; Dahan, F.; Nicodème, F. Structure-Based Description of a Step-by-Step Synthesis of Homo- and Heterodinuclear (4f, 4f′) Lanthanide Complexes. *Inorg. Chem.* **2003**, *42*, 6556–6563. [CrossRef] [PubMed]
24. Lewis, D.J.; Glover, P.B.; Solomons, M.C.; Pikramenou, Z. Purely Heterometallic Lanthanide(III) Macrocycles through Controlled Assembly of Disulfide Bonds for Dual Color Emission. *J. Am. Chem. Soc.* **2011**, *133*, 1033–1043.
25. Zhu, P.; Pan, N.; Li, R.; Dou, J.; Zhang, Y.; Cheng, D.Y.Y.; Wang, D.; Ng, D.K.P.; Jiang, J. Electron-Donating Alkoxy-Group-Driven Synthesis of Heteroleptic Tris(phthalocyaninato) Lanthanide(III) Triple-Deckers with Symmetrical Molecular Structure. *Chem. Eur. J* **2005**, *11*, 1425–1432. [CrossRef]
26. Artizzu, F.; Quochi, F.; Serpe, A.; Sessini, E.; Deplano, P. Tailoring functionality through synthetic strategy in heterolanthanide assemblies. *Inorg. Chem. Front* **2015**, *2*, 213–222.
27. Le Roy, J.J.; Cremers, J.; Thomlinson, I.A.; Slota, M.; Myers, W.K.; Horton, P.H.; Coles, S.J.; Anderson, H.L.; Bogani, L. Tailored homo- and hetero- lanthanide porphyrin dimers: A synthetic strategy for integrating multiple spintronic functionalities into a single molecule. *Chem. Sci.* **2018**, *9*, 8474–8481. [CrossRef]
28. Placidi, M.P.; Villaraza, A.J.L.; Natrajan, L.S.; Sykes, D.; Kenwright, A.M.; Faulkner, S. Synthesis and Spectroscopic Studies on Azo-Dye Derivatives of Polymetallic Lanthanide Complexes: Using Diazotization to Link Metal Complexes Together. *J. Am. Chem. Soc.* **2009**, *131*, 9916–9917. [CrossRef]
29. Natrajan, L.S.; Villaraza, A.J.L.; Kenwright, A.M.; Faulkner, S. Controlled preparation of a heterometallic lanthanide complex containing different lanthanides in symmetrical binding pockets. *Chem. Commun.* **2009**, 6020–6022. [CrossRef]
30. Douib, H.; Flores Gonzalez, J.; Speed, S.; Montigaud, V.; Lefeuvre, B.; Dorcet, V.; Riobé, F.; Maury, O.; Gouasmia, A.; Le Guennic, B.; et al. Modulation of the magnetic and photophysical properties in 3d–4f and 4f–4f′ heterobimetallic complexes involving a tetrathiafulvalene-based ligand. *Dalton Trans.* **2022**, *51*, 16486–16496. [CrossRef]
31. Llunell, M.; Casanova, D.; Cirera, J.; Alemany, P.; Alvarez, S. *SHAPE 2.1*; University of Barcelona: Barcelona, Spain, 2013.
32. Xie, J.; Han, J.; Huang, X.; Jin, C.; Li, L.; Sutter, J.-P. Enhancing the Magnetization Blocking Energy of Biradical-Metal System by Merging Discrete Complexes into One-Dimensional Chains. *Chem. Eur. J* **2023**, *29*, e202203852. [PubMed]
33. Bernot, K.; Pointillart, F.; Rosa, P.; Etienne, M.; Sessoli, R.; Gatteschi, D. Single molecule magnet behaviour in robust dysprosium–biradical complexes. *Chem. Commun.* **2010**, *46*, 6458–6460.
34. Xi, L.; Han, J.; Huang, X.; Li, L. Nitronyl Nitroxide Biradical-Based Binuclear Lanthanide Complexes: Structure and Magnetic Properties. *Magnetochemistry* **2020**, *6*, 48. [CrossRef]
35. Yang, M.; Sun, J.; Guo, J.; Sun, G.; Li, L. Cu–Ln compounds based on nitronyl nitroxide radicals: Synthesis, structure, and magnetic and fluorescence properties. *CrystEngComm* **2016**, *18*, 9345–9356.
36. Wang, X.-L.; Xu, P.-P. Synthesis and Crystal Structure of a Lanthanum-Nitronyl Nitroxide Complex. *Synth. React. Inorg. Met.-Org. Nano-Met. Chem.* **2013**, *43*, 918–921.

37. Ramade, I.; Kahn, O.; Jeannin, Y.; Robert, F. Design and Magnetic Properties of a Magnetically Isolated $Gd^{III}Cu^{II}$ Pair. Crystal Structures of [Gd(hfa)$_3$Cu(salen)], [Y(hfa)$_3$Cu(salen)], [Gd(hfa)$_3$Cu(salen)(Meim)], and [La(hfa)$_3$(H_2O)Cu(salen)] [hfa = Hexafluoroacetylacetonato, salen = *N,N′*-Ethylenebis(salicylideneaminato), Meim = 1-Methylimidazole]. *Inorg. Chem.* **1997**, *36*, 930–936.
38. Xi, L.; Li, H.; Sun, J.; Ma, Y.; Tang, J.; Li, L. Designing Multicoordinating Nitronyl Nitroxide Radical Toward Multinuclear Lanthanide Aggregates. *Inorg. Chem.* **2020**, *59*, 443–451.
39. Han, J.; Jin, C.; Wang, X.; Huang, X.; Song, H.; Xie, J.; Li, L. Magnetic Relaxation in Unique Nitronyl Nitroxide Biradical-Ln–Cu Chains with Ln-bis(NIT)–Cu-bis(NIT)–Ln Units. *Dalton Trans.* **2023**, *52*, 6853–6859.
40. Catala, L.; Le Moigne, J.; Kyritsakas, N.; Rey, P.; Novoa, J.J.; Turek, P. Towards a Better Understanding of the Magnetic Interactions within m-Phenylene α-Nitronyl Imino Nitroxide Based Biradicals. *Chem. Eur. J* **2001**, *7*, 2466–2480. [CrossRef]
41. Aubin, S.M.J.; Sun, Z.; Pardi, L.; Krzystek, J.; Folting, K.; Brunel, L.-C.; Rheingold, A.L.; Christou, G.; Hendrickson, D.N. Reduced Anionic Mn12 Molecules with Half-Integer Ground States as Single-Molecule Magnets. *Inorg. Chem.* **1999**, *38*, 5329–5340.
42. Zheng, Y.Z.; Lan, Y.; Anson, C.E.; Powell, A.K. Anion-perturbed magnetic slow relaxation in planar {Dy_4} clusters. *Inorg. Chem.* **2008**, *47*, 10813–10815. [CrossRef] [PubMed]
43. Dong, H.-M.; Li, H.-Y.; Zhang, Y.-Q.; Yang, E.-C.; Zhao, X.-J. Magnetic Relaxation Dynamics of a Centrosymmetric Dy_2 Single-Molecule Magnet Triggered by Magnetic-Site Dilution and External Magnetic Field. *Inorg. Chem.* **2017**, *56*, 5611–5622. [CrossRef] [PubMed]
44. Gatteschi, D.; Sessoli, R.; Villain, J. *Molecular Nanomagnets*; Oxford University Press: Oxford, UK, 2006.
45. Jing, P.; Xi, L.; Lu, J.; Han, J.; Huang, X.; Jin, C.; Xie, J.; Li, L. Regulating Spin Dynamics of Nitronyl Nitroxide Biradical Lanthanide Complexes through Introducing Different Transition Metals. *Chem.—Asian J.* **2021**, *16*, 793–800. [CrossRef]
46. Xi, L.; Sun, J.; Wang, K.; Lu, J.; Jing, P.; Li, L. Slow magnetic relaxation in Co^{II}–Ln^{III} heterodinuclear complexes achieved through a functionalized nitronyl nitroxide biradical. *Dalton Trans.* **2020**, *49*, 1089–1096. [CrossRef]
47. Wang, K.; Sun, J.; Xi, L.; Lu, J.; Jing, P.; Li, L. Heterometallic Ln–Cu complexes derived from a phenyl pyrimidyl substituted nitronyl nitroxide biradical. *Dalton Trans.* **2019**, *48*, 14383–14389. [CrossRef]
48. Zhou, S.Y.; Li, X.; Li, T.; Tian, L.; Liu, Z.Y.; Wang, X.G. A series of heterospin complexes based on lanthanides and pyridine biradicals: Synthesis, structure and magnetic properties. *RSC Adv.* **2015**, *5*, 17131–17139. [CrossRef]
49. Li, X.; Li, T.; Tian, L.; Liu, Z.Y.; Wang, X.G. A family of rare earth complexes with chelating furan biradicals: Syntheses, structures and magnetic properties. *RSC Adv.* **2015**, *5*, 74864–74873. [CrossRef]
50. Bernot, K.; Bogani, L.; Caneschi, A.; Gatteschi, D.; Sessoli, R. A Family of Rare-Earth-Based Single Chain Magnets: Playing with Anisotropy. *J. Am. Chem. Soc.* **2006**, *128*, 7947–7956. [CrossRef]
51. Kahn, O. *Molecular Magnetism*; Wiley-VCH: Weinheim, Germany, 1993.
52. Sheldrick, G. SHELXT—Integrated space-group and crystal-structure determination. *Acta Crystallogr. A* **2015**, *71*, 3–8. [CrossRef]
53. Sheldrick, G. Crystal structure refinement with SHELXL. *Acta Crystallogr. C* **2015**, *71*, 3–8. [CrossRef]

magnetochemistry

MDPI

Article

Synthesis, X-Ray Crystal Structures, and Magnetic Properties of a Series of Trinuclear Rare-Earth Hepta-Chloride Clusters †

Yingying Pan [1,2], You-Song Ding [1,2,*], Lei Li [1,2] and Zhiping Zheng [1,2,*]

1 Department of Chemistry, Southern University of Science and Technology, Shenzhen 518055, China; 12132777@mail.sustech.edu.cn (Y.P.); 12131184@mail.sustech.edu.cn (L.L.)
2 Key University Laboratory of Rare Earth Chemistry of Guangdong, Southern University of Science and Technology, Shenzhen 518055, China
* Correspondence: dingys@sustech.edu.cn (Y.-S.D.); zhengzp@sustech.edu.cn (Z.Z.)
† Dedicated to Prof. Dai-Zheng Liao on the occasion of his 85th birthday.

Abstract: Organometallic rare-earth complexes have attracted considerable attention in recent years due to their unique structures and exceptional magnetic properties. In this study, we report the synthesis and magnetic characteristics of a family of monopentamethylcyclopentadienyl-coordinated trinuclear rare-earth hepta-chloride clusters $[(Li(THF)(Et_2O))(Cp^*RE)_3(\mu\text{-}Cl)_4(\mu_3\text{-}Cl)_2(\mu_4\text{-}Cl)]$ (**RE_3**: RE =Y, Gd, Tb, Dy, Ho, Er, Tm, Yb, Lu; Cp* = pentamethylcyclopentadienide). These clusters were synthesized by reacting LiCp* with $RECl_3$ in a 1:1 molar ratio within a mixed solvent system (THF: Et_2O = 1:9), resulting in high solubility in common organic solvents such as DCM, THF, and Et_2O. Magnetic studies conducted on these paramagnetic clusters reveal the coexistence of ferromagnetic and antiferromagnetic superexchange interactions in **Gd_3**. Additionally, **Dy_3** exhibits both ferromagnetic and antiferromagnetic intramolecular dipolar interactions. Notably, slow magnetic relaxation was observed in **Dy_3** below 23 K under a zero DC applied field with an energy barrier of 125(6) cm^{-1}.

Keywords: single-molecule magnet; organometallic rare-earth complex; lanthanide chloride cluster; energy barrier; exchange coupling

1. Introduction

Since the pioneering isolation of organometallic rare-earth complexes by Wilkinson and Birmingham in 1954 [1], the synthesis, reactivity, and physical properties of these complexes have attracted considerable attention across various fields due to the distinctive electronic structures of rare-earth ions [2–6]. Notably, significant advancements have been made in this area in recent years. For instance, cyclic multidecker sandwich rare-earth complexes have been successfully isolated and characterized [7]; metal–metal bonding, ligand–metal double bonds, and multicenter chemical bonding involving rare-earth ions have been realized [8–13]; molecular complexes featuring rare-earth ions in unconventional oxidation states (Pr(IV), Pr(V), and Tb(IV)) have been investigated [14–19]; and rare-earth telluride clusters were isolated through an efficient reduction method utilizing KC_8 [20,21]. Certain organometallic rare-earth complexes can be employed for the regio- and stereoselective hydroalkynylation of internal alkynes [22,23]. The complexes $[(C_5Me_4R)Eu(\mu\text{-}BH_4)(THF)_2]_2$ (R = H, Me, Ph, $SiMe_3$, $GeMe_3$, and $P(NMe_2)_2$) serve as potential candidates for luminescence thermometers [24,25]. Some divalent rare-earth complexes also exhibit potential as qubits [26–28]. Furthermore, several organometallic rare-earth complexes demonstrate exceptionally high energy barriers (U_{eff}) and blocking temperatures (T_B) as single-molecule magnet (SMMs) [29–35].

SMMs have emerged as one of the most promising molecular-based advanced functional materials in recent years, owing to their capability to store information within a single molecule. This unique property positions them as leading candidates for ultra-high-density data storage [3,5]. The characteristics of SMMs are typically assessed through U_{eff}, T_B, or hysteresis temperatures (T_H). These parameters are primarily influenced by the magnetic anisotropy of spin centers and the exchange coupling between spins within the molecules [36,37]. Consequently, organometallic rare-earth complexes have opened new avenues for research into SMMs [8,12,29–35,37]. The single-ion anisotropy of lanthanide ions can be significantly enhanced by intercalating Dy(III) ions between compact aromatic rings [29–35] or by utilizing low coordination numbers [38–45]. To facilitate robust exchange couplings, radical ligands are often used to bridge lanthanide ions [46,47]. More recently, metal–metal bonding between lanthanide ions has been successfully achieved, which can also result in significant magnetic exchange coupling in truncated SMMs exhibiting ultrahard magnetism [8,12]. Moreover, the manipulation of intramolecular dipolar interactions has been successfully accomplished in dinuclear rare-earth organometallic complexes by introducing an additional Cl^- ion [48]. To further enhance the exchange couplings in lanthanide complexes, it is essential to develop a comprehensive understanding of the relationships between exchange couplings and structural characteristics [49].

The monopentamethylcyclopentadienyl-coordinated rare-earth fragments {Cp*RE} (where Cp* = pentamethylcyclopentadienide) are of particular interest, as they frequently serve as key components in various intriguing organometallic rare-earth complexes, including chalcogenides [20,21,50], bismuth compounds [51], and hydride clusters of rare-earth elements [52]. The starting materials employed typically consist of a mixture of KCp* and $RECl_3$, or bis-Cp*-coordinated rare-earth complexes, or half-sandwich bis(aminobenzyl) rare-earth complexes [20,21,50–52]. We propose that mono-Cp*-coordinated rare-earth complexes can be atomically precise and cost-effective precursors for synthesizing these clusters. Such mono-Cp*-coordinated rare-earth complexes can be synthesized by reacting MCp* (M = Li, Na, K) and $RECl_3$ in a 1:1 molar ratio (Scheme 1). Previous studies have demonstrated that the hexanuclear complex $[(Cp^*Dy)_6K_4Cl_{16}(THF)_6]$ can be synthesized using KCp* [53]. In contrast, when NaCp* is employed, a dimeric complex $\{Na(\mu\text{-}THF)[Cp^*Gd(THF)]_2(\mu\text{-}Cl)_2\}_2$ is isolated, with $Cp^*GdCl_2(THF)_3$ coexisting in the mother liquor [54]. In this study, we report on the reaction between LiCp* and $RECl_3$ at a 1:1 ratio in a mixed solvent system (THF: Et_2O = 1:9), resulting in the formation of a family of mono-Cp*-coordinated trinuclear rare-earth hepta-chloride clusters $[(Li(THF)(Et_2O))(Cp^*RE)_3(\mu\text{-}Cl)_4(\mu_3\text{-}Cl)_2(\mu_4\text{-}Cl)]$ (**RE$_3$**: RE =Y, Gd, Tb, Dy, Ho, Er, Tm, Yb, Lu). These clusters' structures and magnetic properties were investigated; **Dy$_3$** emerged as a new SMM, exhibiting an energy barrier of 125(6) cm^{-1}.

$$6KCp^* + 6DyCl_3 \rightarrow [(Cp^*Dy)_6K_4Cl_{16}(THF)_6] \quad [53]$$

$$NaCp^* + GdCl_3 \rightarrow [(Cp^*Gd)_4Na_2Cl_6(THF)_6] + Cp^*GdCl_2(THF)_3 \quad [54]$$

$$3LiCp^* + 3RECl_3 \rightarrow [(Cp^*RE)_3LiCl_7(THF)_2(Et_2O)] \quad \text{[This work]}$$

Scheme 1. Synthesis of mono-Cp*-coordinated rare-earth complexes through the reaction of MCp* (where M = Li, Na, K) with $RECl_3$ in a 1:1 molar ratio.

2. Materials and Methods

2.1. Materials and Instruments

All manipulations were conducted using standard Schlenk techniques or within a glovebox under an argon atmosphere. The glassware was dried overnight at 120 °C before use. Diethyl ether (Et_2O, 99+%) (Yonghua Chemical, Changshu, China) was dried over activated alumina and stored over a potassium mirror before utilization. Anhydrous rare-earth chlorides, $RECl_3$, and Cp*Li, were synthesized following the established procedures from the literature [55,56]. All other reagents (AR) were procured from Energy-Chemical, (Shanghai, China) and utilized without further purification. Elemental analyses for carbon (C) and hydrogen (H) were performed usinga Thermo FLASH2000 elemental analyzer (Waltham, MA, USA).

2.2. Syntheses of Complex $\mathbf{RE_3}$

The clusters were synthesized through the reaction of an equimolar mixture of Cp*Li and anhydrous $RECl_3$, which was dispersed in a mixed solvent (THF: Et_2O = 1:9). The resulting mixture was stirred overnight at room temperature. After filtration, the precipitate LiCl (a byproduct) was removed, followed by the evaporation of the solvent under vacuum. The residue was then dissolved in 5 mL of Et_2O. Finally, the slow volatilization of Et_2O yielded crystals of $\mathbf{RE_3}$ suitable for X-ray diffraction after one day, with yields of approximately 25% for RE = Y, Gd, Tb, Dy, Ho, Er, Tm, Yb, and Lu, based on anhydrous $RECl_3$. The elemental analyses calculated (found) the following for $C_{42}H_{71}Cl_7Ln_3LiO_3$: $\mathbf{Y_3}$: C, 43.99 (43.60); H, 6.20 (6.16); $\mathbf{Gd_3}$: C, 37.31 (37.50); H, 5.26 (5.31); $\mathbf{Tb_3}$: C, 37.17 (37.21); H, 5.24 (5.24); $\mathbf{Dy_3}$: C, 36.88 (36.68); H, 5.20 (5.51); $\mathbf{Ho_3}$: C, 36.68 (36.84); H, 5.17 (5.46); $\mathbf{Er_3}$: C, 36.50 (36.45); H, 5.14 (5.44); $\mathbf{Tm_3}$: C, 36.37 (36.15); H, 5.12 (5.42); $\mathbf{Yb_3}$: C, 36.10 (36.23); H, 5.09 (5.39); and $\mathbf{Lu_3}$: C, 36.16 (36.27); H, 5.09 (5.10).

2.3. Crystal Structure Determination

Single-crystal X-ray diffraction (SXRD) studies were conducted using Bruker D8 Venture (Mo $K\alpha$ radiation (λ = 0.71073 Å)) (Bruker, Berlin, Germany) at 100 K for $\mathbf{RE_3}$ (where RE = Gd, Ho-Lu, and Y), and at 200 K for $\mathbf{Tb_3}$ and $\mathbf{Dy_3}$. Using Olex2 (v1.5), the structure was solved with the SHELXT structure solution program using Intrinsic Phasing and refined with the SHELXL refinement package using least squares minimization [57–59]. All hydrogen atoms were placed in calculated, ideal positions and refined as riding on their respective carbon atoms, with displacement parameters also dependent on the parent carbon atom U_{eq} value. The Cambridge Crystallographic Data Centre contains the crystal structures under the following CCDC numbers: 2441413 ($\mathbf{Gd_3}$), 2441414 ($\mathbf{Tb_3}$), 2441415 ($\mathbf{Dy_3}$), 2441416 ($\mathbf{Ho_3}$), 2441417 ($\mathbf{Er_3}$), 2441418 ($\mathbf{Tm_3}$), 2441419 ($\mathbf{Yb_3}$), 2441420 ($\mathbf{Lu_3}$), and 2441421 ($\mathbf{Y_3}$). Crystal data and structure refinement are summarized in Table S1.

2.4. Magnetic Measurements

Crushed polycrystalline samples were sealed with melted eicosane in an NMR tube under vacuum for magnetic measurements. Magnetic susceptibilities were measured with a Quantum Design MPMS-3 SQUID magnetometer (Quantum Design, San Diego, CA, USA) between 2 and 300 K.

2.5. Ab Initio Calculations

OpenMolcas [60] (version: v21.06) was employed to conduct the CASSCF-SO calculations of the electronic structures of $\mathbf{Dy_3}$, utilizing molecular geometries derived from crystallographic analyses. No optimization was performed except for selecting the largest disorder component. Relativistic effects were incorporated using the second-order Douglas–

Kroll–Hess Hamiltonian, and basis sets from the ANO-RCC library were utilized accordingly [61,62]. Specifically, VTZP quality basis sets were applied for Dy atoms, VDZP quality for coordinating C and Cl atoms, and VDZ quality for all other atoms. To conserve disk space and reduce computational demands, the Cholesky decomposition of two-electron integrals was executed with a threshold set at 10^{-8}. The state-averaged CASSCF orbitals corresponding to sextets, quartets, and doublets were optimized with 21, 224, and 490 states, respectively, through the RASSCF module. For constructing and diagonalizing the spin–orbit (SO) coupling Hamiltonian using the RASSI module, 21 sextets, 128 quartets, and 130 doublets were selected. The resulting spin–orbit wavefunctions were decomposed into their corresponding crystal field (CF) wavefunctions. Subsequently, the ground atomic multiplet, associated tensors, energy barrier, and principal magnetic axes for each Dy(III) ion were calculated using SINGLE_ANISO [63]. Additionally, POLY_ANISO was utilized to compute dipolar interactions within $\mathbf{Dy_3}$ [64,65].

3. Results and Discussion

3.1. Syntheses and Structural Characterization

All title $\mathbf{RE_3}$ compounds were synthesized using a consistent procedure, wherein equimolar amounts of Cp*Li and $RECl_3$ were dispersed in a mixed solvent (THF: Et_2O = 1:9) and stirred overnight at room temperature. The byproduct LiCl precipitated out and was subsequently separated by filtration. Notably, the products obtained after solvent removal under vacuum exhibited high solubility in common organic solvents, including DCM, THF, and Et_2O. Single crystals suitable for X-ray diffraction studies were acquired through the slow evaporation of saturated Et_2O solutions of these complexes at room temperature within a glove box.

As revealed by X-ray diffraction, the nine $\mathbf{RE_3}$ isostructural compounds crystallized in the same orthorhombic noncentrosymmetric space group *Pmn*21 (Table S1). The cluster-type molecule (Figure 1a) for $\mathbf{RE_3}$ consists of three {Cp*RE} units, one Li^+, seven Cl^-, and three coordinated solvent molecules, which include two THF and one Et_2O. The three {Cp*RE} fragments and the Li^+ ion are interconnected by four μ_2-Cl, two μ_3-Cl, and one μ_4-Cl (Figure 1b), resulting in a distorted heterometallic tetrahedron (Figure 1c). The μ_4-Cl ion is situated within a seesaw-type polyhedron formed by four metal ions, which have been documented in a limited number of rare-earth clusters [66–70]. It is noteworthy that the structures of these clusters bear resemblance to $[Cp^*LuCl_2]_3[LiCl(THF)_2]$, which was identified as a byproduct during the activation of white phosphorus with organometallic rare-earth complexes [66]. The primary distinction lies in the coordination environments of the Li^+ ions. In this study's $\mathbf{RE_3}$ complex, the Li^+ ion at the vertex of $[Cp^*RE]_3$ is coordinated to one Et_2O molecule and one disordered THF molecule. In contrast, it is coordinated with two THF molecules in $[Cp^*LuCl_2]_3[LiCl(THF)_2]$.

All three RE(III) ions in $\mathbf{RE_3}$ are capped by a Cp*, with RE1 and RE1# coordinated to five Cl^- ligands, while RE2 is associated with four Cl^- ligands and one oxygen atom from THF. The distances between the RE and Cl bonds, ranging from 2.54 Å to 3.00 Å (Tables S2–S10), align well with those reported for other RE–Cl compounds [49,50,53,54]. Similarly, the distances of the RE–C bonds fall within the range of 2.54–2.68 Å (Tables S2–S10), and the distances from the RE(III) ions to the centroid of the coordinated Cp* rings vary from 2.31 Å to 2.36 Å (Tables S2–S10). These measurements also correspond closely with previously documented values for RE-Cp* complexes [47–54]. The three RE ions are interconnected through three bridging Cl atoms, resulting in an interionic distance ranging from 3.92 Å to 4.05 Å (Table S11). Notably, it should be mentioned that RE1 and RE1# are bridged by two μ_3-Cl ligands and one μ_4-Cl ligand; conversely, either pair consisting of (RE1 and RE2) or (RE1# and RE2) is connected via one μ_2-Cl ligand along with one μ_3-Cl ligand and one μ_4-Cl ligand. This distinction results in slightly longer interionic distances for the former case than those observed in the latter. Furthermore, in terms of coordination geometry, the Cp* ligands attached to [$RECl_3RE$] units (Figure 2a) adopt a '*cis* conformation' in cases involving both RE1 and RE1#, whereas they exhibit a '*trans* conformation' when considering pairs such as (RE1 and RE2) or (RE1# and RE2). This variation may influence different magnetic exchange couplings, as discussed further below.

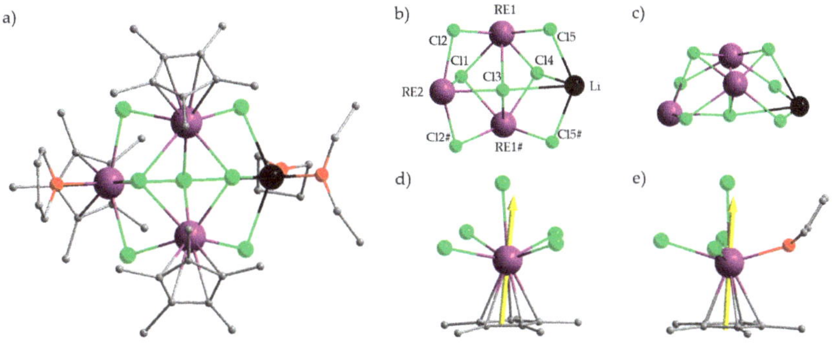

Figure 1. Ball-and-stick representations of **RE$_3$** (**a**) and side views of the RE_3Cl_7 core within **RE$_3$** (**b**,**c**). The local coordination environments for RE1 (**d**) and RE2 (**e**) in **RE$_3$** are illustrated. The yellow arrows indicate the principal magnetic axes of the ground doublets derived from CASSCF-SO calculations for **Dy$_3$**. The color code is as follows: violet, RE; green, Cl; grey, C; and red, O. H atoms are omitted for clarity. Symmetry codes are as follows: #, $1-x$, $+y$, $+z$.

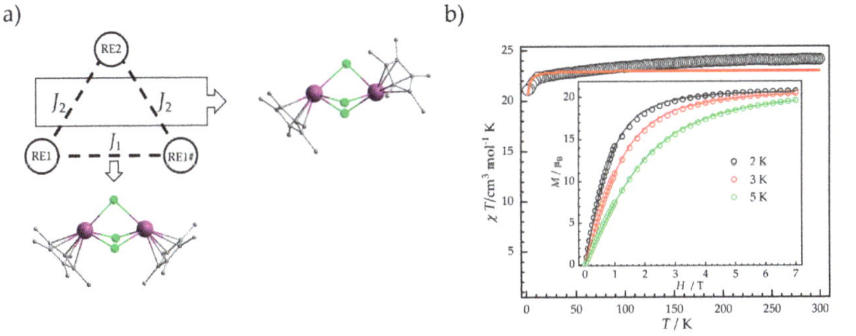

Figure 2. (**a**) The two types of intramolecular magnetic interactions of **RE$_3$** with the corresponding [$RECl_3RE$] units coordinated to Cp* ligands. The symmetry codes are as follows: #, $1-x$, $+y$, $+z$. (**b**) The χT versus T plot for **Gd$_3$** under a 1000 Oe DC field. Inset: the field-dependent magnetization plots at indicated temperatures. The solid lines are the best fits.

3.2. *Static-Field Magnetic Susceptibilities for* **RE$_3$**

To assess the static magnetic properties and intramolecular magnetic interactions in **RE$_3$**, we measured the static-field magnetic susceptibilities under an applied DC field of 1000 Oe across a temperature range of 2–300 K. Additionally, field-dependent magnetic susceptibilities were recorded up to 7 T at low temperatures (Figures 2b and 3). The static-field magnetic susceptibilities for **RE$_3$** exhibited remarkably similar trends (Figures 2b and 3). The experimental χT values at 300 K closely align with the expected values for the corresponding RE(III) ions (Table 1) [71,72]. The decrease in χT values upon cooling for **RE$_3$** (RE = Tb–Yb) can be attributed to the thermal depopulation of Stark levels, anisotropy effects, and/or antiferromagnetic interactions [71,72]. Furthermore, the maximum magnetic moment observed at 2 K and 7 T for **Gd$_3$** corresponds well with the saturation value of

7 μ_B per Gd(III) ion. In contrast, the experimental values for the other **RE_3** compounds (RE = Tb–Yb) are significantly lower than their respective saturation values, indicating a pronounced anisotropy associated with these complexes.

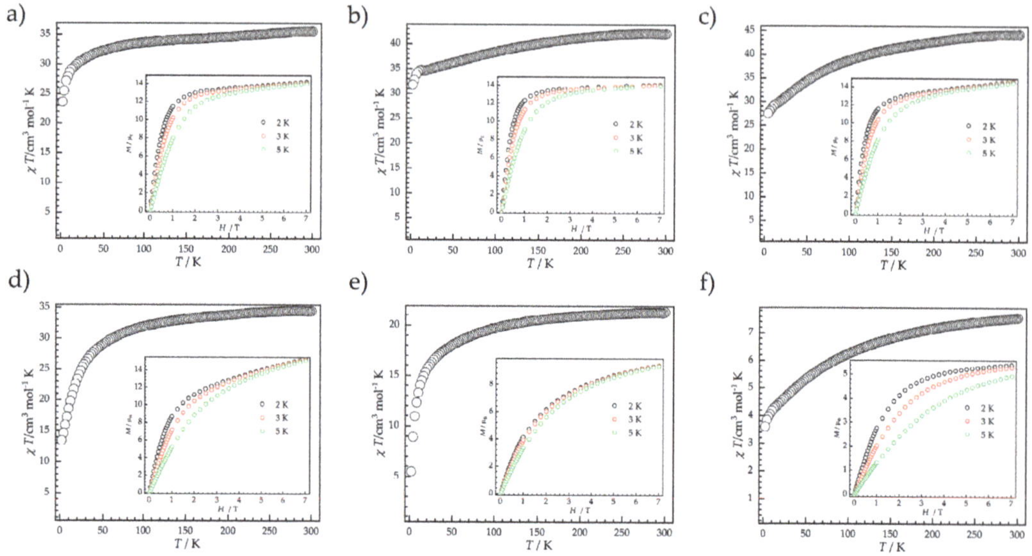

Figure 3. The χT versus T plot for (**a**) **Tb_3**, (**b**) **Dy_3**, (**c**) **Ho_3**, (**d**) **Er_3**, (**e**) **Tm_3**, and (**f**) **Yb_3** under a 1000 Oe DC field. Inset: the field-dependent magnetization plots at indicated temperatures.

Table 1. The total spin momentum (S), total orbit momentum (L), total angular quantum number (J), and g values for the ground states of free RE(III) ions [71,72], χT values, and magnetic moments per RE(III) ion in **RE_3**.

RE(III)	S	L	J	g	**χTValue ($cm^3\ mol^{-1}$ K)**			**Magnetic Moments (μ_B)**	
					Expected	Experimental		Saturated	Experimental
					300 K	300 K	2 K		
Gd	7/2	0	7/2	2	7.87	8.04	7.03	7	6.92
Tb	3	3	6	3/2	11.82	11.86	7.86	9	4.74
Dy	5/2	5	15/2	4/3	14.17	14.03	10.59	10	4.69
Ho	2	6	8	5/4	14.07	14.75	9.17	10	4.86
Er	3/2	6	15/2	6/5	11.48	11.48	4.49	9	5.09
Tm	1	5	6	7/6	7.15	7.11	1.83	7	3.14
Yb	1/2	3	7/2	8/7	2.57	2.52	1.20	4	1.80

Using the PHI program (version 3.1.6), the intramolecular magnetic interactions in **Gd_3** can be assessed by fitting the temperature- and field-dependent magnetic susceptibilities [73]. As previously mentioned, two types of 'conformations' correspond to the different magnetic exchange couplings in **Gd_3** (Figure 2a). Consequently, the Hamiltonian expressed as $\hat{H} = -2J_1(\hat{S}_{Gd1}\cdot\hat{S}_{Gd1\#}) - 2J_2(\hat{S}_{Gd1}\cdot\hat{S}_{Gd2} + \hat{S}_{Gd1\#}\cdot\hat{S}_{Gd2})$ is employed to describe these magnetic exchange couplings. Here, J_1 represents the super-exchange constant for the interaction between Gd1 and Gd1#, while J_2 pertains to the interactions involving either Gd1 with Gd2 or Gd1# with Gd2. The super-exchange constants (J_1 and J_2) derived from this Hamiltonian indicate a ferromagnetic exchange coupling when the value is positive ($J > 0$). Conversely, an antiferromagnetic exchange coupling can be identified if the value is negative. As illustrated in Figure 2b, the optimal fits (solid lines) combined with a variable g-factor in the Zeeman Hamiltonian and the isotropic exchange Hamiltonian mentioned above yielded values of $J_1 = 6.14 \times 10^{-3}$ cm^{-1}, and $J_2 = -1.29 \times 10^{-2}$ cm^{-1} with $g = 1.97$ for **Gd$_3$**. Such weak superexchange interactions among Gd(III) ions are not particularly surprising given that the 4f electrons are confined within contracted 4f orbitals and superexchange interactions mediated by non-radical ligands tend to exhibit inherently weak characteristics [50,51,74].

3.3. *Dynamic Magnetic Susceptibilities for* **Dy$_3$**

The dynamic magnetic properties of **RE$_3$** were examined through AC magnetic susceptibilities measured under a zero applied field at low temperatures (Figures S1–S6). Notably, only **Dy$_3$** exhibited slow magnetic relaxations at these low temperatures, likely attributable to the capacity of Cp* ligands to effectively enhance the single-ion anisotropy of Dy(III) ions. The frequency-dependent AC susceptibilities under a zero applied field were subsequently collected from 2 K to 23 K (Figure 4) and fitted using the generalized Debye model (Figure S7), yielding relaxation times and corresponding α parameters (Table S12). [36,75,76]. The α parameters range from 0.3 to 0.5, indicating a relatively wide distribution of relaxation times. This variation may arise from the differing coordination environments of Dy1 and Dy2 (Figure 1d,e) within **Dy$_3$** and the trimer interactions [74].

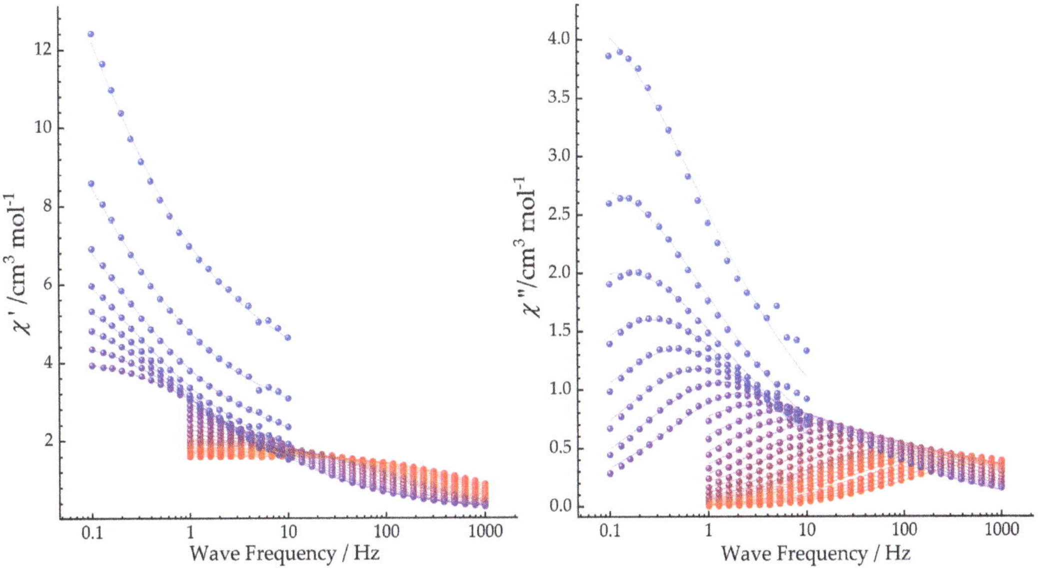

Figure 4. The in-phase (χ', **left**) and out-of-phase (χ'', **right**) frequency-dependent AC susceptibility signals for **Dy$_3$** under a zero DC field, measured from 2 K (blue) to 23 K (red). The solid lines represent the best fit according to the generalized Debye model.

As illustrated in Figure 5a, the relaxation profiles (ln(τ)) for **Dy$_3$** exhibit a rapid increase with rising values of $1/T$ when $1/T$ is less than 0.1 (corresponding to temperatures above 10 K). Subsequently, this increase slows down, ultimately reaching a nearly temperature-independent regime at $1/T = 0.5$ (at a temperature of 2 K). The log–log version for the plot of the magnetic relaxation rate versus the temperature for **Dy$_3$** (inset in Figure 5a) indicates that the Raman process (the linear part) is dominated in the temperature range of 4–15 K [76]. Thus, the relaxation profiles for **Dy3** can be fitted with Equation (1), which combines the Orbach process (U_{eff} is the effective energy barrier, and τ_0 is the attempt frequency of the Orbach process), the Raman process (CT^n), and the quantum tunneling of magnetization (QTM) process (τ_{QTM}). The parameters obtained are as follows: U_{eff} = 125(6) cm^{-1}, τ_0 = 1.4(5) × 10^{-7} s, C = 1.5(3) × 10^{-3} s, n = 4.2(1) and τ_{QTM} = 3.1(2) s. The U_{eff} is lower than the calculated energy gap between the first excited Kramers doublets, and the ground states for the Dy(III) ions in **Dy3**, probably because of the significant contribution of the Raman process to the magnetic relaxations up to 23 K (inset in Figure 5a) and a strong QTM.

$$\tau^{-1} = \tau_0^{-1}\exp(-U_{eff}/T) + CT^n + \tau_{QTM}{}^{-1} \quad (1)$$

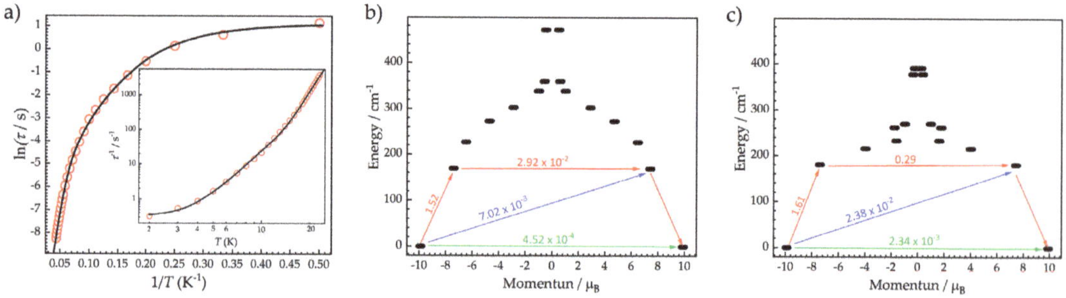

Figure 5. The ln(τ) vs. $1/T$ plots for **Dy3** (inset: log–log version for plot of the magnetic relaxation rate versus the temperature for **Dy3**) (**a**). The solid lines best fit with Equation (1), including the Orbach, Raman, and QTM processes. The electronic states from the CASSCF-SO calculations for Dy1 (**b**) and Dy2 (**c**) in **Dy3**. The thick black lines represent the Kramers doublets as a function of their magnetic moment along the magnetic axis. The red arrows represent the most probable relaxation pathway for the Orbach process. The green lines correspond to the QTM through the ground states. The blue line represents the off-diagonal relaxation process. The numbers at each arrow represent the mean absolute value of the corresponding matrix element of the transition magnetic moment.

3.4. *Ab Initio Calculations for* **Dy3**

The electronic states of the Dy(III) ions in **Dy3** were studied with *OpenMolcas* by complete active space self-consistent field spin–orbit (CASSCF-SO) calculations (Table S13) [60]. For both Dy1 and Dy2 in **Dy3**, the ground states are predominantly characterized by the most magnetic component of $m_J = \pm 15/2$, accounting for over 98%. In contrast, the excited states exhibit various magnetic components (see Tables 2 and 3). The energy gaps between the first excited Kramers doublets and the ground states are 169 cm^{-1} for Dy1 and 180 cm^{-1} for Dy2 (Figure 5b,c). The principal magnetic axes of the ground Kramers' doublets obtained are oriented along the centroid of the coordinated Cp* ring relative to the Dy(III) ion (Figure 1d,e). This orientation indicates that the single-ion anisotropy of the Dy(III) ion is predominantly influenced by the Cp^{*-} ligand, while THF and Cl^- exhibit a similar effect on the single-ion anisotropy of the Dy(III) ion.

The intramolecular dipolar interactions, which play an essential role for the low temperature magnetic property of polynuclear complexes [48,49,77–79], can be calculated accurately using the program POLY_ANISO [63,64]. As shown in Figure 6a, the calculated principal magnetic axes for Dy1 and Dy1# in **Dy3** are oriented along the Dy-μ_4-Cl direction and the Dy-μ_3-Cl direction for Dy2. For Dy1 and Dy1#, the principal magnetic axes are in-plane crossed (Figure 6b), while the principal magnetic axes are out-of-plane crossed for Dy1 and Dy2 or Dy1# and Dy2 (Figure 6c). Such arrangements of the principal magnetic axes cannot minimize the transversal component of the intramolecular dipolar field, so the QTM in **Dy3** remains strong, as mentioned above [48,49,77–79]. The dipolar interaction between Dy1 and Dy1# is antiferromagnetic with an exchange constant of -0.92 cm^{-1}, which is ferromagnetic with an exchange constant of 0.39 cm^{-1} for Dy1 and Dy or Dy1# and Dy2. These values are close to the dipolar interactions in other polynuclear Dy(III) complexes [65].

Table 2. The electronic structure of Dy1 in $\mathbf{Dy_3}$ was calculated with the crystal field parameters obtained from CASSCF-SO at the crystal structure. Each row corresponds to a Kramers doublet. Other components with low contributions are not displayed.

Energy (cm^{-1})	g1	g2	g3	Wavefunction
0.00	1.04×10^{-3}	1.67×10^{-3}	19.81	98.9%$\lvert \pm 15/2\rangle$...
168.911	6.26×10^{-2}	9.52×10^{-2}	17.64	58.2%$\lvert \pm 13/2\rangle$ + 14.5%$\lvert \pm 11/2\rangle$ + 14.5%$\lvert \pm 9/2\rangle$ + 7.9%$\lvert \pm 7/2\rangle$...
227.038	1.60×10^{-2}	2.86×10^{-2}	14.49	37.3%$\lvert \pm 13/2\rangle$ + 17.1%$\lvert \pm 9/2\rangle$ + 16%$\lvert \pm 11/2\rangle$ + 16%$\lvert \pm 7/2\rangle$...
273.431	3.53	4.72	10.07	48.1%$\lvert \pm 11/2\rangle$ + 16.8%$\lvert \pm 5/2\rangle$ + 13.3%$\lvert \pm 7/2\rangle$ + 10.8%$\lvert \pm 3/2\rangle$...
302.717	2.56	5.62	10.56	42.8%$\lvert \pm 9/2\rangle$ + 18.8%$\lvert \pm 7/2\rangle$ + 11.9%$\lvert \pm 11/2\rangle$ + 10.8%$\lvert \pm 3/2\rangle$...
338.500	0.14	2.65	13.20	30.6%$\lvert \pm 5/2\rangle$ + 25.8%$\lvert \pm 1/2\rangle$ + 25.6%$\lvert \pm 7/2\rangle$+8.1%$\lvert \pm 9/2\rangle$...
360.075	0.56	1.97	14.98	39.1%$\lvert \pm 3/2\rangle$ + 20.2%$\lvert \pm 5/2\rangle$ + 16.1%$\lvert \pm 1/2\rangle$ + 10.9%$\lvert \pm 7/2\rangle$...
472.122	2.36×10^{-2}	3.85×10^{-2}	19.77	42.9%$\lvert \pm 1/2\rangle$ + 28.9%$\lvert \pm 3/2\rangle$ + 15.6%$\lvert \pm 5/2\rangle$ + 7.5%$\lvert \pm 7/2\rangle$...

Table 3. The electronic structure of Dy2 in $\mathbf{Dy_3}$ was calculated with the crystal field parameters obtained from CASSCF-SO at the crystal structure. Each row corresponds to a Kramers doublet. Other components with low contributions are not displayed.

Energy (cm^{-1})	g1	g2	g3	Wavefunction
0.00	6.24×10^{-3}	9.61×10^{-3}	19.84	99.5%$\lvert \pm 15/2\rangle$...
180.303	0.80	0.85	16.23	73.3%$\lvert \pm 13/2\rangle$ + 8.8%$\lvert \pm 5/2\rangle$ + 5.8%$\lvert \pm 11/2\rangle$ + 5.6%$\lvert \pm 7/2\rangle$...
214.563	10.20	7.12	2.75	31.6%$\lvert \pm 9/2\rangle$ + 23.4%$\lvert \pm 11/2\rangle$ + 11.2%$\lvert \pm 3/2\rangle$ + 10.7%$\lvert \pm 13/2\rangle$...
233.352	0.68	4.13	8.94	31.4%$\lvert \pm 11/2\rangle$ + 21.6%$\lvert \pm 7/2\rangle$ + 11.9%$\lvert \pm 9/2\rangle$+10.6%$\lvert \pm 13/2\rangle$...
261.853	10.85	7.67	1.04	46.8%$\lvert \pm 7/2\rangle$ + 42.4%$\lvert \pm 9/2\rangle$ + 4.3%$\lvert \pm 5/2\rangle$ +3.3%$\lvert \pm 11/2\rangle$...
270.398	8.77	5.42	0.48	44.4%$\lvert \pm 5/2\rangle$ + 30.6%$\lvert \pm 11/2\rangle$ + 16.9%$\lvert \pm 3/2\rangle$ + 3.6%$\lvert \pm 13/2\rangle$...
378.250	0.58	2.73	14.23	53%$\lvert \pm 3/2\rangle$ + 32.6%$\lvert \pm 5/2\rangle$ + 5%$\lvert \pm 1/2\rangle$ + 4.5%$\lvert \pm 7/2\rangle$...
390.606	0.39	3.26	15.02	68.9%$\lvert \pm 1/2\rangle$ + 11.5%$\lvert \pm 7/2\rangle$ + 8.8%$\lvert \pm 9/2\rangle$ + 8.2%$\lvert \pm 3/2\rangle$...

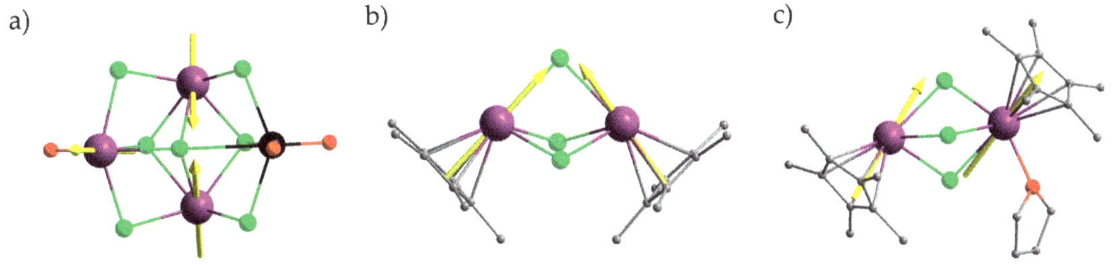

Figure 6. Arrangements of the principal magnetic axes for $\mathbf{Dy_3}$ (**a**), for Dy1 and Dy1# (**b**), for Dy1 and Dy2 or Dy1# and Dy2 (**c**). The yellow arrows represent the principal magnetic axes of the ground doublets obtained from the CASSCF-SO calculations.

4. Conclusions

The synthesis, structure determinations, magnetic properties, and electronic structures of novel trinuclear rare-earth hepta-chloride clusters were reported. These mono-Cp*-coordinated rare-earth complexes obtained by the reaction of LiCp* and $RECl_3$ in a 1:1 ratio can serve as starting materials for more interesting organometallic rare-earth clusters due to their high solubility in common organic solvents. Complex **Dy3** is a novel single-molecule magnet that exhibits a significant contribution from the Orbach process, the Raman process, and QTM. The energy barrier of 125(6) cm^{-1} is primarily attributed to the single-ion anisotropy associated with its Dy(III) ions. Further magnetic studies indicate that the '*cis* conformation' of the Cp* ligands coordinated to the [$RECl_3RE$] units results in ferromagnetic superexchange interactions in **Gd3**, while it leads to antiferromagnetic intramolecular dipolar interactions in **Dy3**. Conversely, the '*trans* conformation' exhibits opposing effects. It remains unclear how the ligands control the exchange interactions. The investigation into the control of exchange interactions in molecular magnetic materials is ongoing, which may pave the way for the development of more advanced molecular magnetic materials.

Supplementary Materials: The following supporting information can be downloaded at: https://www.mdpi.com/article/10.3390/magnetochemistry11050038/s1. Table S1: Crystal data and structure refinement for **RE3** (RE=Gd, Tb, Dy, Ho, Er, Tm, Yb, Lu, Y); Table S2–S10: Selected bond distances (Å) for **Gd3**, **Tb3**, **Dy3**, **Ho3**, **Er3**, **Tm3**, **Yb3**, **Lu3** and **Y3**; Table S11: Selected RE–RE distances (Å) for **RE3**; Table S12: Relaxation fitting parameters obtained using a generalized Debye model for **Dy3** from 23 K to 2 K under zero DC field; Table S13: Ab initio calculated crystal field parameters (B_q^k) for Dy1 and Dy2 in Dy3; Figures S1–S6: Temperature-dependence of the in-phase (χ', top) and out-of-phase (χ'', bottom) AC susceptibility signals under zero DC field by standard AC susceptibility measurements for **Tb3**, **Dy3**, **Ho3**, **Er3**, **Tm3** and **Yb3** at indicated frequencies; Figure S7: Cole–Cole plots using the frequency-dependence AC susceptibility data under the zero DC field for **Dy3** from 2 K (blue) to 23 K (red). The solid lines are the best fits.

Author Contributions: Synthesis and characterization of the materials: Y.P. and L.L.; measurement and analysis of magnetic data, ab initio calculations: Y.-S.D.; supervision and writing—review: Y.-S.D. and Z.Z. All authors have read and agreed to the published version of the manuscript.

Funding: This research was funded by the National Natural Science Foundation of China (22101116, 92261203, and 21971106), the Stable Support Plan Program of Shenzhen Natural Science Fund (20200925161141006), and the Shenzhen Fundamental Research Program (JCYJ20220530115001002 and JCYJ20220818100417037).

Data Availability Statement: The original contributions presented in this study are included in the article/Supplementary Materials. Further inquiries can be directed to the corresponding authors.

Conflicts of Interest: The authors declare no conflicts of interest.

References

1. Wilkinson, G.; Birmingham, J.M. Cyclopentadienyl compounds of Sc, Y, La, Ce and some lanthanide elements. *J. Am. Chem. Soc.* **1954**, *76*, 6210. [CrossRef]
2. Zimmermann, M.; Anwander, R. Homoleptic rare-earth metal complexes containing Ln–C σ-Bonds. *Chem. Rev.* **2010**, *110*, 6194–6259. [CrossRef]
3. Harriman, K.L.M.; Murugesu, M. An organolanthanide building block approach to single-molecule magnets. *Acc. Chem. Res.* **2016**, *49*, 1158–1167. [CrossRef] [PubMed]
4. Nishiura, M.; Guo, F.; Hou, Z. Half-sandwich rare-earth-catalyzed olefin polymerization, carbometalation, and hydroarylation. *Acc. Chem. Res.* **2015**, *48*, 2209–2220. [CrossRef]
5. Day, B.M.; Guo, F.-S.; Layfield, R.A. Cyclopentadienyl ligands in lanthanide single-molecule magnets: One ring to rule them all? *Acc. Chem. Res.* **2018**, *51*, 1880–1889. [CrossRef]
6. Ortu, F. Rare earth starting materials and methodologies for synthetic chemistry. *Chem. Rev.* **2022**, *122*, 6040–6116. [CrossRef]
7. Münzfeld, L.; Gillhuber, S.; Hauser, A.; Lebedkin, S.; Hädinger, P.; Knöfel, N.D.; Zovko, C.; Gamer, M.T.; Weigend, F.; Kappes, M.M.; et al. Synthesis and properties of cyclic sandwich compounds. *Nature* **2023**, *620*, 92–96. [CrossRef]

8. Gould, C.A.; McClain, K.R.; Reta, D.; Kragskow, J.G.C.; Marchiori, D.A.; Lachman, E.; Choi, E.-S.; Analytis, J.G.; Britt, R.D.; Chilton, N.F.; et al. Ultrahard magnetism from mixed-valence dilanthanide complexes with metal-metal bonding. *Science* **2022**, *375*, 198–202. [CrossRef] [PubMed]
9. McClain, K.R.; Kwon, H.; Chakarawet, K.; Nabi, R.; Kragskow, J.G.C.; Chilton, N.F.; Britt, R.D.; Long, J.R.; Harvey, B.G. A trinuclear gadolinium cluster with a three-center one-electron bond and an $S = 11$ ground state. *J. Am. Chem. Soc.* **2023**, *145*, 8996–9002. [CrossRef]
10. Wedal, J.C.; Anderson-Sanchez, L.M.; Dumas, M.T.; Gould, C.A.; Beltrán-Leiva, M.J.; Celis-Barros, C.; Páez-Hernández, D.; Ziller, J.W.; Long, J.R.; Evans, W.J. Synthesis and crystallographic characterization of a reduced bimetallic yttrium ansa-metallocene hydride complex, $[K(crypt)][(\mu\text{-}Cp^{An})Y(\mu\text{-}H)]_2$ ($Cp^{An} = Me_2Si[C_5H_3(SiMe_3)\text{-}3]_2$), with a 3.4 Å yttrium-yttrium distance. *J. Am. Chem. Soc.* **2023**, *145*, 10730–10742. [CrossRef]
11. Wang, Y.; Liang, J.; Deng, C.; Sun, R.; Fu, P.-X.; Wang, B.-W.; Gao, S.; Huang, W. Two-electron oxidations at a single cerium center. *J. Am. Chem. Soc.* **2023**, *145*, 22466–22474. [CrossRef] [PubMed]
12. Kwon, H.; McClain, K.R.; Kragskow, J.G.C.; Staab, J.K.; Ozerov, M.; Meihaus, K.R.; Harvey, B.G.; Choi, E.S.; Chilton, N.F.; Long, J.R. Coercive fields exceeding 30 T in the mixed-valence single-molecule magnet $(Cp^{iPr5})_2Ho_2I_3$. *J. Am. Chem. Soc.* **2024**, *146*, 18714–18721. [CrossRef]
13. Guo, Y.; Jiang, X.-L.; Wu, Q.-Y.; Liu, K.; Wang, W.; Hu, K.-q.; Mei, L.; Chai, Z.-f.; Gibson, J.K.; Yu, J.-p.; et al. 4f/5d hybridization induced single-electron delocalization in an azide-bridged dicerium complex. *J. Am. Chem. Soc.* **2024**, *146*, 7088–7096. [CrossRef] [PubMed]
14. Palumbo, C.T.; Zivkovic, I.; Scopelliti, R.; Mazzanti, M. Molecular Complex of Tb in the +4 Oxidation State. *J. Am. Chem. Soc.* **2019**, *141*, 9827–9831. [CrossRef]
15. Willauer, A.R.; Palumbo, C.T.; Scopelliti, R.; Zivkovic, I.; Douair, I.; Maron, L.; Mazzanti, M. Stabilization of the Oxidation State +IV in Siloxide-Supported Terbium Compounds. *Angew. Chem. Int. Ed.* **2020**, *59*, 3549–3553. [CrossRef]
16. Willauer, A.R.; Palumbo, C.T.; Fadaei-Tirani, F.; Zivkovic, I.; Douair, I.; Maron, L.; Mazzanti, M. Accessing the +IV Oxidation State in Molecular Complexes of Praseodymium. *J. Am. Chem. Soc.* **2020**, *142*, 5538–5542. [CrossRef] [PubMed]
17. Rice, N.T.; Popov, I.A.; Russo, D.R.; Bacsa, J.; Batista, E.R.; Yang, P.; Telser, J.; La Pierre, H.S. Design, Isolation, and Spectroscopic Analysis of a Tetravalent Terbium Complex. *J. Am. Chem. Soc.* **2019**, *141*, 13222–13233. [CrossRef]
18. Xue, T.; Yang, Q.-S.; Li, L.; Chang, X.-Y.; Ding, Y.-S.; Zheng, Z. Supramolecular assemblies of tetravalent terbium complex units: Syntheses, structure, and materials properties. *Chem. Sci.* **2025**, *16*, 6805–6811. [CrossRef] [PubMed]
19. Boggiano, A.C.; Studvick, C.M.; Roy Chowdhury, S.; Niklas, J.E.; Tateyama, H.; Wu, H.; Leisen, J.E.; Kleemiss, F.; Vlaisavljevich, B.; Popov, I.A.; et al. Praseodymium in the formal +5 oxidation state. *Nat. Chem.* **2025**. [CrossRef]
20. Ding, Y.-S.; Jiang, X.-L.; Li, L.; Xu, C.-Q.; Li, J.; Zheng, Z. Atomically precise semiconductor clusters of rare-earth tellurides. *Nat. Synth.* **2024**, *3*, 655–661. [CrossRef]
21. Li, L.; Ding, Y.-S.; Zheng, Z. Lanthanide-based molecular magnetic semiconductors. *Angew. Chem. Int. Ed.* **2024**, *63*, e202410019. [CrossRef] [PubMed]
22. Hao, N.; Jiao, T.; Sun, Z.; Mishra, A.; Zhuo, Q.; Nishiura, M.; Hou, Z.; Cong, X. Regio- and stereoselective hydroalkynylation of internal alkynes with terminal alkynes by half-sandwich rare-earth catalysts. *J. Am. Chem. Soc.* **2025**, *147*, 6149–6161. [CrossRef] [PubMed]
23. Ma, Y.; Mishra, A.; Jiao, T.; Chang, W.; Lou, S.-J.; Nishiura, M.; Cong, X.; Hou, Z. Heteroatom-assisted regio- and stereoselective hydrosilylation of unsymmetric internal alkynes by scandium catalyst. *Angew. Chem. Int. Ed.* **2025**, *64*, e202502665. [CrossRef]
24. Diaz-Rodriguez, R.M.; Gálico, D.A.; Chartrand, D.; Suturina, E.A.; Murugesu, M. Toward opto-structural correlation to investigate luminescence thermometry in an organometallic Eu(II) complex. *J. Am. Chem. Soc.* **2022**, *144*, 912–921. [CrossRef] [PubMed]
25. Diaz-Rodriguez, R.M.; Gálico, D.A.; Chartrand, D.; Murugesu, M. Ligand effects on the emission characteristics of molecular Eu(II) luminescence thermometers. *J. Am. Chem. Soc.* **2024**, *146*, 34118–34129. [CrossRef]
26. Kundu, K.; White, J.R.K.; Moehring, S.A.; Yu, J.M.; Ziller, J.W.; Furche, F.; Evans, W.J.; Hill, S. A 9.2-GHz clock transition in a Lu(II) molecular spin qubit arising from a 3,467-MHz hyperfine interaction. *Nat. Chem.* **2022**, *14*, 392–397. [CrossRef]
27. Smith, P.W.; Hruby, J.; Evans, W.J.; Hill, S.; Minasian, S.G. Identification of an X-Band clock transition in Cp'_3Pr^- enabled by a $4f^25d^1$ configuration. *J. Am. Chem. Soc.* **2024**, *146*, 5781–5785. [CrossRef]
28. Nodaraki, L.E.; Ariciu, A.-M.; Huh, D.N.; Liu, J.; Martins, D.O.T.A.; Ortu, F.; Winpenny, R.E.P.; Chilton, N.F.; McInnes, E.J.L.; Mills, D.P.; et al. Ligand effects on the spin relaxation dynamics and coherent manipulation of organometallic La(II) potential qudits. *J. Am. Chem. Soc.* **2024**, *146*, 15000–15009. [CrossRef]
29. Goodwin, C.A.P.; Ortu, F.; Reta, D.; Chilton, N.F.; Mills, D.P. Molecular magnetic hysteresis at 60 kelvin in dysprosocenium. *Nature* **2017**, *548*, 439–442. [CrossRef]
30. Guo, F.-S.; Day, B.M.; Chen, Y.-C.; Tong, M.-L.; Mansikkamäki, A.; Layfield, R.A. A dysprosium metallocene single-molecule magnet functioning at the axial limit. *Angew. Chem. Int. Ed.* **2017**, *56*, 11445–11449. [CrossRef]

31. Guo, F.-S.; Day, B.M.; Chen, Y.-C.; Tong, M.-L.; Mansikkamäki, A.; Layfield, R.A. Magnetic hysteresis up to 80 kelvin in a dysprosium metallocene single-molecule magnet. *Science* **2018**, *362*, 1400–1403. [CrossRef] [PubMed]
32. Guo, F.-S.; He, M.; Huang, G.-Z.; Giblin, S.R.; Billington, D.; Heinemann, F.W.; Tong, M.-L.; Mansikkamäki, A.; Layfield, R.A. Discovery of a dysprosium metallocene single-molecule magnet with two high-temperature orbach processes. *Inorg. Chem.* **2022**, *61*, 6017–6025. [CrossRef]
33. Jin, P.-B.; Luo, Q.-C.; Gransbury, G.K.; Vitorica-Yrezabal, I.J.; Hajdu, T.; Strashnov, I.; McInnes, E.J.L.; Winpenny, R.E.P.; Chilton, N.F.; Mills, D.P.; et al. Thermally stable terbium(II) and dysprosium(II) bis-amidinate complexes. *J. Am. Chem. Soc.* **2023**, *145*, 27993–28009. [CrossRef] [PubMed]
34. Wang, Y.; Luo, Q.-C.; Zheng, Y.-Z. Organolanthanide single-molecule magnets with heterocyclic ligands. *Angew. Chem. Int. Ed.* **2024**, *63*, e202407016. [CrossRef]
35. Liu, M.; Chen, Y.-C.; Zheng, J.; Peng, X.-H.; Tong, M.-L.; Mansikkamäki, A.; Guo, F.-S. Isolation and single-molecule magnetism of neutral rare-earth(III) sandwich complexes supported by compact aromatic rings. *Sci. China Chem.* **2025**. [CrossRef]
36. Liu, J.-L.; Chen, Y.-C.; Tong, M.-L. Symmetry strategies for high performance lanthanide-based single-molecule magnets. *Chem. Soc. Rev.* **2018**, *47*, 2431–2453. [CrossRef]
37. Chen, Y.-C.; Tong, M.-L. Single-molecule magnets beyond a single lanthanide ion: The art of coupling. *Chem. Sci.* **2022**, *13*, 8716–8726. [CrossRef] [PubMed]
38. Meng, Y.-S.; Xu, L.; Xiong, J.; Yuan, Q.; Liu, T.; Wang, B.-W.; Gao, S. Low-coordinate single-ion magnets by intercalation of lanthanides into a phenol matrix. *Angew. Chem. Int. Ed.* **2018**, *57*, 4673–4676. [CrossRef]
39. Wang, C.; Sun, R.; Chen, Y.; Wang, B.-W.; Wang, Z.-M.; Gao, S. Assembling high-temperature single-molecule magnets with low-coordinate bis(amido) dysprosium unit $[DyN_2]^+$ via Cl–K–Cl linkage. *CCS Chem.* **2020**, *2*, 362–368. [CrossRef]
40. Sun, R.; Wang, C.; Wang, B.-W.; Wang, Z.-M.; Chen, Y.-F.; Tamm, M.; Gao, S. Low-coordinate bis(imidazolin-2-iminato) dysprosium(III) single-molecule magnets. *Inorg. Chem. Front.* **2023**, *10*, 485–492. [CrossRef]
41. Sun, R.; Li, C.-Y.; Huang, Z.-N.; Wang, C.; Chen, Y.; Sun, H.-L.; Wang, Z.-M.; Wang, B.-W.; Gao, S. One is better than two: Improved performance in mono(imidazolin-2-iminato) dysprosium(III) single-molecule magnets featuring short Dy–N bond. *CCS Chem.* 2024. [CrossRef]
42. Emerson-King, J.; Gransbury, G.K.; Whitehead, G.F.S.; Vitorica-Yrezabal, I.J.; Rouzières, M.; Clérac, R.; Chilton, N.F.; Mills, D.P. Isolation of a bent dysprosium bis(amide) single-molecule magnet. *J. Am. Chem. Soc.* **2024**, *146*, 3331–3342. [CrossRef]
43. Benner, F.; Jena, R.; Odom, A.L.; Demir, S. Magnetic hysteresis in a dysprosium bis(amide) complex. *J. Am. Chem. Soc.* **2025**, *147*, 8156–8167. [CrossRef] [PubMed]
44. Yang, Q.-Q.; Wang, Y.-F.; Wang, Y.-X.; Tang, M.-J.; Yin, B. Ab initio prediction of key parameters and magneto-structural correlation of tetracoordinated lanthanide single-ion magnets. *Phys. Chem. Chem. Phys.* **2023**, *25*, 18387–18399. [CrossRef]
45. Wang, Y.-X.; Wang, Y.-F.; Yin, B. Theoretical study of pentacoordinated lanthanide single-ion magnets via ab initio electronic structure calculation. *Magnetochemistry* **2025**, *11*, 3. [CrossRef]
46. Zhang, P.; Luo, Q.-C.; Zhu, Z.; He, W.; Song, N.; Lv, J.; Wang, X.; Zhai, Q.-G.; Zheng, Y.-Z.; Tang, J. Radical-bridged heterometallic single-molecule magnets incorporating four lanthanoceniums. *Angew. Chem. Int. Ed.* **2023**, *62*, e202218540. [CrossRef]
47. Bajaj, N.; Mavragani, N.; Kitos, A.A.; Chartrand, D.; Maris, T.; Mansikkamäki, A.; Murugesu, M. Hard single-molecule magnet behavior and strong magnetic coupling in pyrazinyl radical-bridged lanthanide metallocenes. *Chem* **2024**, *10*, 2484–2499. [CrossRef]
48. Huang, J.-Q.; Chen, Q.-W.; Ding, Y.-S.; Zhu, X.-F.; Wang, B.-W.; Pan, F.; Zheng, Z. Enhancement of single-molecule magnet properties by manipulating intramolecular dipolar interactions. *Adv. Sci.* **2025**, *12*, 2409730. [CrossRef]
49. David, G.; Le Guennic, B.; Reta, D. Promoting exchange coupling in $(Cp^{iPr5})_2Gd_2X_3$ complexes. *Chem. Commun.* **2024**, *60*, 11988–11991. [CrossRef]
50. Li, L.; Xue, T.-J.; Ding, Y.-S.; Zheng, Z. Rare-earth chalcogenidotetrachloride clusters (RE_3ECl_4, RE = Dy, Gd, Y.; E = S, Se, Te): Syntheses and materials properties. *J. Mater. Chem. C* **2024**, *12*, 16506–16514.
51. Zhang, P.; Benner, F.; Chilton, N.F.; Demir, S. Organometallic lanthanide bismuth cluster single-molecule magnets. *Chem.* **2022**, *8*, 717–730. [CrossRef]
52. Shima, T.; Nishiura, M.; Hou, Z. Tetra-, penta-, and hexanuclear yttrium hydride clusters from half-sandwich bis(aminobenzyl) complexes containing various cyclopentadienyl ligands. *Organometallics.* **2011**, *30*, 2513–2524. [CrossRef]
53. Wu, J.; Demeshko, S.; Dechert, S.; Meyer, F. Hexanuclear $[Cp^*Dy]_6$ single-molecule magnet. *Chem. Commun.* **2020**, *56*, 3887–3890. [CrossRef]
54. Shen, Q.; Qi, M.; Lin, Y. Synthesis and X-ray crystal structure of a monopentamethylcyclopentadienyl derivative of gadolinium $\{Na(\mu\text{-THF})[Cp^*Gd(THF)]_2(\mu\text{-Cl})_2\}_2\cdot 6THF$. *J. Organomet. Chem.* **1990**, *399*, 247–254. [CrossRef]
55. Huang, W.; Upton, B.M.; Khan, S.I.; Diaconescu, P.L. Synthesis and characterization of paramagnetic lanthanide benzyl complexes. *Organometallics.* **2013**, *32*, 1379–1386. [CrossRef]

56. Watson, P.L.; Whitney, J.F.; Harlow, R.L. (Pentamethylcyclopentadienyl)ytterbium and -lutetium complexes by metal oxidation and metathesis. *Inorg. Chem.* **1981**, *20*, 3271–3278. [CrossRef]
57. Dolomanov, O.V.; Bourhis, L.J.; Gildea, R.J.; Howard, J.A.K.; Puschmann, H. OLEX2: A complete structure solution, refinement and analysis program. *J. Appl. Cryst.* **2009**, *42*, 339–341. [CrossRef]
58. Sheldrick, G. SHELXT-Integrated space-group and crystal-structure determination. *Acta Crystallogr. A* **2015**, *71*, 3–8. [CrossRef]
59. Sheldrick, G. Crystal structure refinement with SHELXL. *Acta Crystallogr. C* **2015**, *71*, 3–8. [CrossRef]
60. Li Manni, G.; Fdez. Galván, I.; Alavi, A.; Aleotti, F.; Aquilante, F.; Autschbach, J.; Avagliano, D.; Baiardi, A.; Bao, J.J.; Battaglia, S.; et al. The OpenMolcas web: A community-driven approach to advancing computational chemistry. *J. Chem. Theory. Comput.* **2023**, *19*, 6933–6991. [CrossRef]
61. Roos, B.O.; Lindh, R.; Malmqvist, P.-Å.; Veryazov, V.; Widmark, P.-O. Main group atoms and dimers studied with a new relativistic ANO basis set. *J. Phys. Chem. A* **2004**, *108*, 2851–2858. [CrossRef]
62. Roos, B.O.; Lindh, R.; Malmqvist, P.-Å.; Veryazov, V.; Widmark, P.-O. New relativistic ANO basis sets for transition metal atoms. *J. Phys. Chem. A* **2005**, *109*, 6575–6579. [CrossRef] [PubMed]
63. Ungur, L.; Chibotaru, L.F. Ab Initio Crystal Field for Lanthanides. *Chem. Eur. J.* **2017**, *23*, 3708–3718. [CrossRef]
64. Chibotaru, L.F.; Ungur, L.; Soncini, A. The origin of nonmagnetic kramers doublets in the ground state of dysprosium triangles: Evidence for a toroidal magnetic moment. *Angew. Chem. Int. Ed.* **2008**, *47*, 4126–4129. [CrossRef] [PubMed]
65. Ungur, L.; Van den Heuvel, W.; Chibotaru, L.F. Ab initio investigation of the non-collinear magnetic structure and the lowest magnetic excitations in dysprosium triangles. *New J. Chem.* **2009**, *33*, 1224–1230. [CrossRef]
66. Du, S.; Yin, J.; Chi, Y.; Xu, L.; Zhang, W.-X. Dual functionalization of white phosphorus: Formation, characterization, and reactivity of rare-earth-metal cyclo-P_3 complexes. *Angew. Chem. Int. Ed.* **2017**, *56*, 15886–15890. [CrossRef]
67. Dubé, T.; Gambarotta, S.; Yap, G.P.A.; Conoci, S. Preparation and characterization of two mixed-valence samarium octameric clusters. *Organometallics* **2000**, *19*, 115–117. [CrossRef]
68. Gromada, J.; Mortreux, A.; Chenal, T.; Ziller, J.W.; Leising, F.; Carpentier, J.-F. Neodymium alkoxides: Synthesis, characterization and their combinations with dialkylmagnesiums as unique systems for polymerization and block copolymerization of ethylene and methyl methacrylate. *Chem. Eur. J.* **2002**, *8*, 3773–3788. [CrossRef]
69. Dietrich, H.M.; Schuster, O.; Törnroos, K.W.; Anwander, R. Heterobimetallic half-lanthanidocene clusters: Novel mixed tetramethylaluminato/chloro coordination. *Angew. Chem. Int. Ed.* **2006**, *45*, 4858–4863. [CrossRef]
70. Evans, W.J.; Champagne, T.M.; Davis, B.L.; Allen, N.T.; Nyce, G.W.; Johnston, M.A.; Lin, Y.-C.; Khvostov, A.; Ziller, J.W. Structural studies of mono(pentamethylcyclopentadienyl)lanthanide complexes. *J. Coord. Chem.* **2006**, *59*, 1069–1087. [CrossRef]
71. Tang, J.; Zhang, P. A basis for lanthanide single-molecule magnets. In *Lanthanide Single Molecule Magnets*; Springer: Berlin/Heidelberg, Germany, 2015; pp. 1–39.
72. Jiang, S.-D.; Wang, B.-W.; Gao, S. Advances in lanthanide single-ion magnets. In *Molecular Nanomagnets and Related Phenomena*; Gao, S., Ed.; Springer: Berlin/Heidelberg, Germany, 2015; pp. 111–141.
73. Chilton, N.F.; Anderson, R.P.; Turner, L.D.; Soncini, A.; Murray, K.S. PHI: A powerful new program for the analysis of anisotropic monomeric and exchange-coupled polynuclear d- and f-block complexes. *J. Comput. Chem.* **2013**, *34*, 1164–1175. [CrossRef]
74. Han, T.; Ding, Y.-S.; Li, Z.-H.; Yu, K.-X.; Zhai, Y.-Q.; Chilton, N.F.; Zheng, Y.-Z. A dichlorido-bridged dinuclear Dy(III) single-molecule magnet with an effective energy barrier larger than 600 K. *Chem. Commun.* **2019**, *55*, 7930–7933. [CrossRef] [PubMed]
75. Guo, Y.-N.; Xu, G.-F.; Guo, Y.; Tang, J. Relaxation dynamics of dysprosium(III) single-molecule magnets. *Dalton Trans.* **2011**, *40*, 9953–9963. [CrossRef] [PubMed]
76. Reta, D.; Chilton, N.F. Uncertainty estimates for magnetic relaxation times and magnetic relaxation parameters. *Phys. Chem. Chem. Phys.* **2019**, *21*, 23567–23575. [CrossRef] [PubMed]
77. Deng, W.; Du, S.-N.; Ruan, Z.-Y.; Zhao, X.-J.; Chen, Y.-C.; Liu, J.-L.; Tong, M.-L. Aggregation-induced suppression of quantum tunneling by manipulating intermolecular arrangements of magnetic dipoles. *Aggregate* **2024**, *5*, e441. [CrossRef]
78. Ding, Y.-S.; Chen, Q.-W.; Huang, J.-Q.; Zhu, X.-F.; Zheng, Z. Sandwich-type single-molecule magnets complexes of Er(III) with ansa-cyclooctatetraenyl ligands. *Chin. J. Chem.* **2024**, *42*, 3099–3106. [CrossRef]
79. Chen, Q.; Ding, Y.; Zhu, X.; Wang, B.; Zheng, Z. Tuning intramolecular dipole interactions in dinuclear single-molecule magnetic complexes of Er(III) with monosubstituted cyclooctatetraenide ligands. *J. Rare. Earth.* 2024. [CrossRef]

magnetochemistry

MDPI

Article

Multifunctional Synergistic Response Induced by Phase Transition in Molecular Compounds

Xiao-Feng Chen [1], Tao Wang [1], Dan Liao [1], Nan Wu [1], Yan Peng [1,*], Shi-Yong Zhang [2,*] and Zhao-Bo Hu [1,*]

1 Jiangxi Provincial Key Laboratory of Functional Crystalline Materials Chemistry, Jiangxi University of Science and Technology, Ganzhou 341000, China; 6120220189@mail.jxust.edu.cn (X.-F.C.); 15779035038@163.com (T.W.); dan_liao1030@163.com (D.L.); 19972579477@163.com (N.W.)
2 School of Chemistry and Chemical Engineering, Gannan Normal University, Ganzhou 341000, China
* Correspondence: yan.peng@jxust.edu.cn (Y.P.); zhsy1207@126.com (S.-Y.Z.); huzhaobo@smail.nju.edu.cn (Z.-B.H.)

Abstract: Two organic–inorganic materials $(TMAA)_2[CoCl_4]$ (**1**) and $(TMAA)_2[MnCl_4]$ (**2**) (TMAA = N,N,N-trimethyl-1-adamantylammonium hydroxide) were synthesized and characterized. It was found that both compounds exhibit first-order structural phase transition at high-temperature regions. As the temperature approaches the phase transition point, significant abnormal changes were observed in the dielectric properties and $\chi_M T$ values of compounds **1** and **2**. This phenomenon strongly highlights the dielectric bistable and spin bistable properties of compounds **1** and **2**. Further research shows that the dielectric constants of the compounds undergo significant changes upon the application of an external magnetic field, providing strong evidence for the existence of magnetic–dielectric coupling effects within compounds **1** and **2**.

Keywords: multifunctional synergistic response; spin bistability; dielectric; phase transitions

1. Introduction

Bistable materials are special materials whose physical properties can undergo reversible switching between different states when stimulated by external factors [1]. In traditional inorganic bistable materials, inorganic ferromagnets and ceramic ferroelectrics are the most widely studied [2–4]. The direction of magnetization in inorganic ferromagnets can be reversed by an external magnetic field, while the spontaneous polarization direction in ceramic ferroelectrics can be altered by an applied electric field. These bistable characteristics found broad applications in information processing and data storage. Over the past few decades, molecular bistable materials have attracted significant attention and been extensively studied within the scientific community [5–10]. Compared to traditional inorganic materials, molecular materials offer several advantages. First, their process is simpler, their structures are easier to control, and they are more environmentally friendly [11–17]. Second, the diversity of molecular structures and functions greatly expands the scope of bistability [18]. In molecular materials, bistable phenomena are not limited to ferromagnetic and ferroelectric properties, but also switchable magnetism [19,20]. In related studies, molecular bistable materials have gained significant interest from researchers, driven not only by a strong desire to understand the correlation between material properties and structures but also by the high expectation for developing next-generation high-performance functional materials [21], which are promising for applications in sensors [22,23], data storage [24,25], spintronics [26,27], displays [28,29], actuator devices [30,31], and so forth [32–37].

 https://doi.org/10.3390/magnetochemistry11050041

It is well known that the structure determines the physical properties of a material [38,39]. Therefore, changes in magnetic, electrical, optical, and other physical properties of a material are closely related to structural changes [40–44]. External stimuli such as temperature and light can trigger alterations in the spatial or electronic structure of the material, thereby modulating its physical properties [45–49]. Organic–inorganic hybrid materials consist of organic cationic guests and inorganic hosts, which can exhibit characteristics of both organic and inorganic materials [50]. For example, perovskites with the general formulae A_2BX_4 and ABX_3 (where A is an organic ammonium cation, B is a metal ion, and X is a halide ion) [51,52] can display varying magnetic, luminescent, ferroelectric, and dielectric properties depending on the compositions, and may even exhibit multiple physical properties simultaneously. For instance, the compound $(C_6H_5C_2H_3FNH_3)_2[MnCl_4]$ reported by us is capable of controlling fluorescence under a strong magnetic field, demonstrating a magneto-optical effect [52]. Therefore, it is anticipated that the structural phase transitions of organic–inorganic hybrid materials driven by temperature changes could enable the realization of multifunctional bistable effects.

Herein, two examples of inorganic–inorganic hybrid materials $(TMAA)_2[MCl_4]$ (M = Co (**1**) and Mn (**2**)) (TMAA = N,N,N-trimethyl-1-adamantylammonium hydroxide) were synthesized. Both compounds **1** and **2** exhibit a first-order structural phase transition at high temperatures. At the phase transition temperature, both the magnetic and dielectric properties of compounds **1** and **2** show bistable behavior. Moreover, when an in situ magnetic field is applied to compounds **1** and **2**, their dielectric constants undergo significant changes, which strongly suggests the presence of magneto-electric coupling effect in these compounds.

2. Materials and Methods

2.1. Materials

All chemicals and medications were acquired from reagent suppliers of Sigma Aldrich (Sigma Aldrich, St. Louis, MO, USA) and Innochem (Innochem, Beijing, China) without further purification. Powder X-ray diffraction (PXRD) was measured using Rigaku Miniflex 600 equipment (Rigaku Corporation, Tokyo, Japan) at a test angle of 5–50°. The phase purities of the samples employed in subsequent analyses were confirmed through powder X-ray diffraction (PXRD) characterization, which demonstrated excellent consistency with the simulated patterns derived from single-crystal X-ray diffraction data. The Elemental Vario EL analyzer (Elementar Analysensysteme GmbH, Langenselbold, Germany) was used to perform elemental analysis of compounds comprising C, H, and N. The complex dielectric constant ($\varepsilon = \varepsilon' - i\varepsilon''$) was measured using a TH2828A precision LCR meter (Tonghui Electronics Co., Ltd., Changzhou, China) equipped with temperature-controlled capabilities. The DC susceptibilities of compounds **1** and **2** were measured using a Quantum Design SQUID-based MPMS-3-type magnetometer (Quantum Design, Inc., San Diego, CA, USA). Magnetic data of compounds **1** and **2** were recorded in the temperature range of 2.0–430 K with a magnetic field of 1000 Oe. At the same time, in order to study the effect of phase transition on magnetism, the magnetic data changes during cooling and heating were recorded. The magnetization of compound **1** and **2** was recorded in field range of 0–7 T and at corresponding temperatures.

*2.2. Synthesis of $(TMAA)_2[CoCl_4]$ (**1**)*

$CoCl_2·6H_2O$ (0.2379 g, 1 mmol) and TMAA (0.2113 g, 1 mmol) were dissolved in 10 mL of hydrochloric acid and stirred for 10 min. Then, the mixture was added into deionized water (10 mL). Solvent stirring for 10 min followed by filtration volatilization on a heating table at 40 °C gave blue-violet crystals after a few weeks. The yield of

$(TMAA)_2[CoCl_4]$ was 54% based on TMAA. FT-IR (KBr, cm^{-1}): 3852 (s), 3756 (s), 3649 (s), 2915 (s), 2858 (w), 1637 (w), 1555 (w), 1484 (m), 1374 (w), 1300 (w), 1231 (w), 1106 (w), 1036 (m), 948 (m), 838 (m), 568 (w), 456 (w). Calcd (%) for $C_{26}H_{48}Cl_4CoN_2{\cdot}2H_2O$: C, 49.93; H, 8.38; N, 4.48. wt% found: C, 50.06; H, 8.11; N, 4.39.

2.3. *Synthesis of $(TMAA)_2[MnCl_4]$* (**2**)

Green crystals of complex **2** were synthesized in the same way as compound **1**, except that an equivalent amount of $MnCl_2$ was used in place of $CoCl_2{\cdot}6H_2O$. The yield of $(TMAA)_2[MnCl_4]$ was 63% based on TMAA. FT-IR (KBr, cm^{-1}): 3835 (w), 3738 (s), 3617 (s), 2917 (s), 2360 (w), 1920 (w), 1830 (w), 1701 (m), 1483 (s), 1360 (s), 1232 (m), 950 (s), 838 (s), 763 (w), 619 (s), 476 (m). Calcd (%) for $C_{26}H_{48}Cl_4MnN_2{\cdot}H_2O$: C, 51.75; H, 8.36; N, 4.64. wt% found: C, 51.78; H, 8.18; N, 4.57.

2.4. *X-Ray Single-Crystal Structure Determination*

Single-crystal X-ray diffraction data were collected at different temperature using a Bruker D8 QUEST (Bruker Corporation, Billerica, MA, USA) diffractometer with Mo Kα radiation (λ = 0.71073 Å). Initial unit cell parameters were determined and data integration was performed using the APEX II software package. Structural solutions were obtained through intrinsic phasing methods implemented in SHELXT 2019, followed by full-matrix least-squares refinement against F^2 using SHELXL within the Olex2 interface. All non-hydrogen atoms were refined with anisotropic displacement parameters to achieve optimal structural models.

3. Results and Discussion

3.1. *Structural Description*

The structures of compounds **1** and **2** have been reported [52], and they are very similar to each other (Figure 1). The IR spectra and TG analyses were measured as shown in Figures S3 and S4. Therefore, in our detailed discussion of their structures, we will focus solely on compound **1** as a representative example. At a low temperature of 150 K, the crystal structure of compound **1** belongs to the monoclinic system, with the space group $P2_1/n$. The unit cell parameters are as follows: a = 8.9214(3) Å, b = 21.7558(8) Å, c = 14.7144(5) Å, β = 92.685(3)°, and Z = 4, with a unit cell volume V = 2852.81 (18) Å^3. At this temperature, the Co–Cl bond lengths fluctuate between 2.2742 Å and 2.2839 Å, while the Cl–Co–Cl bond angle ranges from 105.06° to 116.16° (Figure 1 and Table S2). At 340 K, some atoms are disordered, with the Co–Cl bond length ranging from 2.2661 Å to 2.2830 and the Cl–Co–Cl bond angle ranging from 104.87° to 116.26° (Table S2). Additionally, temperature-dependent powder X-ray diffraction (PXRD) patterns of compound **1** confirm the occurrence of a phase transition (Figure 2). Although the comparison between the high-temperature and low-temperature phases of compounds **1** and **2** shows no significant changes in the unit cell parameters, the $TMAA^+$ cation exhibits noticeable disorder as the temperature increases. Since the space group remains unchanged during the temperature variation, the phase transition is classified as an isostructural transition, with the intermediate phase assumed to be partially disorder. It is found that the rigid cyclic structure of $TMAA^+$ occupies a relatively large volume, with the terminal carbon atoms positioned at low-energy sites. This restricts atomic movement, allowing only small vibrations within a limited range [52].

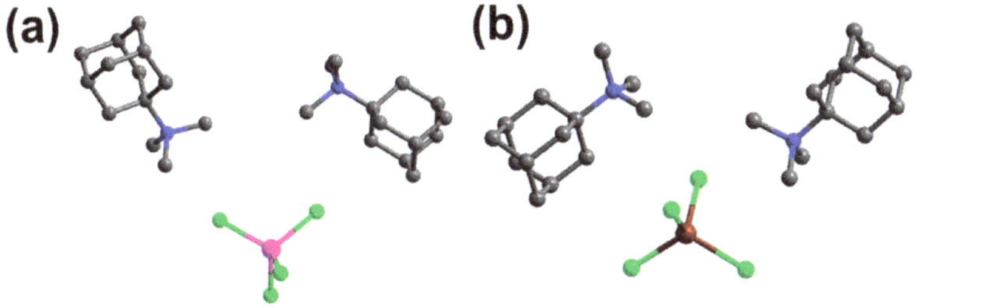

Figure 1. The molecular structures of **1** (**a**) and **2** (**b**). The pink, dark red, bright green blue, and gray spheres represent Co, Mn, Cl, N, and C atoms, respectively.

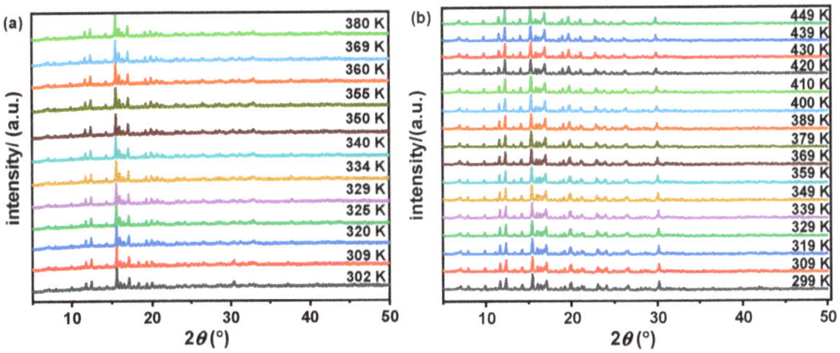

Figure 2. PXRD patterns of compounds **1** (**a**) and **2** (**b**) at variable temperatures.

3.2. Magnetic Properties

Variable-temperature dc magnetic susceptibility measurements were performed on polycrystalline samples of **1** and **2**. We conducted measurements of the molar magnetic susceptibility ($\chi_M T$) for compounds **1** and **2** over a temperature range of 2 to 300 K, under an external DC magnetic field of 1 kOe. As shown in Figure 3a, at room temperature, compound **1** exhibits a $\chi_M T$ value of 3.06 cm^3 K mol^{-1}, which is significantly higher than the theoretical value of 1.875 cm^3 K mol^{-1} for a single Co(II) ion ($S = 3/2$, $g = 2.0$). This discrepancy is attributed to the strong orbital contribution of the Co(II) ion [53,54]. Between 300 and 75 K, the $\chi_M T$ value for compound **1** remains nearly constant as the temperature decreases. Below 75 K, $\chi_M T$ drops sharply, reaching a minimum of 1.7 cm^3 K mol^{-1} at 2 K, which is attributed to the anisotropy of the Co(II) ion. We also measured the magnetic moment (M) as a function of magnetic field strength (H) for compound **1** over the full magnetic field range (0–7 T) at various temperatures (1.8, 2.5, 5.0, and 10 K). At 1.8 K and a magnetic field strength of 7 T, the magnetization of compound **1** reaches a maximum value of 2.35 μ_B, which is considerably lower than the theoretical saturation value of 3.0 μ_B for Co(II) ion (Figure 3a inset), further confirming the strong magnetic anisotropy present in compound **1**. For compound **2**, the $\chi_M T$ curve shows similar behavior to that of compound **1**, but with the key difference that Mn ions do not exhibit magnetic anisotropy. The $\chi_M T$ value for compound **2** at room temperature is 4.375 cm^3 K mol^{-1}, which is close to the theoretical value for Mn(II) ion (Figure 3b). As the temperature decreases, the $\chi_M T$ value for compound **2** remains constant until about 15 K. Below this temperature, $\chi_M T$ decreases rapidly, reaching a minimum value of 4.0 cm^3 K mol^{-1} at 2 K, due to zero-field splitting. At 1.8 K and a magnetic field of 7 T, the magnetization of compound **2** reaches its maximum value of 5.0 μ_B, approaching the saturation value for Mn(II) (Figure 3b inset). The DC magnetic susceptibility data for compounds **1** and **2** were fitted using the Hamiltonian operator in Formula 1 by PHI v3.1.6 software (Figure 3a,b) [55]. The optimal fitting parameters obtained were g = 2.51, D = 14.48 cm^{-1}, E = 3.00 cm^{-1} and TIP = 4.9 × 10^{-4} cm^3 mol^{-1} for compound **1**; g = 1.99, D = 0.23 cm^{-1}, and TIP = 6.59 × 10^{-5} cm^3 mol^{-1} for compound **2**.

$$\hat{H} = D\left[\hat{S}_z^2 - \frac{S(S+1)}{3}\right] + g\mu_B SH \quad (1)$$

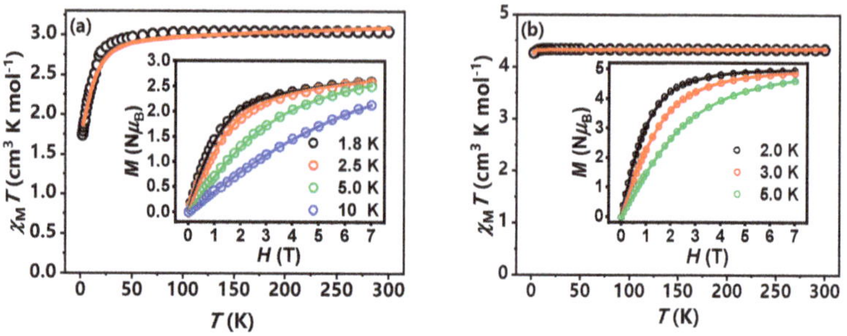

Figure 3. Variable-temperature dc susceptibility data of **1** (**a**) and **2** (**b**). Inset: Variable-temperature, variable-field dc magnetization data collected at different temperatures from 1 to 7 T. Solid lines indicate the best fits with the *PHI* program [55].

In order to investigate the magnetic bistability of compounds **1** and **2**, we measured the $\chi_M T$ vs. T curves of compounds **1** and **2** during the heating and cooling processes. As shown in Figure 4, two pairs of reversible heat anomalies were observed at 329 K/297 K and 359/343 K (heating/cooling, separately) for compound **1**, indicating two phase transitions at 329 K and 359 K for compound **1**. For compound **2**, the $\chi_M T$ remains almost unchanged before the phase transition temperature. When the temperature approaches the phase transition point, there is an abnormal change in the $\chi_M T$ value, which reaches their maximum value at 408 K. As heating continues, the $\chi_M T$ value remains almost unchanged at higher temperatures. Subsequently, as the temperature decreases, there is almost no change in the χ_MT value. During cooling, at 399 K, the minimum values of $\chi_M T$ reach 4.29 cm^3 K mol^{-1}. Then, as the temperature further decreases, the $\chi_M T$ value begins to stabilize, indicating only one phase transitions at 408 K for compound **2**.

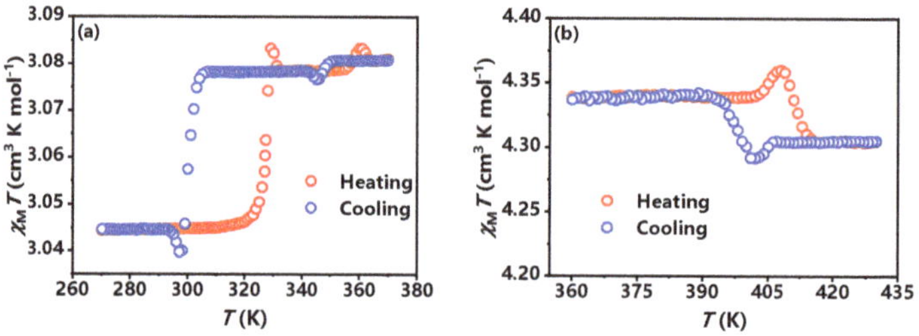

Figure 4. Temperature dependence of $\chi_M T$ vs. T during heating/cooling of **1** (**a**) and **2** (**b**) at high temperatures.

3.3. Dielectric Properties

The thermogravimetric analysis (TGA) curves of compounds **1** and **2** are shown in Figure S4. It can be seen from the figures that both compounds are decomposed at about 525 K, showing excellent thermal stability. The dielectric measurements of compounds **1** and **2** were obtained in the temperature range 300 to 430 K at 100 kHz (Figure S5). The value of ε′ for compound **1** remains at 17.52 at LTP. When the temperature increases to

345 K, the value of ε′ suddenly shows a step increase and reaches 45.91, which indicates an obvious dielectric abnormality. During the cooling process, the change in ε′ value is similar to that for heating process, indicating that compound **1** undergoes a reversible phase transition. For compound **2**, the value of ε′ is 3.19 at LTP, while it is 14.14 at 413.5 K. The abnormal change in ε′ value is attributed to the order–disorder transition of the $[TMAA]^+$ cations. Under an applied magnetic field, the peak of the curves does not change. However, the dielectric constant values decrease. The dielectric constant values decrease with a stronger magnetic field (Figure 5). This kind of physical phenomenon has been reported, but it is still a challenge to give a precise reason. However, it is clear that the magnetic field leads to such behavior probably through Lorentzian force, a slight change in the structure of paramagnetic compounds and/or the Hall effect.

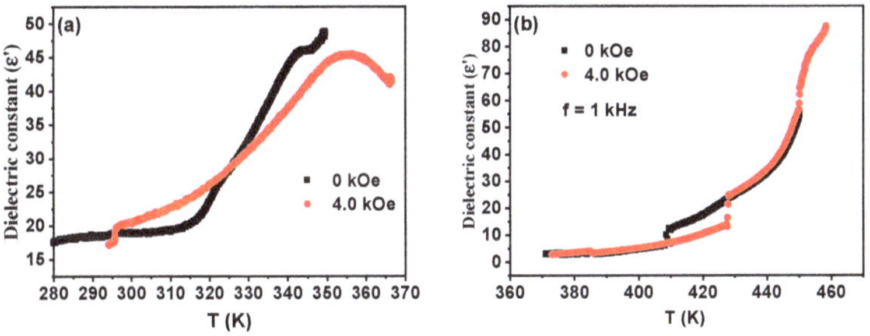

Figure 5. Dielectric curves of compounds **1** (**a**) and **2** (**b**) at different magnetic fields.

4. Conclusions

Two organic–inorganic hybrid materials $(TMAA)_2[CoCl_4]$ (**1**) and $(TMAA)_2[MnCl_4]$ (**2**) were synthesized through the reaction between N,N,N-Trimethyladamantan-1-aminium hydroxide and cobaltous chloride/manganese chloride. Both compounds show a first structure transition at high temperature, owing to the order–disorder character of the phase transition. Both compounds show a magnetic and dielectric response. Compound **1** has two-step spin bistability, while **2** only displayed one-step spin bistability. Both compounds demonstrated magnetoelectric coupling at high temperature. This study demonstrates that organic–inorganic hybrids are multifunctional synergistically responsive materials with the induction of phase transitions in molecular compounds.

Supplementary Materials: The following supporting information can be downloaded at https://www.mdpi.com/article/10.3390/magnetochemistry11050041/s1: Figure S1: Packing picture of compounds **1** (a) and **2** (b); Figure S2: (a) The molecular structures of **1@340** at 340 K. The pink, bright green blue and gray spheres represent Co, Cl, N and C atoms, respectively. (b) Packing picture of compounds **1@340** at 340 K. Red lines indicate hydrogen bonds; Figure S3: The IR spectra of compounds **1** and **2**; Figure S4: TG analysis diagram of compounds **1** (a) and **2** (b). Figure S5. Dielectric curves of compounds **1** (a) and **2** (b) at 100 kHz. Table S1: Crystallographic data and structural refinement parameters for compounds **1**, **1@340** and **2**; Table S2: Bond lengths [Å] and angles [°] for compounds **1**, **1@340** and **2**.

Author Contributions: The manuscript was prepared with the contributions of all authors. X.-F.C.: Synthesis, IR, TG and single-crystal X-ray diffraction studies; T.W.: Dielectric property characterization and synthesis; D.L.: PXRD; N.W.: Magnetic property characterization and writing—original draft; S.-Y.Z.: Magnetic and dielectric property characterization; Y.P. and Z.-B.H.: Careful follow-up and improvement of the manuscript, general idea and plan for the publication. All authors have read and agreed to the published version of the manuscript.

Funding: National Natural Science Foundation of China (22261022, 22161002 and 22105089), Natural Science Foundation of Jiangxi Province (20224BAB214005 and 20242BAB23018), Jiangxi Provincial Key Laboratory of Functional Crystalline Materials Chemistry (2024SSY05161) and Double Thousand Talents Program in Jiangxi Province (jxsq2020101087).

Institutional Review Board Statement: Not applicable.

Informed Consent Statement: Not applicable.

Data Availability Statement: CCDC 2115536–2115539 contain the supplementary crystallographic data for complexes **1** and **2**, respectively. These data can be obtained free of charge via http://www.ccdc.cam.ac.uk/conts/retrieving.html (accessed on 4 May 2025) or from Cambridge Crystallographic Data Centre (12 Union Road, Cambridge CB2 1EZ, UK; fax: (+44) 1223-336-033; or e-mail: deposit@ccdc.cam.ac.uk). The original contributions presented in the study are included in the article and Supplementary Materials; further inquiries can be directed to the corresponding authors.

Acknowledgments: We would like to acknowledge the Nantong University Analytical Testing Center for its support for testing.

Conflicts of Interest: The authors declare no conflicts of interest.

References

1. Sato, O. Dynamic molecular crystals with switchable physical properties. *Nat. Chem.* **2016**, *8*, 644–656. [CrossRef] [PubMed]
2. Wang, J.; Neaton, J.B.; Zheng, H.; Nagarajan, V.; Ogale, S.B.; Liu, B.; Vihland, D.; Vaithyanathan, V.; Schlom, D.G.; Waghmare, U.V.; et al. Epitaxial $BiFeO_3$ Multiferroic Thin Film. *Heterostruct. Sci.* **2003**, *299*, 1719–1722.
3. Fiebig, M.; Lottermoser, T.; Meier, D.; Trassin, M. The evolution of multiferroics. *Nat. Rev. Mater.* **2016**, *1*, 16046. [CrossRef]
4. Feng, Z.W.; Shi, T.; Geng, G.Z.; Li, J.J.; Deng, Z.L.; Kivshar, Y.; Li, X.P. Dual-band polarized upconversion photoluminescence enhanced by resonant dielectric metasurfaces. *eLight* **2023**, *3*, 21. [CrossRef]
5. Sun, B.; Liu, X.F.; Li, X.Y.; Cao, Y.; Yan, Z.; Fu, L.; Tang, N.J.; Wang, Q.; Shao, X.F.; Yang, D.Z.; et al. Reversible Thermochromism and Strong Ferromagnetism in Two-Dimensional Hybrid Perovskites. *Angew. Chem. Int. Ed.* **2020**, *59*, 203–208. [CrossRef]
6. Liu, J.Y.; Zhang, S.Y.; Zeng, Y.; Shu, X.; Du, Z.Y.; He, C.T.; Zhang, W.X.; Chen, X.M. Molecular Dynamics, Phase Transition and Frequency-Tuned Dielectric Switch of an Ionic Co-Crystal. *Angew. Chem. Int. Ed.* **2018**, *130*, 8164–8168. [CrossRef]
7. Jia, Z.H.; Liu, J.Y.; Liu, D.X.; Zhang, S.Y.; Du, Z.Y.; He, C.T.; Zhang, W.X.; Chen, X.M. Four-step thermosensitive dielectric response arising from motionable low-symmetry ammonium confined in deformable supramolecular cages. *J. Mater. Chem. C* **2021**, *9*, 8076–8082. [CrossRef]
8. Wang, C.F.; Wang, N.; Shi, C.; Ye, H.Y.; Zhang, Y.; Miao, L.P. NH_4^+/K^+-substitution-induced C–F–K coordination bonds for designing the highest-temperature hybrid halide double perovskite ferroelastic. *Chin. Chem. Lett.* **2023**, *34*, 107774. [CrossRef]
9. Wang, C.F.; Wang, N.; Liu, L.; Miao, L.P.; Ye, H.Y.; Zhang, Y.; Shi, C. Enantiomeric hybrid high-temperature multiaxial ferroelectrics with a narrow bandgap and high piezoelectricity. *Chin. Chem. Lett.* **2023**, *34*, 108051. [CrossRef]
10. Liu, S.M.; He, L.; Wang, Y.Z.; Shi, P.P.; Ye, Q. Tunable phase transition, band gap and SHG properties by halogen replacement of hybrid perovskites [(thiomorpholinium)PbX_3, X = Cl, Br, I]. *Chin. Chem. Lett.* **2022**, *33*, 1032–1036. [CrossRef]
11. Garcia, V.; Bibes, M. Inside story of ferroelectric memories. *Nature* **2012**, *483*, 279–280. [CrossRef] [PubMed]
12. Harada, J.; Kawamura, Y.; Takahashi, Y.; Uemura, Y.; Hasegawa, T.; Taniguchi, H.; Maruyama, K. Plastic/Ferroelectric Crystals with Easily Switchable Polarization: Low-Voltage Operation, Unprecedentedly High Pyroelectric Performance, and Large Piezoelectric Effect in Polycrystalline Forms. *J. Am. Chem. Soc.* **2019**, *141*, 9349–9357. [CrossRef] [PubMed]
13. Liao, W.Q.; Zeng, Y.L.; Tang, Y.Y.; Peng, H.; Liu, J.C.; Xiong, R.G. Multichannel Control of Multiferroicity in Single Component Homochiral Organic Crystals. *J. Am. Chem. Soc.* **2021**, *143*, 21685–21693. [CrossRef] [PubMed]
14. Liu, K.; Ouyang, B.; Guo, X.J.; Liu, Y. Advances in flexible organic field-effect transistors and their applications for flexible electronics. *Npj Flex. Electron.* **2022**, *6*, 1. [CrossRef]
15. Liu, Y.; Wang, S.; Chen, Z.J.; Xiao, L.X. Applications of ferroelectrics in photovoltaic devices. *Sci. China Mater.* **2016**, *59*, 851. [CrossRef]
16. Tang, Y.Y.; Li, P.F.; Liao, W.Q.; Shi, P.P.; You, Y.M.; Xiong, R.G. Multiaxial Molecular Ferroelectric Thin Films Bring Light to Practical Applications. *J. Am. Chem. Soc.* **2018**, *140*, 8051–8059. [CrossRef]
17. Wei, X.K.; Domingo, N.; Sun, Y.; Balke, N.; Dunin-Borkowski, R.E.; Mayer, J. Progress on Emerging Ferroelectric Materials for Energy Harvesting, Storage and Conversion. *Adv. Energy Mater.* **2022**, *12*, 2201199. [CrossRef]
18. Zhang, W.; Xiong, R.G. Ferroelectric Metal–Organic Frameworks. *Chem. Rev.* **2012**, *112*, 1163–1195. [CrossRef]
19. Horiuchi, S.; Tokura, Y. Organic ferroelectrics. *Nat. Mater.* **2008**, *7*, 357–366. [CrossRef]

20. Gui, L.A.; Chen, J.W.; Zhang, Y.F.; Li, L.H.; Li, J.R.; Hu, Z.B.; Zhang, S.Y.; Zhang, J.L.; Zhang, Z.Y.; Ye, H.Y.; et al. Room-temperature Magnetocapacitance Spanning 97 K Hysteresis in Molecular Material. *Angew. Chem. Int. Ed.* **2025**, *64*, e202416380. [CrossRef]
21. Gui, L.A.; Zhang, Y.F.; Peng, Y.; Hu, Z.B.; Song, Y. Synergetic Responses of Multiple Functions Induced by Phase Transition in Molecular Materials. *ChemPhysChem* **2024**, *25*, e202400297. [CrossRef] [PubMed]
22. Zhang, Z.X.; Zhang, T.; Shi, P.P.; Zhang, W.Y.; Ye, Q.; Fu, D.W. Exploring high-performance integration in a plastic crystal/film with switching and semiconducting behavior. *Inorg. Chem. Front.* **2020**, *7*, 1239–1249. [CrossRef]
23. Gao, Y.F.; Zhang, Z.X.; Zhang, T.; Su, C.Y.; Zhang, W.Y.; Fu, D.W. Regulated molecular rotor in phase transition materials with switchable dielectric and SHG effect. *Mater. Chem. Front.* **2020**, *4*, 3003–3012. [CrossRef]
24. Li, X.Q.; Li, D.; Peng, Y.; Liu, Y.; Wang, J.Q.; Li, L.N.; Yao, Y.P.; Liu, X.T.; Luo, J.H. Exploring a layered iodide perovskite crystal with centimetered dimension for extended spectral polarization-sensitive photodetection. *J. Mater. Chem. C* **2021**, *9*, 9499–9504. [CrossRef]
25. Jia, J.H.; Li, Q.W.; Chen, Y.C.; Liu, J.L.; Tong, M.L. Luminescent single-molecule magnets based on lanthanides: Design strategies, recent advances and magneto-luminescent studies. *Coord. Chem. Rev.* **2019**, *378*, 365–381. [CrossRef]
26. Liu, K.; Zhang, X.J.; Meng, X.X.; Shi, W.; Cheng, P.; Powell, A.K. Constraining the coordination geometries of lanthanide centers and magnetic building blocks in frameworks: A new strategy for molecular nanomagnets. *Chem. Soc. Rev.* **2016**, *45*, 2423–2439. [CrossRef]
27. Zhang, Z.X.; Su, C.Y.; Li, J.; Song, X.J.; Fu, D.W.; Zhang, Y. Ferroelastic Hybrid Bismuth Bromides with Dual Dielectric Switches. *Chem. Mater.* **2021**, *33*, 5790–5799. [CrossRef]
28. Su, C.Y.; Lun, M.M.; Chen, Y.D.; Zhou, Y.C.; Zhang, Z.X.; Chen, M.; Huang, P.Z.; Fu, D.W.; Zhang, Y. Hybrid Optical-Electrical Perovskite Can Be a Ferroelastic Semiconductor. *CCS Chem.* **2022**, *4*, 2009–2019. [CrossRef]
29. Zhang, S.Y.; Shu, X.; Zeng, Y.; Liu, Q.Y.; Du, Z.Y.; He, C.T.; Zhang, W.X.; Chen, X.M. Molecule-based nonlinear optical switch with highly tunable on-off temperature using a dual solid solution approach. *Nat. Commun.* **2020**, *11*, 2752. [CrossRef]
30. Gao, F.F.; Li, X.; Qin, Y.; Li, Z.G.; Guo, T.M.; Zhang, Z.Z.; Su, G.D.; Jiang, C.Y.; Azeem, M.; Li, W.; et al. Dual-Stimuli-Responsive Photoluminescence of Enantiomeric Two-Dimensional Lead Halide Perovskites. *Adv. Opt. Mater.* **2021**, *9*, 2100003. [CrossRef]
31. Liu, X.L.; Li, D.; Zhao, H.X.; Dong, X.W.; Long, L.S.; Zheng, L.S. Inorganic–Organic Hybrid Molecular Materials: From Multiferroic to Magnetoelectric. *Adv. Mater.* **2021**, *33*, 2004542. [CrossRef] [PubMed]
32. Tang, Y.Y.; Wang, Z.X.; Li, P.F.; You, Y.M.; Stroppa, A.; Xiong, R.G. Brilliant triboluminescence in a potential organic–inorganic hybrid ferroelectric: $(Ph_3PO)_2MnBr_2$. *Inorg. Chem. Front.* **2017**, *4*, 154–159. [CrossRef]
33. Jin, W.H.; Lu, H.H.; Zhang, Q.; Qu, D.H. A dual-mode orthogonally tunable fluorescent system covering the whole white light region. *Mater. Chem. Front.* **2020**, *4*, 532–536. [CrossRef]
34. Wang, Z.P.; Zhang, Z.Z.; Tao, L.Q.; Shen, N.N.; Hu, B.; Gong, L.K.; Li, J.R.; Chen, X.P.; Huang, X.Y. Hybrid Chloroantimonates(III): Thermally Induced Triple-Mode Reversible Luminescent Switching and Laser-Printable Rewritable Luminescent Paper. *Angew. Chem. Int. Ed.* **2019**, *58*, 9974–9978. [CrossRef]
35. Li, J.; Xu, C.; Zhang, W.Y.; Shi, P.P.; Ye, Q.; Fu, D.W. Smart and efficient opto-electronic dual response material based on two-dimensional perovskite crystal/thin film. *J. Mater. Chem. C* **2020**, *8*, 1953–1961. [CrossRef]
36. Morita, H.; Tsunashima, R.; Nishihara, S.; Inoue, K.; Omura, Y.; Suzuki, Y.; Kawamata, J.; Hoshino, N.; Akutagawa, T. Ferroelectric Behavior of a Hexamethylenetetramine-Based Molecular Perovskite Structure. *Angew. Chem. Int. Ed.* **2019**, *58*, 9184–9187. [CrossRef]
37. Zhang, Y.F.; Gui, L.A.; Peng, Y.; Hu, Z.B.; Song, Y. Perspective on room temperature and low-field-induced magnetoelectric coupling in molecular complexes. *Dalton Trans.* **2025**, *54*, 881–888. [CrossRef]
38. Clark, J.B.; Hastie, J.W.; Kihlborg, L.H.E.; Metselaar, R.; Thackeray, M.M. Definitions of terms relating to phase transitions of the solid state. *Pure Appl. Chem.* **1994**, *66*, 577–594. [CrossRef]
39. Shi, C.; Han, X.B.; Zhang, W. Structural phase transition-associated dielectric transition and ferroelectricity in coordination compounds. *Coord. Chem. Rev.* **2019**, *378*, 561–576. [CrossRef]
40. Du, Z.Y.; Xu, T.T.; Huang, B.; Su, Y.J.; Xue, W.; He, C.T.; Zhang, W.X.; Chen, X.M. Switchable Guest Molecular Dynamics in a Perovskite-Like Coordination Polymer toward Sensitive Thermoresponsive Dielectric Materials. *Angew. Chem. Int. Ed.* **2015**, *12*, 928–932. [CrossRef]
41. Muhammad; Lim, C.W. From Photonic Crystals to Seismic Metamaterials: A Review via Phononic Crystals and Acoustic Metamaterials. *Arch. Comput. Method. E* **2022**, *29*, 1137–1198. [CrossRef]
42. Nishita, M.; Park, S.Y.; Nishio, T.; Kamizaki, K.; Wang, K.; Takumi, T.; Hashimoto, R.; Otani, H.; Pazour, G.J.; Hsu, V.W.; et al. Ror2 signaling regulates Golgi structure and transport through IFT20 for tumor invasiveness. *Sci. Rep.* **2017**, *7*, 1. [CrossRef] [PubMed]
43. Sen, A.; Swain, D.; Guru Row, T.N.; Sundaresan, A. Unprecedented 30 K hysteresis across switchable dielectric and magnetic properties in a bright luminescent organic-inorganic halide $(CH_6N_3)_2MnCl_4$. *J. Mater. Chem. C* **2019**, *7*, 4838–4845. [CrossRef]

44. Zhang, Y.; Ye, H.Y.; Cai, H.L.; Fu, D.W.; Ye, Q.; Zhang, W.; Zhou, Q.H.; Wang, J.L.; Yuan, G.L.; Xiong, R.G. Switchable Dielectric, Piezoelectric, and Second-Harmonic Generation Bistability in a New Improper Ferroelectric above Room Temperature. *Adv. Mater.* **2014**, *26*, 4515. [CrossRef]
45. Schmid, H. Some symmetry aspects of ferroics and single phase multiferroics. *J. Phys. Condens. Matter.* **2008**, *20*, 434201. [CrossRef]
46. Spaldin, N.A.; Fiebig, M.; Mostovoy, M. The toroidal moment in condensed-matter physics and its relation to the magnetoelectric effect. *J. Phys. Condens. Matter.* **2008**, *20*, 434203. [CrossRef]
47. Yao, H.Q.; Ding, Z.J.; Zhang, S.Y.; Wu, Y.F.; Qiu, X.; Peng, Y.; Ye, H.Y.; Hu, Z.B. High temperature synergistic response of magnetism, dielectricity, and luminescence in a Mn(II)-based molecular organic–inorganic hybrid. *CrystEngComm* **2024**, *26*, 1767–1772. [CrossRef]
48. Yao, H.Q.; Li, L.H.; Zhang, S.Y.; Peng, Y.; Liu, S.J.; Wen, H.R.; Zhang, J.L.; Ye, H.Y.; Song, Y.; Hu, Z.B. Near-Room-Temperature Synergetic Response of Magnetism, Dielectricity and Luminescence Based on $(C_5NH_{13}Cl)_2MnBr_4$. *Chem. Eur. J.* **2023**, *29*, e202300598. [CrossRef]
49. Cheng, Y.; Ruan, H.P.; Peng, Y.; Li, L.H.; Xie, Z.Q.; Liu, L.; Zhang, S.Y.; Ye, H.Y.; Hu, Z.B. Magnetic, dielectric and luminescence synergetic switchable effects in molecular material $[Et_3NCH_2Cl]_2[MnBr_4]$. *Chin. Chem. Let.* **2024**, *35*, 108554. [CrossRef]
50. Zhao, X.H.; Huang, X.C.; Zhang, S.L.; Shao, D.; Wei, H.Y.; Wang, X.Y. Cation-Dependent Magnetic Ordering and Room-Temperature Bistability in Azido-Bridged Perovskite-Type Compounds. *J. Am. Chem. Soc.* **2013**, *135*, 16006–16009. [CrossRef]
51. Du, S.N.; Su, D.; Ruan, Z.Y.; Zhou, Y.Q.; Deng, W.; Zhang, W.X.; Sun, Y.; Liu, J.L.; Tong, M.L. Reversible Switchability of Magnetic Anisotropy and Magnetodielectric Effect Induced by Intermolecular Motion. *Angew. Chem. Int. Ed.* **2022**, *61*, e202204700. [CrossRef] [PubMed]
52. Shao, T.; Ren, R.Y.; Huang, P.Z.; Ni, H.F.; Su, C.Y.; Fu, D.W.; Xie, L.Y.; Lu, H.F. Metal ion modulation triggers dielectric double switching and green fluorescence in A_2MX_4-type compounds. *Dalton Trans.* **2022**, *51*, 2005–2011. [CrossRef] [PubMed]
53. Hu, Z.B.; Gui, L.A.; Li, L.H.; Xiao, T.T.; Hand, A.T.; Tin, P.; Ozerov, M.; Peng, Y.; Ouyang, Z.; Wang, Z.; et al. Co^{II} Single-ion Magnet and its Multi-dimensional Aggregations: Influence of the Structural Rigidity on Magnetic Relaxation Process. *Chin. Chem. Lett.* **2025**, *36*, 109600. [CrossRef]
54. Li, L.H.; Peng, Y.; Zhang, S.; Xiao, T.; Song, Y.; Ye, H.Y.; Zhang, J.; Zhang, Y.Q.; Wang, Z.; Hu, Z.B. Subtle changes Effect on Slow Relaxation Behaviour of Co(II) Ions Based Metal Organic Frameworks. *Cryst. Growth Des.* **2023**, *23*, 2099–2105. [CrossRef]
55. Chilton, N.F.; Anderson, R.P.; Turner, L.D.; Soncini, A.; Murray, K.S. PHI: A powerful new program for the analysis of anisotropic monomeric and exchange-coupled polynuclear d- and f-block complexes. *J. Comput. Chem.* **2013**, *34*, 1164–1175. [CrossRef]

magnetochemistry

MDPI

Article

Two Dy_2 Zero-Field Single-Molecule Magnets Derived from Hydrazone Schiff Base-Bridging Ligands and 1,3-Di(2-pyridyl)-1,3-propanedione †

Cai-Ming Liu

Beijing National Laboratory for Molecular Sciences, Center for Molecular Science, Institute of Chemistry, Chinese Academy of Sciences, Beijing 100080, China; cmliu@iccas.ac.cn

† Dedicated to Prof. Dai-Zheng Liao on the occasion of his 85th birthday.

Abstract: Two hydrazone Schiff base-bridging ligands with different heterocycles {2-[(*E*)-(5-chloro-2-hydroxyphenyl)methylidene]diazanyl}(pyrazine-2-yl)methanone (H_2L_{Schiff}-1) and (*E*)-*N′*-(2-hydroxy-3-methoxybenzylidene)nicotinohydrazide (H_2L_{Schiff}-2) together with 1,3-di(2-pyridyl)-1,3-propanedione (Hdpp) were chosen to construct two new Dy_2 complexes, $[Dy_2(L_{Schiff}\text{-}1)_2(DMF)_2(dpp)_2]\cdot 0.5DMF$ (**1**) and $[Dy_2(L_{Schiff}\text{-}2)_2(DMF)_2(dpp)_2]\cdot 2DMF$ (**2**). Although the $[N_2O_6]$ coordination spheres are observed for the Dy^{3+} ions in **1** and **2**, their coordination configurations have some differences (both the biaugmented trigonal prism and the Snub diphenoid J84 in **1** and only the biaugmented trigonal prism in **2**). Magnetic research revealed that both **1** and **2** possess ferromagnetic interactions between two Dy^{3+} ions and perform as zero-field single-molecule magnets, with U_{eff}/k values of 49.7 K at 0 Oe for **1** and 151.8 K at 0 Oe for **2**. This work suggests that the heterocycle groups (pyrazine vs. pyridine) on the hydrazone Schiff base-bridging ligands have effects on the SMM properties of **1** and **2**.

Keywords: single-molecule magnet; Dy_2 complex; ferromagnetic interaction; hydrazone Schiff base; heterocycles′ effect; 1,3-di(2-pyridyl)-1,3-propanedione

1. Introduction

Single-molecule magnets (SMMs) are an important branch of molecular magnets, and their magnetic bistable properties at the nanoscale have potential applications in high-density information storage, quantum computing, and other related fields [1]. Since lanthanide ions, especially the dysprosium (III) ion, naturally have significant magnetic anisotropy and multiple unpaired electrons (corresponding to large ground-state spin values), lanthanide SMMs have become the most developed SMMs in recent years [2]. Among them, the energy barrier (U_{eff}) value and magnetic blocking temperature (T_B) of the SMM containing a single lanthanide ion, that is, the single-ion magnet (SIM), can reach a high level [3–8], but the sensitivity of metal–organic compounds to air will affect the application of this type of SMMs; while the SMM containing multiple lanthanide ions, requires consideration of the consistency of the magnetic axis orientation of each lanthanide ion [9], its U_{eff} value and T_B value are more difficult to make a big breakthrough. There is no doubt that the magnetic axis orientation consistency of the Ln(III) binuclear SMM is the easiest to achieve among the SMMs with multiple lanthanide (III) ions, which can be achieved through the ferromagnetic interaction between two lanthanide (III) ions [10–29]. Notably, the ferromagnetic interaction can partially or completely quench quantum tunneling of magnetization (QTM), activating SMM behavior at zero dc field [11].

However, the magnetic interactions between 4f and 4f electrons are usually difficult; even if magnetic exchange occurs, it often exhibits weak antiferromagnetic coupling.

Hydrazone Schiff base-bridging ligands are commonly used to construct Ln SMMs [30,31]; they have been used to construct some ferromagnetically coupled Dy_2 SMMs [10–14,32–38], the magnetic properties of which may be affected by several factors. For example, the coordination anion [32], and the substituent on the terminal ligands [33] are the most common influencing factors. From the perspective of molecular engineering, heterocycles on hydrazone Schiff base-bridging ligands are also important influencing factors. Recently, we used two hydrazone Schiff bases (Scheme 1) with different heterocycles (pyrazine vs. pyridine) as bridging ligands and both 1,3-di(2-pyridyl)-1,3-propanedione (Hdpp) (Scheme 1) and DMF as terminal ligands to construct two new Dy_2 complexes: $[Dy_2(L_{Schiff}\text{-}1)_2(DMF)_2(dpp)_2]\cdot0.5DMF$ {H_2L_{Schiff}-1 = {2-[(*E*)-(5-chloro-2-hydroxyphenyl)methylidene]diazanyl}(pyrazine-2-yl)methanone} (**1**) and $[Dy_2(L_{Schiff}\text{-}2)_2(DMF)_2(dpp)_2]\cdot2DMF$ {H_2L_{Schiff}-2 = (*E*)-*N*′-(2-hydroxy-3-methoxybenzylidene) nicotinohydrazide} (**2**), which exhibit ferromagnetic interactions and perform zero-field SMM properties, with U_{eff}/k values of 49.7 K and 151.8 K at 0 Oe for **1** and **2**, respectively, suggesting that the heterocycles on hydrazone Schiff base-bridging ligands have an effect on the magnetic properties of Dy_2 SMMs.

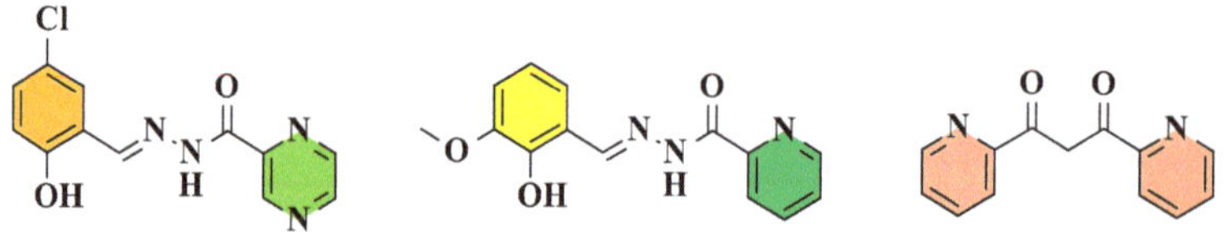

Scheme 1. Hydrazone Schiff bases (H_2L_{Schiff}-1 and H_2L_{Schiff}-2) and 1,3-di(2-pyridyl)-1,3-propanedione (Hdpp).

2. Results

2.1. Crystal Structures

Dy_2 complexes **1** and **2** are crystallized in *P*-1 space group and $P2_1/c$ space group, respectively. As shown in Figure 1, there are two crystallographically independent molecules in **1**, each consisting of two Dy^{3+} ions, two $L_{Schiff}\text{-}1^{2-}$ bridging ligands, two dpp^- terminal ligands, and two coordinated DMF molecules. In addition, the lattice of **1** contains DMF solvent molecules. Notably, both independent molecules are centrosymmetric; consequently, the anisotropy axes of the two ferromagnetic Dy^{3+} ions in **1** have to be parallel to each other, which may be beneficial to its SMM properties. The asymmetric unit is composed of one Dy^{3+} ion, one $L_{Schiff}\text{-}1^{2-}$ bridging ligand, one dpp^- terminal ligand, and one coordinated DMF molecule. In each molecule the two $L_{Schiff}\text{-}1^{2-}$ bridging ligands chelate two Dy^{3+} ions from opposite positions to form a rough plane; a dpp^- terminal ligand and a DMF molecule then coordinate each dysprosium (III) ion from the upper and lower directions of the plane, and a $[N_2O_6]$ coordination sphere is formed, which is coordinated by one $N_{pyrazine}$ atom and O_{acyl} atom from one $L_{Schiff}\text{-}1^{2-}$ ligand, one N_{imine} atom, one O_{acyl} atom and one O_{phenol} atom from the other $L_{Schiff}\text{-}1^{2-}$ ligand, two O atoms from the dpp^- terminal ligand, and one O_{DMF} atom. Further analysis with the SHAPE 2.1 software [39] revealed that the coordination configurations of Dy1 and Dy2 are the biaugmented trigonal prism and Snub diphenoid J84, respectively, with CShM values of 1.818 for the C_{2v} symmetry and 20.565 for the D_{2d} symmetry (Table S1). Notably, the pyridine N atom in the dpp^- terminal ligand does not participate in the coordination, unlike some SMMs containing lanthanide (III) ions [40,41]. The Dy1–Dy1$^{\#1}$ ($^{\#1}$ $1 - x, 2 - y, 1 - z$) distance (3.942 Å) is a little longer than the Dy1-Dy1[#2] ([#2] $-x, 1-y, 2-z$) distance (3.932 Å). The mean distance of Dy1–O (2.324 Å) is comparable with that of Dy2–O (2.321 Å), and the average distance of Dy1–N (2.541 Å) is comparable with that of Dy2–N (2.541 Å) (Table 1). Notably, the latest CCDC retrieval revealed that **1** represents the first complex derived from the new H_2L_{Schiff}-1 ligand.

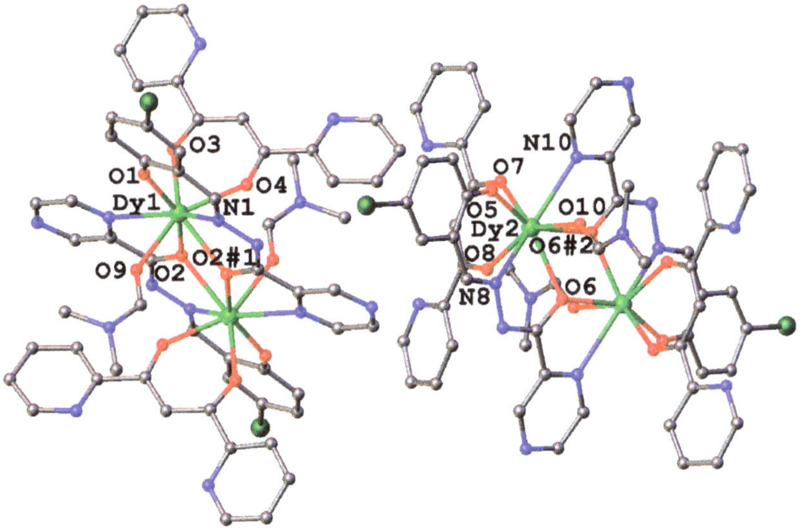

Figure 1. Crystal structure of **1**, all H atoms and solvent molecules are omitted for clarity. Symmetry code: [#1] $1-x, 2-y, 1-z$; [#2] $-x, 1-y, 2-z$.

Table 1. Selected bond lengths (Å) and angles (°) of **1**.

Dy1–O4	2.303(6)	Dy1–O1 [#1]	2.191(6)
Dy1–O3	2.355(6)	Dy1–O2	2.352(5)
Dy1–O2 [#1]	2.385(5)	Dy1–O9	2.360(6)
Dy1–N3	2.579(7)	Dy1–N1 [#1]	2.502(7)
Dy2–O5	2.197(6)	Dy2–O8	2.302(6)
Dy2–O6 [#2]	2.340(6)	Dy2–O6	2.387(5)
Dy2–O7	2.352(6)	Dy2–O10	2.351(6)
Dy2–N10	2.584(7)	Dy2–N8	2.497(7)
O4–Dy1–O3	71.4(2)	O4–Dy1–O9	147.9(2)
O3–Dy1–O9	140.7(2)	Dy1–O2–Dy1 [#1]	112.7(2)
O8–Dy2–O7	71.4(2)	O8–Dy2–O10	148.5(2)
O10–Dy2–O7	140.0(2)	Dy2 [#2]–O6–Dy2	112.6(2)

[#1] $1-x, 2-y, 1-z$; [#2] $-x, 1-y, 2-z$.

As shown in Figure 2, the main structure of **2** is composed of two Dy^{3+} ions, two L_{Schiff}-2^{2-} bridging ligands, two dpp^- anions, and two coordinated DMF molecules. Additionally, **2** contains some DMF solvent molecules in its lattice as well. Different from **1**, only one crystallographically independent molecule exists in **2**; there is a symmetry center in this molecule as well, and the asymmetric unit is composed of one Dy^{3+} ion, one L_{Schiff}-2^{2-} bridging ligand, one dpp^- anion, and one coordinated DMF molecule. Similar to **1**, the two L_{Schiff}-2^{2-} bridging ligands in **2** coordinate two Dy^{3+} ions to form a rough plane; and a dpp^- anion and a DMF molecule then finish the coordination of the Dy^{3+} ion, generating a $[N_2O_6]$ coordination sphere. Each Dy^{3+} cation is coordinated by one N_{pydine} atom and

O_{acyl} atom of one L_{Schiff}-2^{2-} ligand, one N_{imine} atom, one O_{acyl} atom and one O_{phenol} atom of the other L_{Schiff}-2^{2-} ligand, two O atoms of the dpp^- anion, and one O_{DMF} atom. The coordination configuration of Dy1 in **2** is the biaugmented trigonal prism, with a CShM value of 2.247 for the C_{2v} symmetry (Table S2). The Dy1–Dy1$^{\#3}$ ($^{\#3}$ $1-x, 1-y, 1-z$) distance in **2** (3.885 Å) is slightly shorter than the Dy–Dy distance in **1** (3.942 Å and 3.932 Å) (Tables 1 and 2). The mean distance of Dy1–O in **2** (2.315 Å) is slightly shorter than those of Dy–O in **1** (2.324 Å and 2.321 Å), and the average distance of Dy1–N in **2** (2.525 Å) is a little smaller than those of Dy–N in **1** (2.541 Å and 2.541 Å) (Tables 1 and 2). However, the Dy1–O3–Dy1$^{\#3}$ bond angle in **2** {113.95(8)°} is slightly larger than the corresponding Dy–O–Dy bond angles in **1** {112.7(2)° and 112.6(2)°} (Tables 1 and 2).

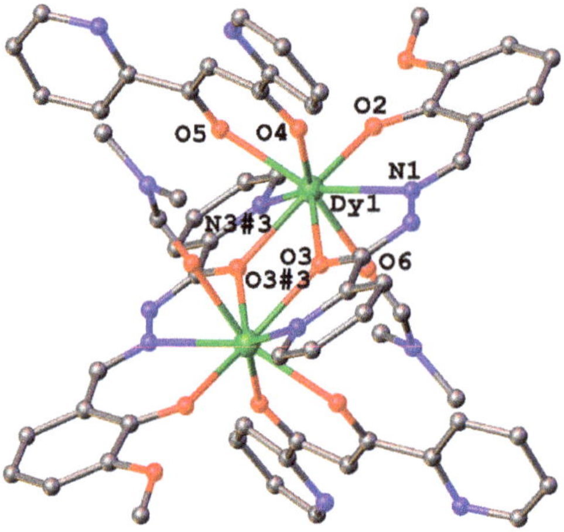

Figure 2. Crystal structure of **2**, all H atoms and solvent molecules are omitted for clarity. Symmetry code: $^{\#3}$ $1-x, 1-y, 1-z$.

Table 2. Selected bond lengths (Å) and angles (°) of **2**.

Dy1–O4	2.326(2)	Dy1–O2	2.176(2)
Dy1–O5	2.361(2)	Dy1–O3	2.312(2)
Dy1–O3 $^{\#3}$	2.322(2)	Dy1–O6	2.392(2)
Dy1–N1	2.509(2)	Dy1–N3 $^{\#3}$	2.541(2)
O4–Dy1–O5	71.04(7)	O4–Dy1–O6	149.21(8)
O5–Dy1–O6	139.50(8)	Dy1–O3–Dy1 $^{\#3}$	113.95(8)

$^{\#3}$ $1-x, 1-y, 1-z$.

2.2. Magnetic Properties

The thermal variation of $\chi_M T$ in the range of 2–300 K for **1** and **2** at 1000 Oe is shown in Figure 3. The $\chi_M T$ values at 300 K are 28.36 cm^3 K mol^{-1} for **1** and 28.37 cm^3 K mol^{-1} for **2**, both are very close to the theoretical value of the two separated Dy^{3+} ions (28.34 cm^3 K mol^{-1}). Their $\chi_M T$ values all decrease slowly with the decrease in temperature, reaching minimums of 27.04 cm^3 K mol^{-1} at 42 K for **1** and 28.12 cm^3 K mol^{-1} at 55 K for **2**, and then increasing sharply to 40.32 cm^3 K mol^{-1} for **1** and 41.67 cm^3 K mol^{-1} for **2** at 2 K. Such a trend of the $\chi_M T$–T curve is caused by two factors, the depopulation of M_j levels of the Dy^{3+} ion and the intramolecular ferromagnetic interaction between the two Dy^{3+} ions [10–28]. More and more ferromagnetic Dy_2 SMMs based on hydrazine Schiff base ligands have been explored (Table 3), in which the Dy···Dy distance, the Dy–O bond distance and the Dy–O–Dy angle of the Dy_2O_2 core are important parameters that determine whether a ferromagnetic interaction is formed or not, because these three factors have a subtle regulation of the overlap between the magnetic orbitals of the two Dy^{3+} ions, thus affecting their magnetic interactions. As can be seen from Table 3, the ferromagnetic interaction can be formed in the range of 3.864–3.992 Å of the Dy···Dy distance, 2.317–2.380 Å of the average Dy–O bond distance, and 108.9–115.1° of the Dy–O–Dy angle for the eight-coordinate sphere. However, in the case of a nine-coordinate sphere [42], the ferromagnetic interaction may not be formed when the average Dy–O bond length is greater than 2.400 Å (Table 3). The field-variable magnetization of **1** and **2** at different low temperatures was then determined, and their M–H/T curves at 2–6 K do not coincide together (Figures S1 and S2), suggesting possible magnetic anisotropy.

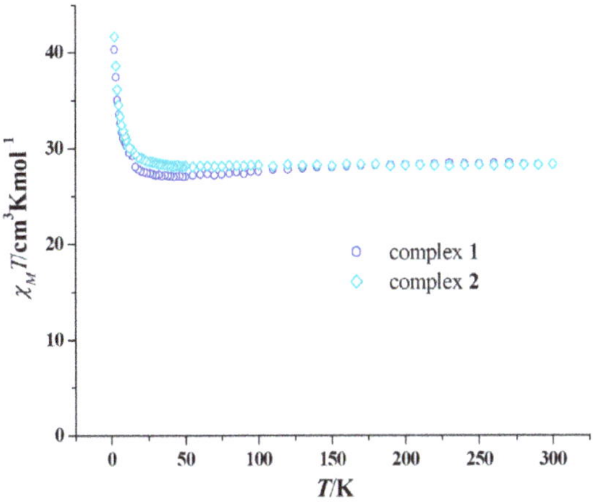

Figure 3. Plots of $\chi_M T$ versus T of **1** and **2** measured under a 1000 Oe dc field.

Table 3. Main structural parameters in the Dy_2O_2 core and nature of the magnetic interactions for some Dy_2 SMMs with hydrazine Schiff base ligands.

Complex	Magnetic Interaction	Dy···Dy (Å)	Dy···O (Å) [a]	Dy···Dy (°)	Coordination Sphere	Ref.
$[Dy_2(ovph)_2Cl_2(MeOH)_3]\cdot sol$	ferromagnetic	3.864	2.333	112.3, 111.5	N_2O_6, $N_2O_3Cl_2$	[11]
$[Dy_2(L1)_2(NO_3)_2(CH_3OH)_2]\cdot sol$	ferromagnetic	3.923	2.327	114.9	N_2O_6	[12]
$Dy_2(L2)_2(DMF)_2(NO_3)_2$	ferromagnetic	3.869	2.319	113.0	N_2O_6	[32]
$Dy_2(L2)_2(DMF)_2(AcO)_2$	ferromagnetic	3.961	2.348	115.1	N_2O_6	[32]
$[Dy_2(TTA)_2(L3)_2(CH_3OH)_2]\cdot sol$	ferromagnetic	3.992	2.374	114.5	N_2O_6	[33]
$[Dy_2(tfa)_2(L3)_2(CH_3OH)_2]$	ferromagnetic	3.911	2.363	111.7	N_2O_6	[33]
$[Dy_2(R\text{-}L)_2(L4)_2(DMA)_2]$	ferromagnetic	3.879	2.380	109.4, 108.9	N_2O_6	[34]
$[Dy_2(S\text{-}L)_2(L5)_2(H_2O)(MeOH)]\cdot sol$	ferromagnetic	3.901	2.369	111.3, 110.4	N_2O_6	[35]
$[Dy_2(D\text{-}tfc)_2(L6)_2(H_2O)_2]\cdot sol$	ferromagnetic	3.902	2.357	111.6, 111.9	N_2O_6	[36]
$[Dy_2(D\text{-}tfc)_2(L7)_2(H_2O)_2]\cdot sol$	ferromagnetic	3.899	2.352	111.9, 112.1	N_2O_6	[36]
$[Dy_2(D\text{-}pfc)_2(L6)_2(DMF)_2]$	ferromagnetic	3.906	2.333	113.4, 113.3	N_2O_6	[36]
$[Dy_2(D\text{-}pfc)_2(L7)_2(H_2O)_2]\cdot sol$	ferromagnetic	3.918	2.353	113.2, 112.2	N_2O_6	[36]
$[Dy_2(L_{Schiff}\text{-}1)_2(DMF)_2(dpp)_2]\cdot sol$	ferromagnetic	3.942, 3.932	2.366	112.7, 112.6	N_2O_6	this work
$[Dy_2(L_{Schiff}\text{-}2)_2(DMF)_2(dpp)_2]\cdot sol$	ferromagnetic	3.885	2.317	113.95	N_2O_6	this work
$[Dy_2(NO_3)_2(H_2L8)_2]\cdot NO_3\cdot sol$		3.976	2.414	110.5, 111.0	N_3O_6	[42]

[a] Average values. H_2ovph: pyridine-2-carboxylic acid [(2-hydroxy-3-methoxyphenyl)methylene] hydrazide; H_2L1: N′-((2-hydroxy-1-naphthyl)methylene) picolinohydrazide; H_2L2: N′-(2-hydroxy-5-methylphenyl)-pyrazine-2-carbohydrazide; H_2L3: N′-(2-hydroxy-5-methylphenyl)-pyrazine-2-carbohydrazide; HTTA: 2-thenoyltrifluoroacetone; Htfa: trifluoroacetylacetone; H_2L4: (*E*)-*N*′-(2-hydroxybenzylidene)-3-aminopyrazine-2-carbohydrazide; *R*-HL: (*R*)-(+)-chlocyphos; H_2L5: (*E*)-*N*′-(5-fluoro-2-hydroxybenzylidene)pyrazine-2-carbohydrazide; *S*-HL: (*S*)-5,5′,6,6′,7,7′,8,8′-octahydro-1,1′-bi-2-naphthyl phosphate; H_2L6: (*E*)-*N*′-(2-hydroxy-3-methoxybenzylidene)pyrazine-2-carbohydrazide; *D*-Htfc: (+)-3-trifluoroacetyl camphor; H_2L7: (*E*)-*N*′-(2-hydroxy-3-methoxybenzylidene)nicotinohydrazide; *D*-Hpfc: 3-(perfluorobutyryl)-(+)/(−)-camphor; H_4L8: 6-((bis(2-hydroxyethyl)amino)-N′-(2hydroxybenzylidene)picolinohydrazide).

The measurement of ac magnetic susceptibility revealed that both **1** and **2** are zero-field SMMs. As shown in Figure 4a, at 0 Oe the χ'' versus T curves at 50–1399 Hz of **1** show clear frequency dependence, and peaks can be observed in the range of 250–1399 Hz. On the other hand, the χ'' versus ν curves at 0 Oe of **1** also display frequency dependence, and peaks can be observed in the range of 2.0–6.5 K (Figure 4b). The ln(τ)–$1/T$ plot based on these χ'' versus ν curves is a curve rather than a straight line (Figure 4c), suggesting that the magnetic relaxation of **1** includes other processes such as the quantum tunneling of magnetization (QTM), the Raman process in addition to the Orbach process; therefore, this curve was fitted with $\tau^{-1} = \tau_{QTM}{}^{-1} + CT^n + \tau_0{}^{-1}\exp(-U_{eff}/kT)$, giving τ_{QTM} = 0.0145 s, n = 2.46, C = 1.33 s^{-1} K$^{-2.46}$, U_{eff}/k = 49.7 K and $\tau_0 = 7.9 \times 10^{-6}$ s. The τ_0 value of **1** (7.9×10^{-6} s) is normal for the Dy(III) SMMs (10^{-6} s–10^{-11} s) [2], and its n value of 2.46 is also within the normal range ($2 < n < 9$). Since the crystal structure of **1** contains two types of coordination configurations of the Dy^{3+} ions, its Cole–Cole curves at 0 Oe could be fitted with the sum of two modified Debye functions [43–46] (Figure 4d), giving α_1 values of 0.276–0.385, and α_2 values of 0.001–0.108. Moreover, no hysteresis loop can be formed at 1.9 K for **1** (Figure S3). The magnetic axes of **1** could be calculated with the electrostatic model implemented in the Magellan program [47]; as expected, the two magnetic axes in each independent molecule are arranged in perfect parallel (Figure S4).

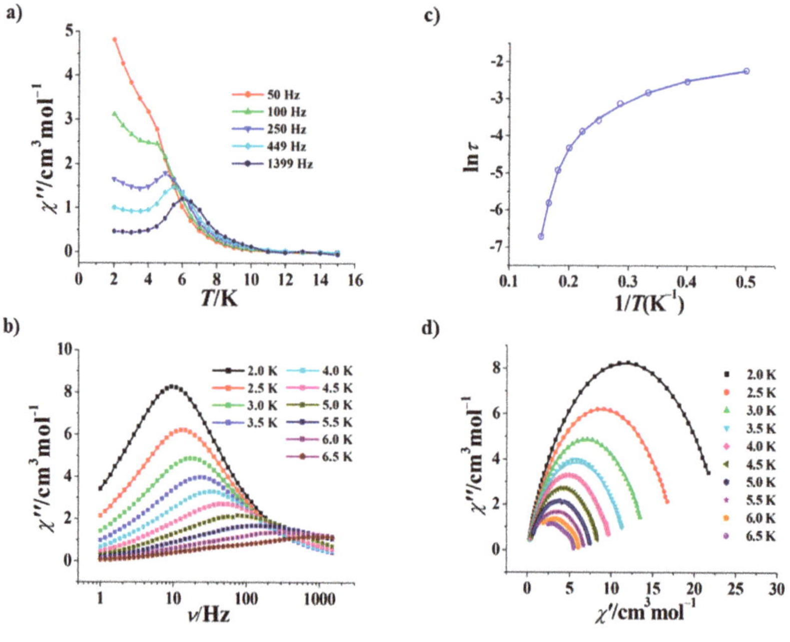

Figure 4. χ'' versus T curves for **1** (H_{dc} = 0 Oe) (**a**); χ'' versus ν curves at 2.0 to 6.5 K for **1** (H_{dc} = 0 Oe) (**b**); plot of ln(τ) versus $1/T$ for **1** (H_{dc} = 0 Oe), the solid line represents the best fitting with QTM + Raman + Orbach (**c**); Cole–Cole curves at 2.0 to 6.5 K for **1** (H_{dc} = 0 Oe), the solid lines represent the best fitting with the sum of two modified Debye functions [43–46] (**d**).

We also investigated the effect of 1500 Oe, a commonly used dc magnetic field, on the magnetic relaxation of **1**. As shown in Figure 5a, at 1500 Oe the χ'' versus T curves of **1** show more clear frequency dependence, and peaks can be observed in a wider range of 10–1399 Hz with respect to 0 Oe. The χ'' versus ν curves at 1500 Oe of **1** are also frequency-dependent, but peaks can be observed in a narrower range of 3.0–6.5 K with respect to 0 Oe (Figure 5b). The ln(τ)–$1/T$ plot at 1500 Oe of **1** is still a curve (Figure S5), which was fitted using $\tau^{-1} = AT + CT^n + \tau_0^{-1}\exp(-U_{eff}/kT)$, giving $A = 0.892$, $n = 5.91$, $C = 0.00352\ s^{-1}\ K^{-5.91}$, $U_{eff}/k = 87.6$ K and $\tau_0 = 2.4 \times 10^{-9}$ s. Similar to other zero-field Dy_2 SMMs [30,32–34], the dc magnetic field (here is 1500 Oe) can partially or completely inhibit QTM, resulting in an increase in the U_{eff}/k value of **1** (from 49.7 K to 87.6 K) and a decrease in the τ_0 value (from 7.9×10^{-6} s to 2.4×10^{-9} s).

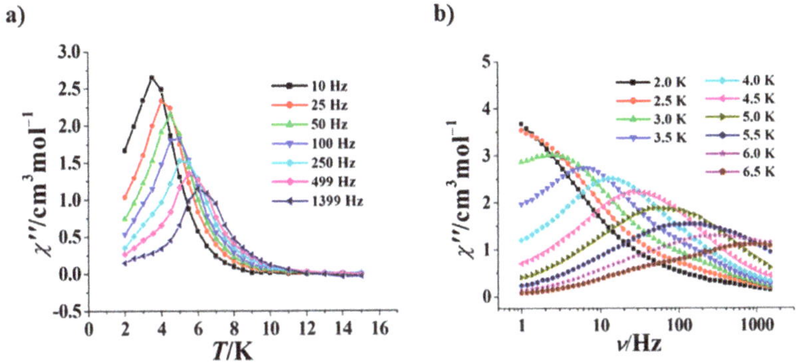

Figure 5. χ'' versus T curves for **1** (H_{dc} = 1500 Oe) (**a**); χ'' versus ν curves at 2.0 to 6.5 K for **1** (H_{dc} = 1500 Oe) (**b**).

The χ'' versus T curves at 0 Oe of **2** show obvious frequency dependence (Figure 6a), and peaks appear in the range of 10–1399 Hz, which is larger than that of **1** (250–1399 Hz). On the other hand, the χ'' versus ν curves at 0 Oe of **2** are also frequency-dependent (Figure 6b), and peaks appear in the range of 5–15 K; the peak temperatures of **2** are obviously higher than those of **1** at 0 Oe. The ln(τ)–1/T plot derived from the χ'' versus ν curves of **2** is a curve as well (Figure 6c), which could be fitted with $\tau^{-1} = CT^n + \tau_0^{-1}\exp(-U_{eff}/kT)$, yielding $n = 5.56$, $C = 0.00014\ s^{-1}K^{-5.56}$, $U_{eff}/k = 151.8$ K, and $\tau_0 = 5.2 \times 10^{-8}$ s. The τ_0 value at 0 Oe of **2** (5.2×10^{-8} s) is smaller than that of **1** (7.9×10^{-6} s); the U_{eff}/k value at 0 Oe of **2** (151.8 K) is comparable with those of other ferromagnetic Dy_2 SMMs based on hydrazine Schiff base ligands and achiral β-diketonate ligands [33] or homochial β-diketonate ligands [36], which is about three times that of **1** (49.7 K). The Cole–Cole curves at 0 Oe of **2** could be fitted by a generalized Debye model [48,49] (Figure 6d), giving small α values in the range of 0.00051–0.01061, suggesting narrow relaxation time distribution. In addition, at 1.9 K no hysteresis loop may be observed for **2** (Figure S6). Due to the presence of a center of symmetry in the molecule, the two magnetic axes of **2** are also perfectly parallel (Figure S7).

Similarly, we also studied the effect of 1500 Oe dc field on the magnetic relaxation of **2**. As shown in Figure 7a, at 1500 Oe the χ'' versus T curves of **2** possess obvious frequency dependence, and peaks may be observed in the same range of 10–1399 Hz as at 0 Oe. The χ'' versus ν curves at 1500 Oe of **2** are frequency-dependent as well, but peaks appear in a narrower range of 6–15 K with respect to 0 Oe (Figure 7b). The ln(τ)–1/T plot at 1500 Oe of **2** is also a curve (Figure S8), which could also be fitted with $\tau^{-1} = AT + CT^n + \tau_0^{-1}\exp(-U_{eff}/kT)$, calculating $A = 0.211$, $n = 5.95$, $C = 0.00006\ s^{-1}\ K^{-5.95}$, $U_{eff}/k = 157.9$ K, and $\tau_0 = 5.1 \times 10^{-8}$ s. Similar to **1**, the dc magnetic field of 1500 Oe may partially or completely inhibit QTM, which results in an increase in the U_{eff}/k value of **2** (from 151.8 K to 157.9 K) and a decrease in the τ_0 value (from 5.2×10^{-8} s to 5.1×10^{-8} s). However, the variation amplitude of **2** is smaller than that of **1**, indicating that the QTM effect of **2** is much weaker than that of **1**.

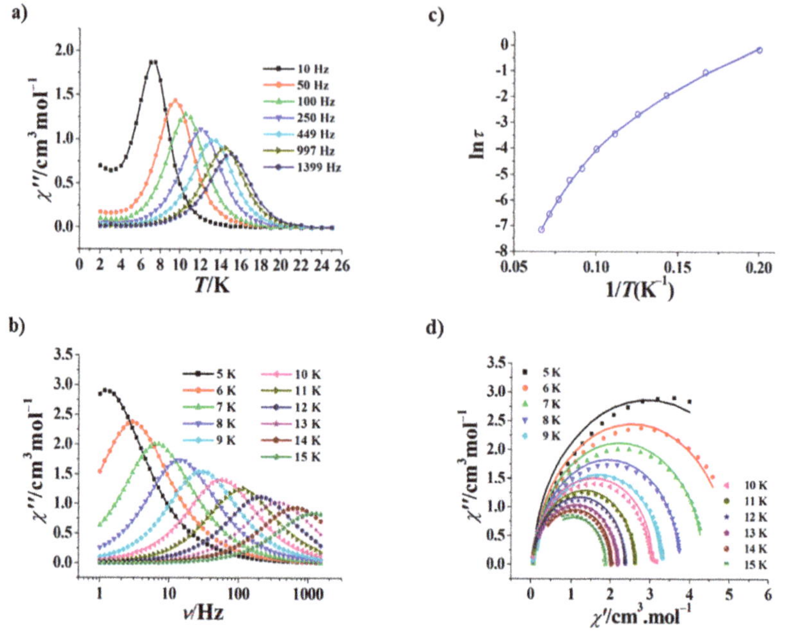

Figure 6. χ'' versus T curves for **2** (H_{dc} =0 Oe) (**a**); χ'' versus ν curves at 5 to 15 K for **2** (H_{dc} =0 Oe) (**b**); plot of $\ln(\tau)$ versus $1/T$ for **2** (H_{dc} = 0 Oe), the solid line represents the best fitting with Raman + Orbach (**c**); Cole–Cole curves at 5 to 15 K for **2** (H_{dc} = 0 Oe), the solid lines represent the best fitting with a generalized Debye model [48,49] (**d**).

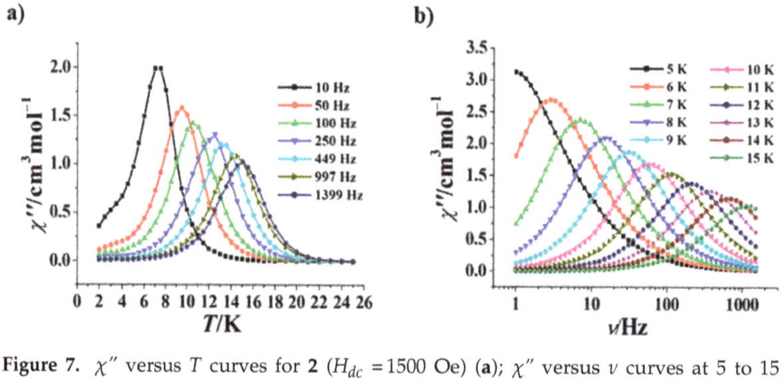

Figure 7. χ'' versus T curves for **2** (H_{dc} =1500 Oe) (**a**); χ'' versus ν curves at 5 to 15 K for **2** (H_{dc} =1500 Oe) (**b**).

It can be seen that although both **1** and **2** have a $[N_2O_6]$ coordination sphere, the change in heterocyclic groups (pyrazine vs. pyridine) on the hydrazone Schiff base-bridging ligand can lead to significant changes in the SMM properties of **1** and **2** (e.g., magnetic relaxation, QTM, energy barrier value, etc.). This is related to the difference in their coordination configurations.

3. Conclusions

In summary, two new Dy_2 zero-field SMMs have been assembled using two hydrazone Schiff base-bridging ligands with different heterocycles (pyrazine vs. pyridine) as the bridging ligand and 1,3-di(2-pyridyl)-1,3-propanedione as the terminal ligand. The ferromagnetic interactions between two Dy^{3+} ions exist in both complexes. As expected, they have different U_{eff}/k values (49.7 K and 151.8 K at 0 Oe for **1** and **2**, respectively). This work demonstrates that the Dy_2 SMMs' magnetic properties can be adjusted by the change in the heterocycle groups (pyrazine vs. pyridine) on the hydrazone Schiff base-bridging ligands. Such molecular engineering provides a good way to design lanthanide SMMs with better performance.

4. Materials and Methods

4.1. General Remarks

The C, H, N elemental analyses were carried out on a Thermo Flash EA1112 elemental analyzer (Thermo Fisher Scientific, Waltham, MA, USA). The infrared spectra were performed on a Bruker VERTEX 70v spectrophotometer (Bruker, Karlsruhe, Germany) with pressed KBr disk. The magnetic susceptibility measurements were measured on a Quantum Design MPMS-XL5 SQUID magnetometer (Quantum Design, San Diego, CA, USA). Diamagnetic corrections were estimated using Pascal's constants for all constituent atoms of **1** and **2**.

4.2. Preparation of H_2L_{Schiff}-1

A mixture of 5-chlorosalicylaldehyde (1565.7 mg, 10 mmol) and pyrazinoic acid hydrazide (1381.3 mg, 10 mmol) was added into 60 mL of MeOH; the mixture was refluxed for 3 h to form a light yellow precipitate, which was naturally cooled and then collected by filtration. Yield: 90%. Anal. Calcd. (%) for $C_{12}H_9ClN_4O_2$ (H_2L_{Schiff}-1): C, 52.09; H, 3.28; N, 20.25%. Found: C, 52.01; H, 3.32; N, 20.17%.

4.3. Preparation of **1**

H_2L_{Schiff}-1 (66.9 mg, 0.25 mmol), 1,3-di(2-pyridyl)-1,3-propanedione (56.6 mg, 0.25 mmol), $LiOH{\cdot}H_2O$ (31.5 mg, 0.75 mmol), and $Dy(ClO_4)_3{\cdot}6H_2O$ (142.2 mg, 0.25 mmol) were added into 15 mL of DMF, this mixture was stirred for 10 h to form yellow precipitates, which were dissolved after adding 20 mL of methanol and 20 mL of CH_2Cl_2 to obtain a yellow solution, and yellow single crystals were grown after slowly volatilizing the solvent. Yield: 55% based on Dy. Anal. Calcd. (%) for $C_{57.5}H_{47.5}Cl_2Dy_2N_{14.5}O_{10.5}$ (**1**): C, 45.87; H, 3.18; N, 13.49%. Found: C, 45.81; H, 3.22; N, 13.43%. IR (KBr): ν = 3422(b,w), 3052(w), 3000(w), 2929(w), 2889(w), 1659(s), 1606(s), 1553(s), 1524(s), 1507(s), 1470(s), 1399(s), 1365(m), 1340(m), 1311(w), 1284(w), 1255(w), 1240(w), 1171(w), 1156(w), 1112(w), 1088(w), 1050(w), 1032(w), 1007(w), 994(w), 973(w), 948(w), 906(w), 875(w), 861(w), 828(w), 808(w), 774(w), 752(m), 716(m), 684(w), 653(w), 610(w), 553(w), 534(w), 506(w), 474(w), 451(w), 429(w) cm^{-1}.

4.4. Preparation of **2**

H_2L_{Schiff}-2 (64.8 mg, 0.25 mmol), 1,3-di(2-pyridyl)-1,3-propanedione (56.6 mg, 0.25 mmol), $LiOH \cdot H_2O$ (31.5 mg, 0.75 mmol), and $Dy(ClO_4)_3{\cdot}6H_2O$ (142.2 mg, 0.25 mmol) were added into 5 mL of DMF; this mixture was stirred for 10 h to form yellow precipitates, which were dissolved after adding 15 mL of methanol and 20 mL of CH_2Cl_2 to obtain a yellow solution, and yellow single crystals were grown after slowly volatilizing the solvent. Yield: 45% based on Dy. Anal. Calcd. (%) for $C_{66}H_{68}Dy_2N_{14}O_{14}$ (**2**): C, 49.35; H, 4.27; N, 12.21%. Found: C, 49.25; H, 4.31; N, 12.15%. IR (KBr): ν = 3415(b,w), 3131(w), 3045(w), 2998(w), 2930(w), 2898(w), 2826(w), 1656(s), 1603(s), 1555(s), 1521(s), 1505(s), 1473(s), 1453(s), 1402(s), 1343(m), 1315(w), 1286(w), 1235(m), 1212(m), 1163(w), 1102(w), 1082(w), 1045(w), 1015(w), 993(w), 968(w), 949(w), 919(w), 861(w), 806(w), 740(m), 718(w), 694(w), 668(w), 635(w), 610(w), 587(w), 536(w), 493(w), 475(w), 456(w), 426(w) cm^{-1}.

4.5. X-Ray Crystallography

The intensity data collection of **1** and **2** was curried on a Rigaku MM007HF diffractometer (Rigaku, Tokyo, Japan) with Mo-Kα radiation (λ = 0.71073 Å) at 170 K. The structures were solved using the olex2.solve structure solution program and refined using the ShelXL-2018 refinement package. All non-hydrogen atoms were refined anisotropically, and the hydrogen atoms were refined as riding atoms. Selected crystal data and structural refinement parameters for **1** and **2** are listed in Table 4.

Table 4. Crystal data and structural refinement parameters for **1** and **2**.

	1	**2**
formula	$C_{57.5}H_{47.5}Cl_2Dy_2N_{14.5}O_{10.5}$	$C_{66}H_{68}Dy_2N_{14}O_{14}$
F_W	1505.50	1606.34
crystal system	triclinic	monoclinic
space group	*P-1*	$P2_1/c$
a [Å]	13.6737(3)	13.5044(2)
b [Å]	15.5510(4)	16.6323(2)
c [Å]	17.1825(4)	15.5855(2)
α [°]	71.194(2)	90
β [°]	68.498(2)	107.308(2)
γ [°]	67.717(2)	90
V [Å^3]	3073.36(14)	3342.13(8)
Z	2	2
ρ_{calc} [g cm^{-3}]	1.627	1.596
μ [mm^{-1}]	2.569	2.294
T [K]	170	170
λ [Å]	0.71073	0.71073
reflections collected	73,968	47,513
unique reflections	10,125	5572
observed reflections	8949	5247
parameters	808	438
GoF	1.181	1.075
R_1 [$I \geq 2\sigma(I)$]	0.0605	0.0242
wR_2 [$I \geq 2\sigma(I)$]	0.1933	0.0639
CCDC	2454030	2454031

Supplementary Materials: The following supporting information can be downloaded at: https://www.mdpi.com/article/10.3390/magnetochemistry11070058/s1, Table S1, Dy (III) ion geometry analysis by SHAPE 2.1 software for **1**; Table S2, Dy (III) ion geometry analysis by SHAPE 2.1 software for **2**; Figure S1, *M versus H/T* plots at 2–6 K of **1**; Figure S2, *M* versus *H/T* plots at 2–6 K of **2**; Figure S3, Hysteresis loop for **1** at 1.9 K; Figure S4, Magnetic axes of the Dy^{3+} ions in **1** calculated using an electrostatic model; Figure S5, Plot of lnτ versus $1/T$ for **1** (H_{dc} = 1500 Oe); Figure S6, Hysteresis loop for **2** at 1.9 K; Figure S7, Magnetic axes of the Dy^{3+} ions in **2** calculated using an electrostatic model; Figure S8, plot of lnτ versus $1/T$ for **2** (H_{dc} = 1500 Oe). CCDC 2454030-2454031 contain the supplementary crystallographic data for **1** and **2**. These data can be obtained free of charge via https://www.ccdc.cam.ac.uk/, or from the Cambridge Crystallographic Data Center, 12 Union Road, Cambridge CB2 1EZ, UK; e-mail: deposit@ccdc.cam.ac.uk.

Funding: This research was funded by the National Natural Science Foundation of China (22271289 and 21871274).

Institutional Review Board Statement: Not applicable.

Informed Consent Statement: Not applicable.

Data Availability Statement: The original contributions presented in this study are included in the article/Supplementary Materials. Further inquiries can be directed to the corresponding author.

Conflicts of Interest: The author declares no conflicts of interest.

References

1. Sessoli, R.; Gatteschi, D.; Caneschi, A.; Novak, M.A. Magnetic Bistability in a Metal-Ion Cluster. *Nature* **1993**, *365*, 141–143. [CrossRef]
2. Woodruff, D.N.; Winpenny, R.E.P.; Layfield, R.A. Lanthanide Single-Molecule Magnets. *Chem. Rev.* **2013**, *113*, 5110–5148. [CrossRef] [PubMed]
3. Jiang, S.-D.; Wang, B.-W.; Sun, H.-L.; Wang, Z.-M.; Gao, S. An Organometallic Single-Ion Magnet. *J. Am. Chem. Soc.* **2011**, *133*, 4730–4733. [CrossRef]
4. Ding, Y.-S.; Chilton, N.F.; Winpenny, R.E.P.; Zheng, Y.-Z. On Approaching the Limit of Molecular Magnetic Anisotropy: A Near-Perfect Pentagonal Bipyramidal Dysprosium(III) Single-Molecule Magnet. *Angew. Chem. Int. Ed.* **2016**, *55*, 16071–16074. [CrossRef]
5. Goodwin, C.A.P.; Ortu, F.; Reta, D.; Chilton, N.F.; Mills, D.P. Molecular magnetic hysteresis at 60 kelvin in dysprosocenium. *Nature* **2017**, *548*, 439–442. [CrossRef]
6. Guo, F.-S.; Day, B.M.; Chen, Y.-C.; Tong, M.-L.; Mansikkamäki, A.; Layfield, R.A. Magnetic hysteresis up to 80 kelvin in a dysprosium metallocene single-molecule magnet. *Science* **2018**, *362*, 1400–1403. [CrossRef]
7. Canaj, A.B.; Dey, S.; Martí, E.R.; Wilson, C.; Rajaraman, G.; Murrie, M. Insight into D_{6h} Symmetry: Targeting Strong Axiality in Stable Dysprosium(III) Hexagonal Bipyramidal Single-Ion Magnets. *Angew. Chem. Int. Ed.* **2019**, *58*, 14146–14151. [CrossRef]
8. Zhu, Z.; Zhao, C.; Feng, T.; Liu, X.; Ying, X.; Li, X.-L.; Zhang, Y.-Q.; Tang, J. Air-Stable Chiral Single-Molecule Magnets with Record Anisotropy Barrier Exceeding 1800 K. *J. Am. Chem. Soc.* **2021**, *143*, 10077–10082. [CrossRef]
9. Liu, C.-M.; Zhang, D.-Q.; Hao, X.; Zhu, D.-B. Arraying Octahedral {Cr_2Dy_4} Units into 3D Single-Molecule-Magnet-Like Inorganic Compounds with Sulfate Bridges. *Inorg. Chem.* **2018**, *57*, 6803–6806. [CrossRef]
10. Lin, P.-H.; Burchell, T.J.; Clérac, R.; Murugesu, M. Dinuclear Dysprosium(III) Single-Molecule Magnets with a Large Anisotropic Barrier. *Angew. Chem. Int. Ed.* **2008**, *47*, 8848–8851. [CrossRef]
11. Guo, Y.-N.; Xu, G.-F.; Wernsdorfer, W.; Ungur, L.; Guo, Y.; Tang, J.; Zhang, H.-J.; Chibotaru, L.F.; Powell, A.K. Strong Axiality and Ising Exchange Interaction Suppress Zero-Field Tunneling of Magnetization of an Asymmetric Dy_2 Single-Molecule Magnet. *J. Am. Chem. Soc.* **2011**, *133*, 11948–11951. [CrossRef] [PubMed]
12. Zou, L.; Zhao, L.; Chen, P.; Guo, Y.-N.; Guo, Y.; Li, Y.-H.; Tang, J. Phenoxido and alkoxido-bridged dinuclear dysprosium complexes showing single-molecule magnet behavior. *Dalton Trans.* **2012**, *41*, 2966–2971. [CrossRef] [PubMed]
13. Jin, C.; Li, X.-L.; Liu, Z.; Mansikkamäki, A.; Tang, J. An investigation into the magnetic interactions in a series of Dy_2 single-molecule magnets. *Dalton Trans.* **2020**, *49*, 10477–10485. [CrossRef] [PubMed]
14. Zhang, L.; Jung, J.; Zhang, P.; Guo, M.; Zhao, L.; Tang, J.; Le Guennic, B. Site-Resolved Two-Step Relaxation Process in an Asymmetric Dy_2 Single-Molecule Magnet. *Chem. Eur. J.* **2016**, *22*, 1392–1398. [CrossRef]
15. Gao, F.; Zhang, Y.-Q.; Sun, W.; Liu, H.; Chen, X. Syntheses, structures and magnetic properties of macrocyclic Schiff base-supported homodinuclear lanthanide complexes. *Dalton Trans.* **2018**, *47*, 11696–11704. [CrossRef]
16. Ghosh, S.; Mandal, S.; Singh, M.K.; Liu, C.-M.; Rajaraman, G.; Mohanta, S. Experimental and theoretical exploration of magnetic exchange interactions and single-molecule magnetic behaviour of bis(η^1:η^2:μ^2-carboxylate)Gd^{III}_2/Dy^{III}_2 systems. *Dalton Trans.* **2018**, *47*, 11455–11469. [CrossRef]
17. Zhang, J.; Zhang, H.; Chen, Y.; Zhang, X.; Li, Y.; Liu, W.; Dong, Y. A series of dinuclear lanthanide complexes with slow magnetic relaxation for Dy_2 and Ho_2. *Dalton Trans.* **2016**, *45*, 16463–16470. [CrossRef]
18. Liu, C.-M.; Zhang, D.-Q.; Hao, X.; Zhu, D.-B. Simultaneous assembly of mononuclear and dinuclear dysprosium(III) complexes behaving as single-molecule magnets in a one-pot hydrothermal synthesis. *Sci. China Chem.* **2017**, *60*, 358–365. [CrossRef]
19. Katoh, K.; Aizawa, Y.; Morita, T.; Breedlove, B.K.; Yamashita, M. Elucidation of Dual Magnetic Relaxation Processes in Dinuclear Dysprosium(III) Phthalocyaninato Triple-Decker Single-Molecule Magnets Depending on the Octacoordination Geometry. *Chem. Eur. J.* **2017**, *23*, 15377–15386. [CrossRef]
20. Wang, Y.-L.; Han, C.-B.; Zhang, Y.-Q.; Liu, Q.-Y.; Liu, C.-M.; Yin, S.-G. Fine-Tuning Ligand to Modulate the Magnetic Anisotropy in a Carboxylate-Bridged Dy_2 Single-Molecule Magnet System. *Inorg. Chem.* **2016**, *55*, 5578–5584. [CrossRef]
21. Guo, M.; Zhang, Y.-Q.; Zhu, Z.; Tang, J. Dysprosium Compounds with Hula-Hoop-like Geometries: The Influence of Magnetic Anisotropy and Magnetic Interactions on Magnetic Relaxation. *Inorg. Chem.* **2018**, *57*, 12213–12221. [CrossRef] [PubMed]
22. Lin, P.; Sun, W.; Yu, M.; Li, G.; Yan, P.; Murugesu, M. An unsymmetrical coordination environment leading to two slow relaxation modes in a Dy_2 single-molecule magnet. *Chem. Commun.* **2011**, *47*, 10993–10995. [CrossRef] [PubMed]

23. Ren, M.; Bao, S.-S.; Hoshino, N.; Akutagawa, T.; Wang, B.; Ding, Y.-C.; Wei, S.; Zheng, L.-M. Solvent Responsive Magnetic Dynamics of a Dinuclear Dysprosium Single-Molecule Magnet. *Chem. Eur. J.* **2013**, *19*, 9619–9628. [CrossRef] [PubMed]
24. Lu, Y.-B.; Jiang, X.-M.; Zhu, S.-D.; Du, Z.-Y.; Liu, C.-M.; Xie, Y.-R.; Liu, L.-X. Anion Effects on Lanthanide(III) Tetrazole-1-acetate Dinuclear Complexes Showing Slow Magnetic Relaxation and Photofluorescent Emission. *Inorg. Chem.* **2016**, *55*, 3738–3749. [CrossRef]
25. Kong, M.; Feng, X.; Wang, J.; Zhang, Y.-Q.; Song, Y. Tuning magnetic anisotropy via terminal ligands along the Dy$\cdots$Dy orientation in novel centrosymmetric [Dy2] single molecule magnets. *Dalton Trans.* **2021**, *50*, 568–577. [CrossRef]
26. Liu, C.-M.; Sun, R.; Wang, B.-W.; Wu, F.; Hao, X.; Shen, Z. Homochiral Ferromagnetic Coupling Dy_2 Single-Molecule Magnets with Strong Magneto-Optical Faraday Effects at Room Temperature. *Inorg. Chem.* **2021**, *60*, 12039–12048. [CrossRef]
27. Tu, H.-R.; Sun, W.-B.; Li, H.-F.; Chen, P.; Tian, Y.-M.; Zhang, W.-Y.; Zhang, Y.-Q.; Yan, P.-F. Complementation and joint contribution of appropriate intramolecular coupling and local ion symmetry to improve magnetic relaxation in a series of dinuclear Dy_2 single-molecule magnets. *Inorg. Chem. Front.* **2017**, *4*, 499–508. [CrossRef]
28. Zhou, Y.-Q.; Deng, W.; Du, S.-N.; Chen, Y.-C.; Ruan, Z.-Y.; Wu, S.-G.; Liu, J.-L.; Tong, M.-L. An alkoxyborate-bridging Dy_2 single-molecule magnet with ferromagnetic coupling. *Inorg. Chem. Front.* **2024**, *11*, 1061–1069. [CrossRef]
29. Faraonov, M.A.; Martynov, A.G.; Polovkova, M.A.; Khasanov, S.S.; Gorbunova, Y.G.; Tsivadze, A.Y.; Otsuka, A.; Yamochi, H.; Kitagawa, H.; Konarev, D.V. Single-Molecule Magnets Based on Heteroleptic Terbium(III) Trisphthalocyaninate in Solvent-Free and Solvent-Containing Forms. *Magnetochemistry* **2023**, *9*, 36. [CrossRef]
30. Liu, C.-M. Assembling 4f and 3d–4f clusters as single-molecule magnets by automatic fixation of atmospheric CO_2. *Dalton Trans.* **2025**, *54*, 9850–9855. [CrossRef]
31. Zhang, P.; Guo, Y.-N.; Tang, J. Recent advances in dysprosium-based single molecule magnets: Structural overview and synthetic strategies. *Coord. Chem. Rev.* **2013**, *257*, 1728–1763. [CrossRef]
32. Liu, C.-M.; Hao, X.; Zhu, D.-M.; Zhang, Y.-Q. Effect of coordinated anions on ferromagnetically coupled Dy_2 zero-field single-molecule magnets. *Dalton Trans.* **2024**, *53*, 6120–6127. [CrossRef] [PubMed]
33. Wang, Y.-J.; Wu, D.-F.; Gou, J.; Duan, Y.-Y.; Li, L.; Chen, H.-H.; Gao, H.-L.; Cui, J.-Z. Modulation of the properties of dinuclear lanthanide complexes through utilizing different β-diketonate co-ligands: Near-infrared luminescence and magnetization dynamics. *Dalton Trans.* **2020**, *49*, 2850–2861. [CrossRef]
34. Liu, C.-M.; Zou, H.-H.; Zhang, Y.-Q.; Hao, X.; Ren, X.-M. Homochiral Dy_2 Schiff Base Complexes Based on $(R)/(S)$-Chlocyphos: Single-Molecule Magnet Behaviours, Proton Conductivities and MCD Effects. *Chin. J. Chem.* **2025**, *43*, 1051–1058. [CrossRef]
35. Liu, C.-M.; Hao, X.; Zhang, Y.-Q. Homochiral Dy_2 zero-field single-molecule magnets derived from axial chiral ligands $(R)/(S)$-octahydro-1,1′-bi-2-naphthyl phosphate. *Dalton Trans.* **2025**, *54*, 4159–4166. [CrossRef]
36. Liu, C.-M.; Sun, R.; Hao, X.; Li, X.-L.; Wang, B.-W. Homochiral Dy_2 single-molecule magnets with strong magneto-optical Faraday effects and strong third-harmonic generation. *Inorg. Chem. Front.* **2024**, *11*, 3296–3308. [CrossRef]
37. Cen, P.; Liu, X.; Zhang, Y.-Q.; Ferrando-Soria, J.; Xie, G.; Chen, S.; Pardo, E. Modulating magnetic dynamics through tailoring the terminal ligands in Dy_2 single-molecule magnets. *Dalton Trans.* **2020**, *49*, 808–816. [CrossRef]
38. Liu, M.-J.; Yuan, J.; Tao, J.; Zhang, Y.-Q.; Liu, C.-M.; Kou, H.-Z. Rhodamine Salicylaldehyde Hydrazone Dy(III) Complexes: Fluorescence and Magnetism. *Inorg. Chem.* **2018**, *57*, 4061–4069. [CrossRef]
39. Casanova, D.; Llunell, M.; Alemany, P.; Alvarez, S. The Rich Stereochemistry of Eight-Vertex Polyhedra: A Continuous Shape Measures Study. *Chem. Eur. J.* **2005**, *11*, 1479–1494. [CrossRef]
40. Caporale, C.; Sobolev, A.N.; Phonsri, W.; Murray, K.S.; Swain, A.; Rajaraman, G.; Ogden, M.I.; Massi, M.; Fuller, R.O. Lanthanoid pyridyl-β-diketonate 'triangles'. New examples of single molecule toroics. *Dalton Trans.* **2020**, *49*, 17421–17432. [CrossRef]
41. Mao, P.-D.; Yao, N.-T.; Sun, H.-Y.; Yan, F.-F.; Zhang, Y.-Q.; Meng, Y.-S.; Liu, T. Design of Heterometallic $\{Ln^{III}\text{-}M^{V}\}$ (Ln = Dy, Er; M = W, Mo) Molecular Nanomagnets: Protonation Induced Structural Diversification. *Cryst. Growth Des.* **2022**, *23*, 450–464. [CrossRef]
42. Kumar, P.; Swain, A.; Acharya, J.; Li, Y.; Kumar, V.; Rajaraman, G.; Colacio, E.; Chandrasekhar, V. Synthesis, Structure, and Zero-Field SMM Behavior of Homometallic Dy2, Dy4, and Dy6 Complexes. *Inorg. Chem.* **2022**, *61*, 11600–11621. [CrossRef] [PubMed]
43. Guo, Y.-N.; Xu, G.-F.; Gamez, P.; Zhao, L.; Lin, S.-Y.; Deng, R.; Tang, J.; Zhang, H.-J. Two-Step Relaxation in a Linear Tetranuclear Dysprosium(III) Aggregate Showing Single-Molecule Magnet Behavior. *J. Am. Chem. Soc.* **2010**, *132*, 8538–8539. [CrossRef] [PubMed]
44. Liu, C.-M.; Zhang, D.-Q.; Zhu, D.-B. A single-molecule magnet featuring a parallelogram $[Dy_4(OCH_2-)_4]$ core and two magnetic relaxation processes. *Dalton Trans.* **2013**, *42*, 14813–14818. [CrossRef]
45. Langley, S.K.; Chilton, N.F.; Moubaraki, B.; Murray, K.S. Single-Molecule Magnetism in Three Related $\{Co_2^{III}Dy_2^{III}\}$-Acetylacetonate Complexes with Multiple Relaxation Mechanisms. *Inorg. Chem.* **2013**, *52*, 7183–7192. [CrossRef]
46. Jin, Y.-S.; Liu, C.-M.; Zhang, Y.-Q.; Kou, H.-Z. Trinuclear Dy(III) Single- Molecule Magnets with Two-Step Relaxation. *Chin. J. Chem.* **2023**, *41*, 2641–2647. [CrossRef]

47. Chilton, N.F.; Collison, D.; McInnes, E.J.L.; Winpenny, R.E.P.; Soncini, A. An electrostatic model for the determination of magnetic anisotropy in dysprosium complexes. *Nat. Commun.* **2013**, *4*, 2551. [CrossRef]
48. Cole, K.S.; Cole, R.H. Dispersion and Absorption in Dielectrics I. Alternating Current Characteristics. *J. Chem. Phys.* **1941**, *9*, 341–351. [CrossRef]
49. Aubin, S.M.; Sun, Z.; Pardi, L.; Krzysteck, J.; Folting, K.; Brunel, L.-J.; Rheingold, A.L.; Christou, G.; Hendrickson, D.N. Reduced Anionic Mn_{12} Molecules with Half-Integer Ground States as Single-Molecule Magnets. *Inorg. Chem.* **1999**, *38*, 5329–5340. [CrossRef]

magnetochemistry

MDPI

Article

Synthesis, Crystal Structures and Magnetic Properties of Lanthanide Complexes with Rhodamine Benzoyl Hydrazone Ligands †

Lin Miao [1], Dong-Mei Zhu [2], Cai-Ming Liu [3], Yi-Quan Zhang [2,*] and Hui-Zhong Kou [1,*]

1 Engineering Research Center of Advanced Rare Earth Materials (Ministry of Education), Department of Chemistry, Tsinghua University, Beijing 100084, China; ml19@tsinghua.org.cn
2 Ministry of Education Key Laboratory of NSLSCS, School of Physical Science and Technology, Nanjing Normal University, Nanjing 210023, China; 221001007@njnu.edu.cn
3 Beijing National Laboratory for Molecular Sciences, Center for Molecular Science, Institute of Chemistry, Chinese Academy of Sciences, Beijing 100190, China; cmliu@iccas.ac.cn
* Correspondence: zhangyiquan@njnu.edu.cn (Y.-Q.Z.); kouhz@mail.tsinghua.edu.cn (H.-Z.K.)
† Dedicated to Prof. Dai-Zheng Liao on the occasion of his 85th birthday.

Abstract: Given the outstanding magnetic characteristics of lanthanide ions, the development of mononuclear or multinuclear lanthanide complexes becomes imperative. Previous research showed that a series of mononuclear Dy(III) complexes of rhodamine benzoyl hydrazone Schiff base ligands exhibit remarkable single-molecule magnetic properties and fluorescence. In this study, we used analogous ligands to synthesize lanthanide complexes $[Dy(HL^1\text{-o})(NO_3)_2(CH_3OH)_2]NO_3\cdot CH_3OH$ (complex **1·MeOH**) and tetranuclear complexes $[Ln_4(L^1\text{-c})_2(L^2)_2(\mu_3\text{-}OH)_2(NO_3)_2(CH_3OH)_4](NO_3)_2\cdot 2CH_3CN\cdot 5CH_3OH\cdot 2H_2O$ (Ln = Dy, complex **2**; Ln = Gd, complex **3**). Magnetic susceptibility measurements show that **1·2H$_2$O** is a single-molecule magnet, **2** shows slow magnetic relaxation and **3** is a magnetic cooling material with the magnetic entropy change of 9.81 J kg^{-1} K^{-1} at 2 K and 5 T. The theoretical calculations on **1·MeOH** indicate that it shows good magnetic anisotropy with the calculated energy barrier of 194.6 cm^{-1}.

Keywords: single-molecule magnets; magnetocaloric effect; lanthanide; rhodamine; Schiff base

1. Introduction

Single-molecule magnets (SMMs) are a type of molecular complex that exhibits slow magnetic relaxation at the single molecular level [1–5]. They have unique electronic and magnetic structures, making them promising candidates for next-generation data storage devices [6,7]. Due to the potential for high-density data storage, the demand for SMM has increased significantly. However, the main challenge of developing SMMs for practical applications lies in their low effective energy barriers (U_{eff}) and blocking temperatures (T_{B}), which limits their stability and applicability. In recent years, there have been many efforts to improve the performance of SMMs, including using lanthanide ions with suitable symmetry (D_{4h}, D_{5h}, D_{6h} for Dy(III)) and ligand fields to enhance the magnetic anisotropy required for SMMs [8,9]. For example, the lanthanide complexes of the cyclopentadiene series have excellent SMM properties, constantly breaking the record of U_{eff} [10–12]. Currently, the two highest U_{eff} are held by the binuclear Dysprosium cyclopentadiene complexes $(Cp^iPr_5)_2Dy_2I_3$ and a mononuclear complex

$[Dy\{N(Si^iPr_3)[Si(^iPr)_2C(CH_3)=CHCH_3]\}\{N(Si^iPr_3)(Si^iPr_2Et)\}][Al\{OC(CF_3)_3\}_4]$ with the effective energy barrier of 1631 cm^{-1} and 1843 cm^{-1}, respectively [13,14].

The research on bifunctional SMMs is currently a hot topic, including chiral SMMs and fluorescent SMMs [15–22]. Combining fluorescence enhances our understanding of SMM properties and broadens their application prospects. For instance, fluorescence-labeled SMMs allow for tracking of their locations and behaviors in a solution or solid, while the f–f emission peak positions reflect the energy level splitting of the lanthanide ion, from which the energy barrier value can be obtained [19]. The fluorescence of lanthanide SMMs can take advantage of the intrinsic fluorescence of 4f ions, and can also be induced by incorporating fluorescent ligands, such as spiropyran, rhodamine, tetraethylene and pyrene chromophores [17]. The studies on fluorescent lanthanide SMMs reveal that the choice of ligands can significantly influence the magnetic and optical properties of the complexes.

Our group has been interested in complexes of rhodamine ligands. We have found that rhodamine Fe(II) complexes [23,24] show the coupling effect of spin crossover and fluorescence, and the rhodamine dysprosium SMMs show improved magnetic properties by incorporating the phenolic hydroxyl group [25–29]. All the above complexes are mononuclear species, as the used tridentate chelate ligands tend to coordinate with the same central metal ion. Here we extend our research to polynuclear lanthanide complexes. We choose the ligand HL^1 (Scheme 1a) with a methoxy group at the ortho position of the phenolic hydroxyl group, conducive to the formation of multinuclear structures [30]. We further utilize the compartmental hexadentate ligand H_2L^2 (Scheme 1b) to coordinate with different lanthanide metal ions. Herein, we report the synthesis, crystal structures and magnetic properties of three lanthanide complexes, i.e., mononuclear $[Dy(HL^1\text{-o})(NO_3)_2(CH_3OH)_2]NO_3\cdot CH_3OH$ (complex **1·MeOH**) and tetranuclear $[Ln_4(L^1\text{-c})_2(L^2)_2(\mu_3\text{-OH})_2(NO_3)_2(CH_3OH)_4](NO_3)_2\cdot 2CH_3CN\cdot 5CH_3OH\cdot 2H_2O$ (Ln = Dy, complex **2**; Ln = Gd, complex **3**).

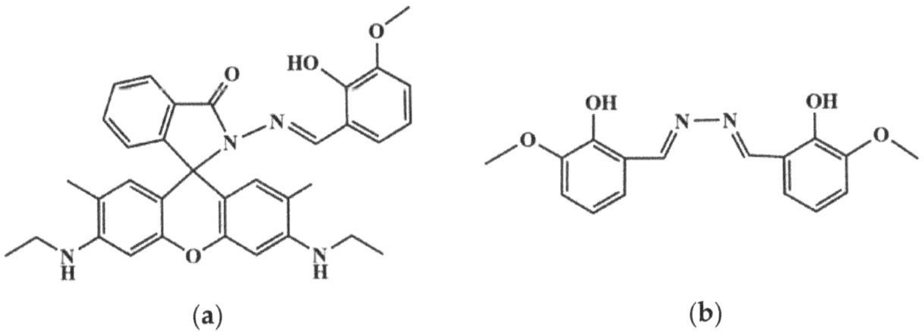

Scheme 1. Structures of the ligands HL^1 (**a**) and H_2L^2 (**b**).

2. Materials and Methods

2.1. Synthesis

All reagents (Sigma Aldrich, St. Louis, MO, USA) used in this study were commercially available and used without further purification. The ligands HL^1 and H_2L^2 were synthesized by method from the literature [28,29]. All syntheses were performed under ambient conditions at room temperature.

2.1.1. Synthesis of Complexes **1·MeOH** and **1·2H₂O**

$Dy(NO_3)_3\cdot 5H_2O$ (0.2 mmol, 112.5 mg) and HL^1 (0.2 mmol, 86.5 mg) were stirred in methanol (10 mL) to obtain a red solution. Well-shaped red crystals of **1·MeOH** suitable for single-crystal structural determination can be obtained by evaporation of the above solution at room temperature for a week. Anal. Calcd for **1·MeOH** ($C_{37}H_{46}N_7O_{16}Dy$): H,

4.60; C, 44.12; N, 9.73. Found: H, 4.55; C, 42.75; N, 9.79. The measured data are consistent with that for $[Dy(HL^1\text{-o})(NO_3)_2(CH_3OH)_2]NO_3\cdot 2H_2O$ ($C_{36}H_{46}N_7O_{17}Dy$, **1·2H₂O**): H, 4.58; C, 42.76; N, 9.70.

2.1.2. Synthesis of Complex **2**

Mixing $Dy(NO_3)_3\cdot 5H_2O$ (0.5 mmol, 216 mg) and HL^1 (0.2 mmol, 112.5 mg) in methanol (5 mL) and acetonitrile (5 mL) gave a red solution, to which H_2L^2 (0.2 mmol, 60 mg) in CH_2Cl_2 (2 mL) and Et_3N (100 μL, 2.16 mmol) were added successively. The solution turned yellow and became turbid. After filtration, the filtrate was put into a glass tube and covered with ether for slow diffusion. After a week, yellow crystals can be obtained for X-ray single-crystal diffraction analysis. Anal. Calcd for **2** ($C_{113}H_{142}N_{18}O_{41}Dy_4$): H, 4.680; C, 44.376; N, 8.244. Found: H, 4.343; C, 43.577; N, 7.926.

2.1.3. Synthesis of Complex **3**

The synthetic method is similar to that of complex **2**, except that $Dy(NO_3)_3\cdot 5H_2O$ is replaced with $Gd(NO_3)_3\cdot 6H_2O$ (0.5 mmol, 225 mg). High-quality yellow single crystals were obtained. Anal. Calcd for **3** ($C_{113}H_{142}N_{18}O_{41}Gd_4$): H, 4.712; C, 44.683; N, 8.301. Found: H, 4.511; C, 43.001; N, 7.699.

2.2. Physical Measurements

The IR spectra with a KBr tablet were obtained on WQF-510A FTIR equipment (Beifen-Ruili, Beijing, China) with a sweeping interval of 2 cm^{-1}. The elemental analyses were performed on an Elementar Vario Cario Erballo analyzer (Elementar, Langenselbold, Germany). The powder X-ray diffraction (PXRD) measurements were recorded on a Rigaku Miniflex 600 X-ray diffractometer (Rigaku Oxford Diffraction, Tokyo, Japan) at room temperature with a sweeping speed of 10°/min. The UV-vis spectra were measured with a U-3900 spectrophotometer (Hitachi, Tokyo, Japan) between 280 and 600 nm. The fluorescence spectra were recorded on a Lengguang F98 fluorescence spectrophotometer (Lengguang, Shanghai, China). Single-crystal X-ray data were collected on a Rigaku SuperNova, Dual, Cu at zero, AtlasS2 (Rigaku Oxford Diffraction, Tokyo, Japan). The structure was solved by the program SHELXT-2013 and refined by a full matrix least-squares method based on F^2 using SHELXL-2014/7 method. Hydrogen atoms were added geometrically and refined using a riding model. Temperature- and field-dependent magnetic susceptibility measurements were carried out on a Quantum Design MPMS-XL5 SQUID magnetometer (Quantum Design, San Diego, CA, USA). The experimental magnetic susceptibilities were corrected for the diamagnetism of the constituent atoms (Pascal's tables).

3. Results and Discussion

According to the previous reports, the rhodamine-based ligand shows interesting ring-opened and ring-closed properties under different acid-base conditions (Scheme 2). Upon addition of $Ln(NO_3)_3$, the reactant solution turned red, indicating formation of ring-opened HL^1-o. Red single crystals of **1·MeOH** suitable for single-crystal X-ray diffraction were obtained from the resulting red solution. Yellow single crystals of **2** and **3** are obtained when HL^1 and H_2L^2 react with $Ln(NO_3)_3$ in the presence of Et_3N, showing that the ring-closed ligand $(L^1\text{-c})^-$ is present in the tetranuclear complexes.

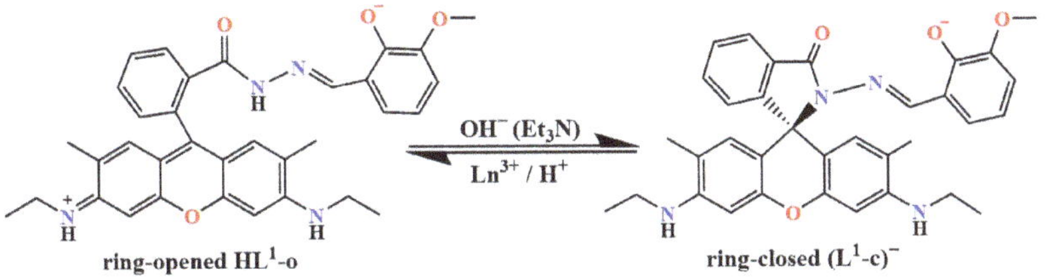

Scheme 2. Transformation process of ring-opened HL^1-o to ring-closed $(L^1\text{-c})^-$.

3.1. Crystal Structures

High-quality single crystals of **1·MeOH**, **2** and **3** were used for X-ray single crystal diffraction measurements. Since complexes **2** and **3** are isomorphic, the structure of complex **2** is described in detail as a representative. The crystallographic data are given in Table S1, and the selected bond distances and bond angles are listed in Tables S2 and S3. Hydrogen bonding for **1·MeOH**, **2** and **3** is listed in Tables S4–S6.

The crystal data show that complex **1·MeOH** crystallizes in the triclinic space group of $P-1$. It consists of the $[Dy(HL^1\text{-o})(NO_3)_2(CH_3OH)_2]^+$ cation (Figure 1a) and a NO_3^- charge-balancing anion. Dy(III) has a typical O_8N nine-coordination, with two chelating nitrates, two methanol and one ring-opened ligand HL^1-o participating in the coordination. Calculated by SHAPE software (version 2.1), the local symmetry of the Dy(III) ion is close to C_s (Table S7), and the deviation parameter is 1.359. The coordination pattern is shown in the Figure S1. The bond length for Dy1–N1 is 2.508(3) Å. The shortest Dy–O bond length is 2.192(3) Å (Dy1–O1), and the longest Dy–O bond length is 2.498(3) Å for Dy1–O7. This indicates that the short $Dy\text{–}O_{phenoxy}$ bond is dominant. The ligand HL^1-o is in a ring-opened form (Figure S2), with only three atoms available for coordination, while the oxygen in the methoxyl group is free. It is worth noting that the hydrogen on the phenol of the ligand is removed, but the nitrogen atom (N2) of the carbonyl group is protonated. The hydrazide C9–N2–N1 bond angle is 116.71°, which is a clear proof of the presence of hydrogen on N2. Otherwise, the C–N–N bond angle is smaller than 110° due to the stronger electrostatic repulsion of two bond pairs by lone pair on N2. This results in the entire neutral ligand, HL^1-o. This hydrogen (H22) attached to N2 forms a significant hydrogen bond with the free nitrate (O11), with an O11···H22 distance of 1.970 Å. Interestingly, the nitrate (O13) also forms another hydrogen bond with the free methanol molecule (O14), with an O13···H14A distance of 1.954 Å (Figure S3). The shortest intermolecular Dy···Dy distance is 6.637 Å. The short intermolecular Dy···Dy distance may lead to appreciable intermolecular magnetic interactions. Complex **1·MeOH** also forms a supramolecular one-dimensional chain structure via π-π stacking (Figure S4). The interplane distance between adjacent oxyanthracene planes is 3.311 Å, and the center···center distance is 3.692 Å.

The crystal data for complex **2** show that it crystallizes in the monoclinic space group of $C2/c$. It comprises a C_2 rotation axis-related $[Dy_4(L^1\text{-c})_2(L^2)_2(\mu_3\text{-OH})_2(NO_3)_2(CH_3OH)_4]^{2+}$ cation (Figure 1b), with two charge-balancing NO_3^- anions. The Dy_4O_6 core is formed by the μ_3-OH^- and phenoxyl oxygen bridges (Figure S5). There are two different coordination environments of Dy(III) ions within the Dy_4 molecule, as shown in Figure S6. Dy1 exhibits nine-coordination, and Dy2 is an eight-coordinate. Calculated by SHAPE software, Dy1 is close to the local symmetry of C_{4v} (Table S8) with the deviation parameter is 0.375; Dy2 is close to the local symmetry of D_{4d} (Table S9), and the deviation parameter is 1.216. The coordination pattern is shown in the Figure S6. The distance between Dy1···Dy2 is 3.538 Å, and between Dy1···Dy1A is 3.904 Å. In addition, Dy1···Dy2A is 4.327 Å and Dy2···Dy2A is 6.757 Å. The four Dy(III) ions are not coplanar (Figure S7), forming a saddle-shaped Dy_4

quadrangle with a dihedral angle of 23.0°. The coordination patterns of the ligands are shown in Figure S8. The ligand $(L^1\text{-c})^-$ is in a ring-closed form, and the four coordination atoms of $(L^1\text{-c})^-$ are all involved in coordination, bridging Dy1 and Dy2. Ligand $(L^2)^{2-}$ is a hexadentate ligand, but only five atoms participate in the coordination and bridge three Dy ions (Figure S8). The N–N single bond in the middle of the ligand $(L^2)^{2-}$ undergoes rotation, and the dihedral angle between the two benzene planes is 65.76°. The shortest Dy1–O bond length is 2.328(6) Å, and the longest is 2.442(6) Å with the average Dy1–O bond length of 2.395 Å. The shortest Dy2–O bond length is 2.202(8) Å, and the longest Dy2–O bond length is 2.501(7) Å. In addition, solvent molecules including two CH_3CN, five CH_3OH and two H_2O molecules crystallize in the lattice of the crystal. The nearest intermolecular Dy···Dy distance for complex 2 is 9.454 Å, large enough to exclude any intermolecular magnetic interaction.

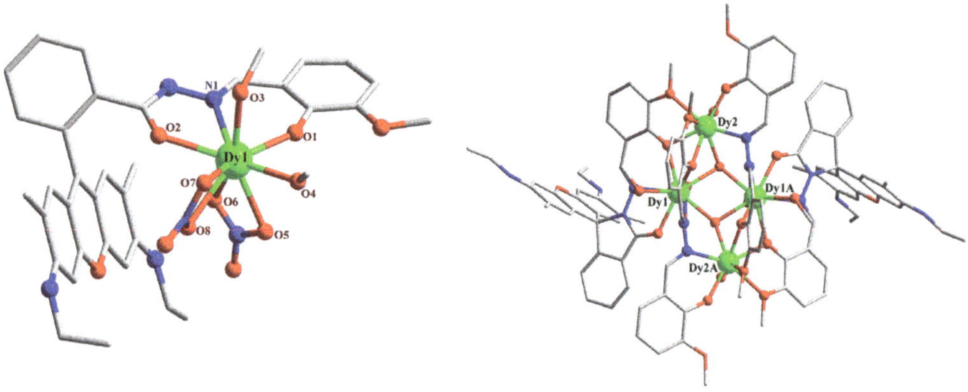

(**a**) (**b**)

Figure 1. The crystal structures of complexes **1·MeOH** (**a**) and **2** (**b**). The lattice solvents and hydrogen atoms are omitted for clarity.

3.2. Characterizations

The PXRD data for the three complexes show that the XRD patterns of complexes **2** and **3** are in good agreement with that simulated from single-crystal data; however, the data for **1·2H₂O** shows an inconformity with that of complex **1·MeOH**, further confirming the partial desolvation and hydration of the crystals (Figure S9). The infrared spectra of complexes **1·2H₂O**, **2** and **3** and the ligands HL^1 and H_2L^2 show that the peak at 1693 cm^{-1} for HL^1 disappears (Figure S10), which may be due to the fact that the C=O bond of ligand HL^1 participates in the coordination in the enol form, which weakened the vibration peak of C=O. The strong peaks at 1625 cm^{-1} of HL^1 and 1616 cm^{-1} of complex **1·2H₂O** may be assigned to the stretching vibration of C=N bond of the ligand. The peak of complex **1·2H₂O** at 1368 cm^{-1} is the characteristic peak of nitrate ion compared with ligand HL^1. For complexes **2** and **3**, the spectra are very similar (Figure S10). The strong peaks at 1090 cm^{-1} of ligand HL^1 and 1070 cm^{-1} of ligand H_2L^2 may be attributed to the methoxyl group. Upon coordination of this oxygen atom, the peak intensities undergo a notable weakening, particularly evident in the IR spectra of complexes **2** and **3**. The emergence of new peaks at 1382 cm^{-1} and 1380 cm^{-1} in complexes **2** and **3**, respectively, can be attributed to the asymmetric vibration of nitrate.

UV-Vis spectra of complexes **1·2H₂O**, **2** and **3** in ethanol (10^{-5} mol/L) are shown in Figure S11. Complex **1·2H₂O** is pink in EtOH, and complexes **2** and **3** are nearly colorless in EtOH. The absorption peaks of ligand HL^1 and complex **1·2H₂O** at 302 nm and 306 nm should be due to the ligand-to-ligand charge transfer (LLCT) of the ligand, while the new absorption peak at 534 nm for complex **1·2H$_2$O** indicates that ligand HL^1 is in the ring-opened form. The oxaanthracene group in HL^1-o has a planar conjugated rigid structure, which can cause π-π^* transition. Complexes **2** and **3**, however, have no obvious absorption peaks in the visible region, which is consistent with the presence of ring-closed HL^1-c.

The fluorescence spectra of the complexes in ethanol (10^{-5} mol/L) are shown in Figures S12 and S13. The fluorescence emission spectrum (λ_{ex} = 354 nm) of complex **1** shows a maximum emission at 569 nm (Figure 2), indicating that the ring-opened rhodamine ligand emits green light upon excitation. This fluorescence behavior is consistent with that of complexes with ring-opened rhodamine ligand. When excited by the light of 378 nm, the emission spectra of complexes **2** and **3** are obtained. It can be found that complex **2** has almost no fluorescence, and complex **3** still has weak fluorescence at 549 nm, presumably due to the presence of trace ring-opened ligand.

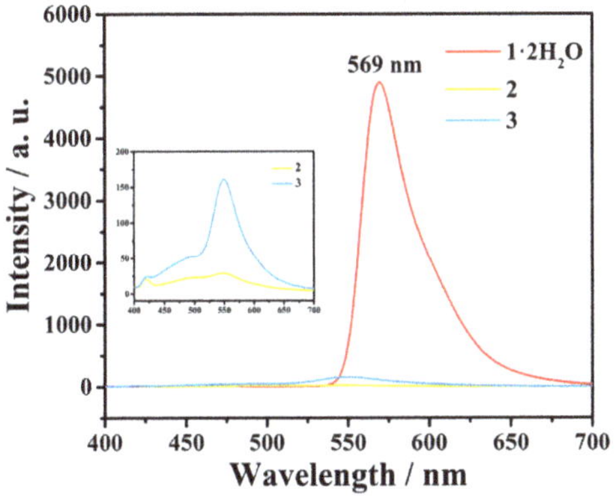

Figure 2. Fluorescence spectra (λ_{ex} = 354 nm) for complexes **1·2H$_2$O**, **2** and **3** in ethanol (10 μM).

3.3. Magnetic Measurements

The direct current (DC) and alternating current (AC) magnetic susceptibilities of complexes **1·2H$_2$O**, **2** and **3** were measured on a MPMS-XL5 SQUID magnetometer. Figure 3 shows the temperature dependence of $\chi_M T$ for the three complexes at 2–300 K under 1000-Oe field. At room temperature, the maximum $\chi_M T$ values are 14.00 cm^3 K mol^{-1}, 56.21 cm^3 K mol^{-1} and 31.36 cm^3 K mol^{-1}, which are close to the theoretical values of a Dy(III) ion (14.18 cm^3 K mol^{-1}, $S = 5/2$, $L = 5$, $g_J = 4/3$), four Dy(III) ions (56.72 cm^3 K mol^{-1}) and four Gd(III) ions (31.52 cm^3 K mol^{-1}, $S = 7/2$, $L = 0$, $g = 2.0$), respectively. For complex **1·2H$_2$O**, the $\chi_M T$ value first decreases slowly with the decrease in temperature from 300 K to 18 K, reaching a minimum of 12.27 cm^3 K mol^{-1} at 18 K. Afterwards, the $\chi_M T$ value rapidly increases to 16.29 cm^3 K mol^{-1} at 2 K. The increase in low-temperature magnetic susceptibility may be attributed to intermolecular ferromagnetic interactions in complex **1·2H$_2$O**. The shortest distance between Dy···Dy of 6.637 Å should be responsible for the intermolecular magnetic interaction. A similar situation occurs to a 1D Dy(III) coordination polymer, which has stronger intermolecular ferromagnetic coupling, and the shortest intermolecular Dy···Dy distances are even greater than that of complex **1·MeOH** [31]. For complex **2**, $\chi_M T$ value decreases slowly with the decrease in temperature to 100 K, and then quickly to 12 K before $\chi_M T$ slightly increases to 52.40 cm^3 K mol^{-1} at 2 K. The increase of $\chi_M T$ below 12 K may be related to the intramolecular Dy···Dy ferromagnetic interaction via the μ_3-OH$^-$ and phenoxyl bridges. As for complex **3**, $\chi_M T$ decreases monotonically throughout the entire temperatures. From 300 K to 10 K, the $\chi_M T$ value decreases slowly, and then rapidly from 10 K to 2 K, reaching 20.74 cm^3 K mol^{-1}, which suggests that there is an intramolecular antiferromagnetic interaction between the tetranuclear Gd(III) ions in the complex. Magnetic fitting of tetranuclear gadolinium complex **3** was carried out by MagPack software (version 1.0) [32], and the dark blue solid curve was obtained. The Hamiltonian operator expression based on the molecular structure (Figure S14) is as follows:

$$\hat{H} = -2J_1(\hat{S}_{Gd1}\hat{S}_{Gd2} + \hat{S}_{Gd1A}\hat{S}_{Gd2A}) - 2J_2(\hat{S}_{Gd1}\hat{S}_{Gd2A} + \hat{S}_{Gd1A}\hat{S}_{Gd2}) - 2J_3\hat{S}_{Gd1}\hat{S}_{Gd1A} \quad (1)$$

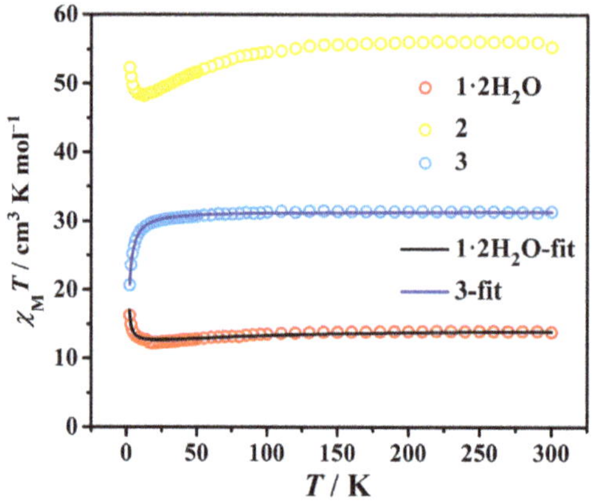

Figure 3. Temperature dependence of $\chi_M T$ for complexes **1·2H$_2$O**, **2** and **3** at 2–300 K under 1000 Oe field. The solid lines represent the fitting results.

It was found that the fitting curve was in good agreement with the experimental data with the fitting parameters of $J_1 = J_2 = -0.022$ cm^{-1}, $J_3 = -0.036$ cm^{-1} and $g = 2.000$, which also indicated that there was weak intramolecular Gd···Gd antiferromagnetic coupling.

The field-dependent magnetization of complexes **1·2H$_2$O**, **2** and **3** in the temperature range of 2–10 K was measured, as shown in Figures 4, S15 and S16. The magnetization of complex **3** increases with the increase in magnetic field intensity at 2 K, and increases rapidly at 0 T to 3 T, and gradually approaches the maximum magnetization of 28.4 Nβ at 5 T (Figure S16). When the temperature increases, the maximum magnetization value decreases. At 10 K, the magnetization of complex **3** is far from saturated at 5 T. Contrary to complexes **1·2H$_2$O** and **2** (Figure S15), complex **3** basically reached saturation at a temperature of 2 K and 5 T. The theoretical saturation value of four independent Gd(III) ions is 28 Nβ, which is consistent with the maximum experimental saturation value, indicating that the four Gd(III) ions in complex **3** have negligibly weak magnetic interaction. Magnetic fitting of complex **3** (Figure S16) shows that the experimental M-H curve at 2 K is lower than the calculated Brillouin curve of non-interacting S_{Gd} spin, which also indicates that there is weak antiferromagnetic coupling.

Some gadolinium compounds have demonstrated exceptionally outstanding magnetocaloric properties, showcasing remarkable potential in various advanced technological applications [33–42]. We further characterized the magnetocaloric effect (MCE) of complex **3**. Using $-\Delta S_m = nR\ln(2S+1)$ ($R \approx 8.314$ J mol^{-1} K^{-1}), the theoretical magnetic entropy of four independent Gd(III) can be calculated to be 8.32R (22.77 J kg^{-1} K^{-1}). We used the Maxwell Equation (2):

$$\Delta S_m(T)_{\Delta H} = \int \left[\frac{\partial M(T,H)}{\partial T}\right]_H dH \quad (2)$$

The magnetic entropy change values under different temperatures and different external magnetic fields can be calculated (Figure 5). At 2 K and $\Delta H = 5$ T, the value of $-\Delta S_m$

is 9.81 J kg^{-1} K^{-1}, which is attributed to the low Gd(III) content (20.71%) that reduces the density of active magnetic moments and thereby limits the cumulative magnetic response, resulting in an unsatisfactory MCE. Moreover, the presence of weak antiferromagnetic coupling within the system also causes the actual value of $-\Delta S_m$ to be smaller than the theoretical value.

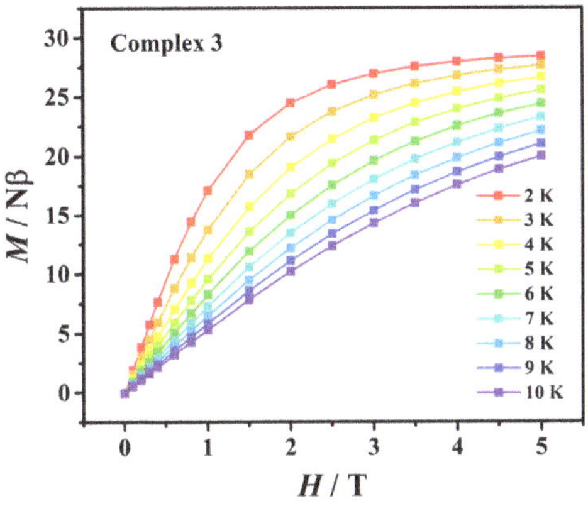

Figure 4. Magnetization curves of complex **3**.

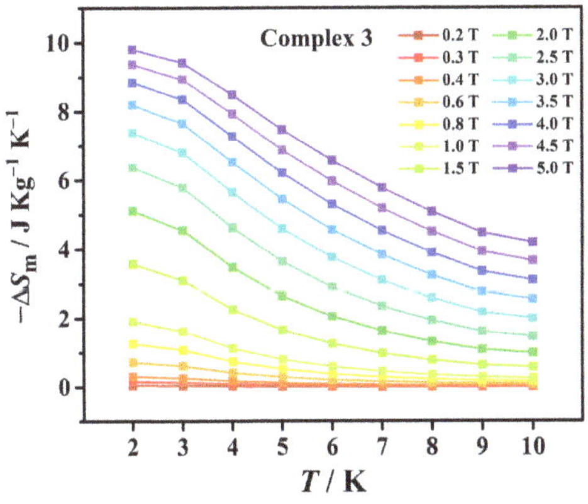

Figure 5. Magnetic entropy change in temperature of complex **3** under different external magnetic fields.

The AC magnetic susceptibility measurements were performed on **1·2H$_2$O** and **2** to study the slow magnetic relaxation. Figure S17 shows temperature dependence of out-of-phase AC susceptibilities (χ''_M) for **1·2H$_2$O** under 0- and 2000-Oe DC field with the frequency of 997 Hz. Below 10 K, no obvious peak is observed in the χ''_M curve under 0 Oe, suggesting a strong quantum tunneling of the magnetization (QTM) effect during the magnetization relaxation process. After applying a 2000-Oe DC field, the peak appears at 4 K, indicating that an external field can effectively suppresses the low-temperature QTM.

In order to find an exact external magnetic field, field dependence of χ''_M for complex **1·2H$_2$O** was measured at 4 K and 997 Hz (Figure S18). With the increase in external magnetic field, the χ''_M first increases rapidly, and then decreases slowly, and the optimal external magnetic field corresponds to the maximum value under 600 Oe.

The AC magnetic susceptibilities of complex **1·2H$_2$O** are measured under 600 Oe (Figure 6). As shown in Figure 6b, with the increase in frequency, the peak of the χ''_M

curve of AC magnetic susceptibility moves to the high temperature region, which is a typical feature of SMMs. As shown in Figure 6c, the Cole–Cole curves in the range of 2.0–4.2 K are approximately semicircular, which is well fitted by the generalized Debye model. The fitting parameters are listed in Table S10, and the α value is in the range of 0.13 to 0.095, indicating that the relaxation time distribution is relatively narrow and has a single relaxation. The lnτ vs. T^{-1} plot shows a certain curvature, indicating that there are multiple relaxation paths. The plot was well fitted with the formula, including Raman and direct processes:

$$\tau^{-1} = AT + CT^{n} \tag{3}$$

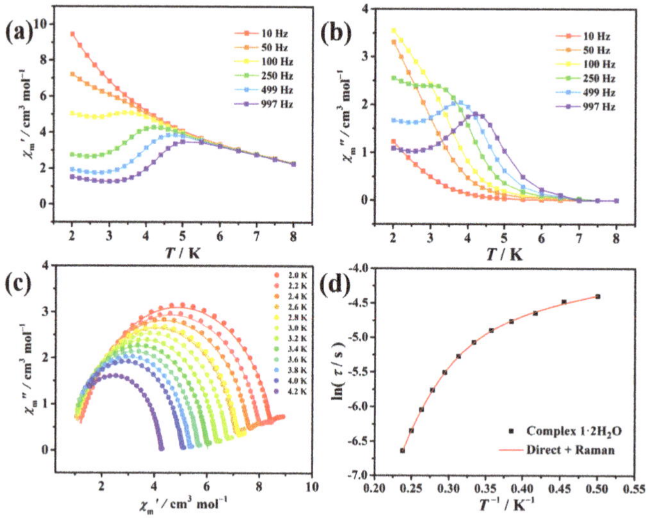

Figure 6. The AC magnetic susceptibility data of complex **1·2H$_2$O** under 600 Oe DC field: (**a**) the χ'_M of AC magnetic susceptibility changes with temperature at different frequencies; (**b**) the χ''_M of AC magnetic susceptibility with temperature at different frequencies; (**c**) the χ''_M of AC magnetic susceptibility is plotted against the χ'_M at different temperatures (Cole–Cole) and fitted by the generalized Debye model; (**d**) lnτ is plotted against T^{-1}, and the red line is the fitting result.

As can be seen from Figure 6d, the fitted curve can well match the data points with the fitting parameters of A = 39.6(3) s^{-1} K^{-1}, n =8.0(1) and C = 0.007(1) s^{-1} K$^{-8.0}$. The n value is in the normal range of 2–9 for Dy(III) SMMs. The inclusion of Orbach relaxation cannot give a satisfactory fitting result, and nor can it only include the Raman process.

Figure S19 shows the χ'_M and χ''_M of AC magnetic susceptibility with temperature at different frequencies of complex **2** under 0 Oe. Within the test temperature range, no obvious peak is observed in the χ''_M curve under 0 Oe, suggesting a strong QTM effect during the relaxation process. The AC plots of complexes **1·2H$_2$O** and **2** are markedly different, which may be due to the fact that complex **1·2H$_2$O**, as a mononuclear Dy complex with one sole short Dy-O$_{phenoxy}$ bond, exhibits strong magnetic anisotropy. In contrast, complex **2**, being a saddle-shaped tetranuclear Dy species, shows diminished magnetic anisotropy due to the cancellation of the anisotropy of four Dy(III) centers.

3.4. Theoretical Calculations

Complete-active-space self-consistent field (CASSCF) calculations on complex **1·MeOH** (Figure S20) on the basis of X-ray determined geometry have been carried out with the OpenMolcas [43] and the SINGLE_ANISO [44–46] programs (see the Supporting Information for details). The calculated energy levels, g (g_x, g_y, g_z) tensors and the predominant m_J values of the lowest eight Kramers doublets (KDs) of complex **1·MeOH** are shown in Table S11. The predominant m_J components for the lowest eight KDs of **1·MeOH** are shown in Table S12, where the ground KD is mostly composed of $m_J = \pm 15/2$. The first excited KD of **1·MeOH** is composed of $m_J = \pm 13/2$, which leads to the large transversal magnetic moment in its first excited KD (Figure 7). The calculations have been validated by fitting the $\chi_M T$ data for **1·2H$_2$O** with the intermolecular coupling constant of $zJ' = 0.07$ cm^{-1} (Figure 3).

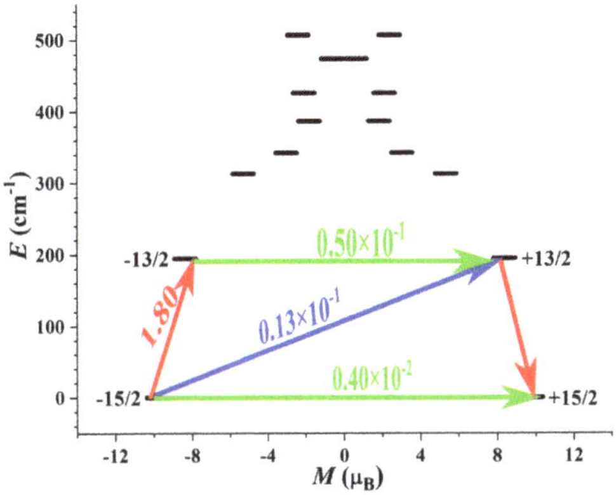

Figure 7. Magnetization blocking barrier of complex **1·MeOH**. The path shown by the red arrows represents the most probable path for magnetic relaxation in the corresponding compound. The numbers at each arrow stand for the mean absolute value of the corresponding matrix element of transition magnetic moment.

The corresponding magnetization blocking barrier of complex **1·MeOH** is shown in Figure 7, where the transversal magnetic moment in the ground KD is close to 10^{-2} μ_B, and thus the QTM in the ground KD is suppressed at low temperature. The transversal magnetic moment in its first excited state of **1·MeOH** is 0.5×10^{-1} μ_B, and therefore allowing the fast thermal-assisted QTM (TA-QTM) in its first excited KD. Thus, the calculated energy barrier is 194.6 cm^{-1}. Due to the unfavorable effects of anharmonic phonons, Raman magnetic relaxation, QTM, et al. on the energy barrier, the experimental effective energy barrier U_{eff} is usually smaller than the calculated one [47–51].

The calculated crystal-field (CF) parameters B (k, q) and corresponding weights are given in Table S13, where the weights of the axial CF parameters B(2,0) and B(4,0) for **1·MeOH** are 20.70% and 7.73%, respectively, which exceed the non-axial components, indicating a perfectly axial symmetry for **1·MeOH**. The main magnetic axis on Dy(III) ion of 1 is indicated in Figure 8a. In addition, the main magnetic axis is also calculated by MAGELLAN software (version 1.0) [52], and the obtained results are shown in Figure 8b. It can also be observed that the direction of the main magnetic axis is almost identical with the direction of the phenol–oxygen bond, showing the shortest Dy-O bond is dominant in magnetic anisotropy.

It has been documented that the magnetic anisotropy is the key factor affecting the SMM behavior, while the magnetic anisotropy is related to the axiality and coordination geometry of Dy(III). Complex **1·MeOH** has a single phenol–oxygen bond with a short distance of 2.193 Å, dictating the orientation of the magnetic axis and anisotropy. However, the calculated energy barrier for **1·MeOH** is 194.6 cm^{-1}, obviously smaller than that (221 cm^{-1}) of a similar Dy(III) complex [Dy(L^R)(H$_2$O)$_4$(MeCN)](ClO$_4$)$_3$·2H$_2$O·MeCN (L^R = salicylaldehyde rhodamine 6G hydrazone) with a single short Dy-O$_{phenoxy}$ bond of 2.175 Å [26]. Careful examination of the crystal structure shows that both complexes have eight-coordinate Dy(III) and different coordination geometry: C_s for complex **1** and

D_{2d} for the latter. Higher coordination geometry and a shorter Dy-O$_{phenoxy}$ bond usually favors higher calculated energy barrier for the latter. In addition, the coordination of nitrate anion in the equatorial positions attenuates the magnetic anisotropy of **1·MeOH**. The under-barrier relaxation might be related to the weathering of the crystals, causing subtle variations in the crystal-field environment that can significantly influence the magnetic properties.

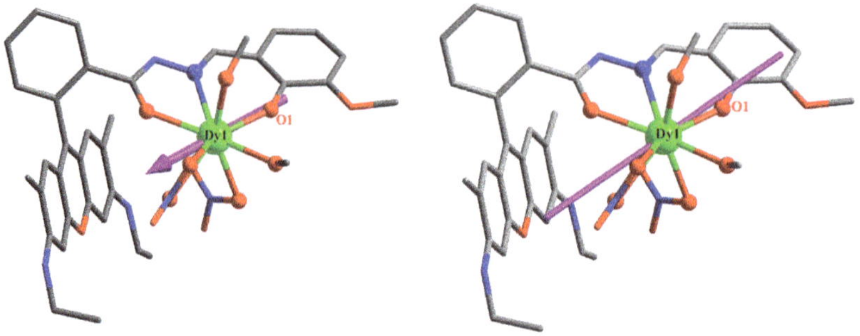

(**a**) (**b**)

Figure 8. Calculated orientation of the local main magnetic axis on Dy(III) of complex **1·MeOH** in the ground KD. (**a**) Ab initio, (**b**) MAGELLAN.

4. Conclusions

In this work, we have developed new Schiff base ligands HL^1 and H_2L^2 to obtain a series of Ln(III) complexes from mononuclear to tetranuclear. The mononuclear complex **1·MeOH** features a ring-opened ligand upon crystallization, and it emits a strong green fluorescence emission in ethanol. Magnetic characterizations show that complex **1·2H₂O** is a typical field-induced SMM with relaxations under-barrier [53]. The ab initio calculations have revealed that complex **1·MeOH** exhibits high magnetic anisotropy. There is an intramolecular ferromagnetic coupling in the Dy_4 complex (**2**), whereas complex **3** is weakly antiferromagnetic. Complex **3** can be used as a magnetic refrigeration material. This work reveals that the careful selection of ligands and the rational control of coordination modes can lead to the development of fluorescent lanthanide single-molecule magnets (Ln-SMMs) and a magnetic refrigeration material. This approach, as demonstrated in recent studies, allows for the fine-tuning of magnetic properties, which is crucial for advancing applications in fields such as information storage, quantum computing and biomedical detection.

Supplementary Materials: The following supporting information can be downloaded at: https://www.mdpi.com/article/10.3390/magnetochemistry11080068/s1, Figure S1. The polyhedral structures around the Dy(III) coordination in complex **1·MeOH**. Figure S2. Coordination mode of ligand HL^1-o in complex **1·MeOH**. Figure S3. Hydrogen bonding in complex **1·MeOH**. Figure S4. One-dimensional chain structure via π-π stacking in complex **1·MeOH**. Figure S5. Dy_4O_6 core skeleton of complex **2**. Figure S6. The polyhedral structures around the Dy(III) coordination in complex **2**. Figure S7. Top and side views of the coordination polyhedron of the tetranuclear Dy(III) surrounding structures in complex **2**. Figure S8. Coordination mode of ligand $(L^1\text{-c})^-$ and $(L^2)^{2-}$ in complex **2**. Figure S9. PXRD pattern for complexes **1·2H₂O**, **2** and **3**. Figure S10. Infrared spectra of the ligand and complexes **1·2H₂O**, **2** and **3**. Figure S11. UV-Vis spectra of the ligands and complexes **1·MeOH**, **2** and **3**. Figure S12. Fluorescence excitation spectra of complex **1·2H₂O**. Figure S13. Fluorescence excitation spectra of complexes **2** and **3**. Figure S14. Magnetic coupling pathways in complex **3**. Figure S15. Magnetization curves of complex **1·2H₂O** (a) and **2** (b). Figure S16. Magnetization curves and Brillouin curve at 2 K of complex **3**. Figure S17. Temperature dependence of χ''_M for complex

$1{\cdot}2H_2O$ under 0 and 2000 Oe DC field. Figure S18. Plots of χ''_M for complex $1{\cdot}2H_2O$ at 4 K and 997 Hz. Figure S19. The AC magnetic susceptibility data of complex **2** under 0 Oe: the χ'_M (a) and χ''_M (b) of AC magnetic susceptibility with temperature at different frequencies. Figure S20. Calculated complete structure of complex **1·MeOH**. H atoms are omitted for clarify. Figure S21. Magnetization data at 3.0 K for $1{\cdot}2H_2O$. The solid line stands for the fit data. Table S1. Crystal data and refinement parameters for complexes **1·MeOH**, **2** and **3**. Table S2. Selected bond lengths (Å) and angles (°) for complex **1·MeOH**. Table S3. Selected bond lengths (Å) and angles (°) for complexes **2** and **3**. Table S4. Hydrogen bonding for complex **1·MeOH**. Table S5. Hydrogen bonding for complex **2**. Table S6. Hydrogen bonds for complex **3**. Table S7. The results of coordination geometric configurations evaluated by SHAPE software for complex **1·MeOH**. Table S8. The results of coordination geometric configurations evaluated by SHAPE software for Dy1 of complex **2**. Table S9. The results of coordination geometric configurations evaluated by SHAPE software for Dy2 of complex **2**. Table S10. Cole-Cole plots fitting parameters for complex $1{\cdot}2H_2O$. Table S11. Calculated energy levels (cm^{-1}), g (g_x, g_y, g_z) tensors and predominant m_J values of the lowest eight Kramers doublets (KDs) of complex **1·MeOH** using CASSCF/RASSI-SO with the OpenMolcas. Table S12. Wave functions with definite projection of the total moment $|m_J>$ for the lowest eight KDs of complex **1·MeOH**. Table S13. Calculated crystal-field parameters B (k, q) and corresponding weights for complex **1·MeOH** using CASSCF/RASSI-SO with the OpenMolcas. References [43,54–58] are cited in the Supplementary Materials.

Author Contributions: Conceptualization and formal analysis, L.M. and C.-M.L.; ab initio calculations, D.-M.Z.; writing—original draft preparation, L.M. and D.-M.Z.; writing—review and editing, supervision, project administration and funding acquisition, H.-Z.K. and Y.-Q.Z. All authors have read and agreed to the published version of the manuscript.

Funding: This research was funded by the National Natural Science Foundation of China, grant numbers 22271171 and 21971142.

Institutional Review Board Statement: Not applicable.

Informed Consent Statement: Not applicable.

Data Availability Statement: The original contributions presented in this study are included in the article/Supplementary Materials. Further inquiries can be directed to the corresponding authors.

Conflicts of Interest: The authors declare no conflicts of interest.

Abbreviations

The following abbreviations are used in this manuscript:

SMMs	Single-molecule magnets
PXRD	Powder X-ray diffraction
LLCT	Ligand-to-ligand charge transfer
DC	Direct current
AC	Alternating current
MCE	Magnetocaloric effect
χ''_M	Out-of-phase AC susceptibilities
QTM	Quantum tunneling of the magnetization
CASSCF	Complete-active-space self-consistent field
KDs	Kramers doublets
TA-QTM	Thermal-assisted QTM
CF	Crystal-field

References

1. Caneschi, A.; Gatteschi, D.; Sessoli, R.; Barra, A.L.; Brunel, L.C.; Guillot, M. Alternating current susceptibility, high field magnetization, and millimeter band EPR evidence for a ground S = 10 state in $[Mn_{12}O_{12}(CH_3COO)_{16}(H_2O)_4]\cdot 2CH_3COOH\cdot 4H_2O$. *J. Am. Chem. Soc.* **1991**, *113*, 5873–5874. [CrossRef]
2. Sessoli, R.; Gatteschi, D.; Caneschi, A.; Novak, M.A. Magnetic bistability in a metal-ion cluster. *Nature* **1993**, *365*, 141–143. [CrossRef]
3. Aubin, S.M.J.; Wemple, M.W.; Adams, D.M.; Tsai, H.-L.; Christou, G.; Hendrickson, D.N. Distorted $Mn^{IV}Mn^{III}_3$ Cubane Complexes as Single-Molecule Magnets. *J. Am. Chem. Soc.* **1996**, *118*, 7746–7754. [CrossRef]
4. Friedman, J.-R.; Sarachik, M.-P. Macroscopic Measurement of Resonant Magnetization Tunneling in High-Spin Molecules. *Phys. Rev. Lett.* **1996**, *76*, 3830–3833. [CrossRef]
5. Leuenberger, M.N.; Loss, D.J.N. Quantum computing in molecular magnets. *Nature* **2001**, *410*, 789–793. [CrossRef]
6. Sorace, L.; Benelli, C.; Gatteschi, D. Lanthanides in Molecular Magnetism: Old Tools in a New Field. *Chem. Soc. Rev.* **2011**, *40*, 3092–3104. [CrossRef]
7. Troiani, F.; Affronte, M. Molecular Spins for Quantum Information Technologies. *Chem. Soc. Rev.* **2011**, *40*, 3119–3129. [CrossRef]
8. Wang, J.; Li, Q.-W.; Wu, S.-G.; Chen, Y.-C.; Wan, R.-C.; Huang, G.-Z.; Liu, Y.; Liu, J.-L.; Reta, D.; Giansiracusa, M.; et al. Opening Magnetic Hysteresis by Axial Ferromagnetic Coupling: From Mono-Decker to Double-Decker Metallacrown. *Angew. Chem. Int. Ed.* **2021**, *60*, 5299–5306. [CrossRef]
9. Wu, S.-G.; Ruan, Z.-Y.; Huang, G.-Z.; Zheng, J.-Y.; Vieru, V.; Taran, G.; Wang, J.; Chen, Y.-C.; Liu, J.-L.; Ho, L.; et al. Field-induced oscillation of magnetization blocking barrier in a holmium metallacrown single-molecule magnet. *Chem* **2021**, *7*, 982–992. [CrossRef]
10. Jiang, S.-D.; Wang, B.-W.; Sun, H.-L.; Wang, Z.-M.; Gao, S. An Organometallic Single-Ion Magnet. *J. Am. Chem. Soc.* **2011**, *133*, 4730–4733. [CrossRef]
11. Goodwin, C.; Ortu, F.; Reta, D.; Chilton, N.; Mills, D. Molecular magnetic hysteresis at 60 kelvin in dysprosocenium. *Nature* **2017**, *548*, 439–442. [CrossRef] [PubMed]
12. Guo, F.-S.; Day, B.; Chen, Y.-C.; Tong, M.-L.; Mansikkamäki, A.; Layfield, R. Magnetic hysteresis up to 80 kelvin in a dysprosium metallocene single-molecule magnet. *Science* **2018**, *362*, 1400–1403. [CrossRef]
13. Gould, C.; McClain, R.; Reta, D.; Kragskow, J.; Marchiori, D.; Lachman, E.; Choi, E.-S.; Analytis, J.; Britt, D.; Chilton, N.; et al. Ultrahard Magnetism from Mixed-Valence Dilanthanide Complexes with Metal-Metal Bonding. *Science* **2022**, *375*, 198–202. [CrossRef]
14. Emerson-King, J.; Gransbury, G.K.; Atkinson, B.E.; Blackmore, W.J.A.; Whitehead, G.F.S.; Chilton, N.F.; Mills, D.P. Soft magnetic hysteresis in a dysprosium amide–alkene complex up to 100 kelvin. *Nature* **2025**, *643*, 125–129. [CrossRef]
15. Hojorat, M.; Sabea, H.-A.; Norel, L.; Bernot, K.; Roisnel, T.; Gendron, F.; Guennic, B.; Trzop, E.; Collet, E.; Long, J.; et al. Hysteresis Photomodulation via Single-Crystal-to-Single-Crystal Isomerization of a Photochromic Chain of Dysprosium Single Molecule Magnets. *J. Am. Chem. Soc.* **2020**, *142*, 931–936. [CrossRef]
16. Bünzli, J.-C. Benefiting from the Unique Properties of Lanthanide Ions. *Acc. Chem. Res.* **2006**, *39*, 53–61. [CrossRef]
17. Jia, J.-H.; Li, Q.-W.; Chen, Y.-C.; Liu, J.-L.; Tong, M.-L. Luminescent single-molecule magnets based on lanthanides: Design strategies, recent advances and magneto-luminescent studies. *Coord. Chem. Rev.* **2019**, *378*, 365–381. [CrossRef]
18. Ruiz, J.; Mota, A.; Rodríguez-Diéguez, A.; Titos, S.; Herrera, J.; Ruiz, E.; Cremades, E.; Costes, J.; Colacio, E. Field and Dilution Effects on the Slow Relaxation of a Luminescent DyO_9 Low-Symmetry Single-Ion Magnet. *Chem. Commun.* **2012**, *48*, 7916–7918. [CrossRef]
19. Bi, Y.; Chen, C.; Zhao, Y.-F.; Zhang, Y.-Q.; Jiang, S.-D.; Wang, B.-W.; Han, J.-B.; Sun, J.-L.; Bian, Z.-Q.; Wang, Z.-M.; et al. Thermostability and Photoluminescence of Dy(III) Single-Molecule Magnets Under a Magnetic Field. *Chem. Sci.* **2016**, *7*, 5020–5031. [CrossRef]
20. Zeng, M.; Zhou, Z.-Y.; Wu, X.-R.; Liu, C.-M.; Kou, H.-Z. Assembly of a Heterotrimetallic Zn_2Dy_2Ir Pentanuclear Complex toward Multifunctional Molecular Materials. *Inorg. Chem.* **2022**, *61*, 14275–14281. [CrossRef]
21. Liu, K.; Zhang, X.-J.; Meng, X.-X.; Shi, W.; Cheng, P.; Powell, A. Constraining the Coordination Geometries of Lanthanide Centers and Magnetic Building Blocks in Frameworks: A New Strategy for Molecular Nanomagnets. *Chem. Soc. Rev.* **2016**, *45*, 2423–2439. [CrossRef]
22. Wang, X.-Q.; Geng, Y.-W.; Wang, Z.; Xie, C.; Han, T.; Cheng, P. Two-Dimensional Metal–Organic Framework with HighPerformance Single-Molecule Magnets as Nodes Showing Magnetic Coercivity Photomodulation. *J. Am. Chem. Soc.* **2025**, *147*, 18044–18053. [CrossRef]
23. Yuan, J.; Wu, S.-Q.; Liu, M.-J.; Sato, O.; Kou, H.-Z. Rhodamine 6G-Labeled Pyridyl Aroylhydrazone Fe(II) Complex Exhibiting Synergetic Spin Crossover and Fluorescence. *J. Am. Chem. Soc.* **2018**, *140*, 9426–9433. [CrossRef]
24. Yuan, J.; Liu, M.-J.; Wu, S.-Q.; Zhu, X.; Zhang, N.; Sato, O.; Kou, H.-Z. Substituent effects on the fluorescent spin-crossover Fe(II) complexes of rhodamine 6G hydrazones. *Inorg. Chem. Front.* **2019**, *6*, 1170–1176. [CrossRef]

25. Liu, M.-J.; Yuan, J.; Tao, J.; Zhang, Y.-Q.; Liu, C.-M.; Kou, H.-Z. Rhodamine Salicylaldehyde Hydrazone Dy(III) Complexes: Fluorescence and Magnetism. *Inorg. Chem.* **2018**, *57*, 4061–4069. [CrossRef]
26. Liu, M.-J.; Wu, S.-Q.; Li, J.-X.; Zhang, Y.-Q.; Sato, O.; Kou, H.-Z. Structural Modulation of Fluorescent Rhodamine-Based Dysprosium(III) Single-Molecule Magnets. *Inorg. Chem.* **2020**, *59*, 2308–2315. [CrossRef]
27. Liu, M.-J.; Fu, Z.-Y.; Sun, R.; Yuan, J.; Liu, C.-M.; Zou, B.; Wang, B.-W.; Kou, H.-Z. Mechanochromic and Single-Molecule Magnetic Properties of a Rhodamine 6G Dy(III) Complex. *ACS Appl. Electron. Mater.* **2021**, *3*, 1368–1374. [CrossRef]
28. Miao, L.; Liu, M.-J.; Ding, M.-M.; Zhang, Y.-Q.; Kou, H.-Z. A Dy(III) Fluorescent Single-Molecule Magnet Based on a Rhodamine 6G Ligand. *Inorganics* **2021**, *9*, 51. [CrossRef]
29. Miao, L.; Liu, M.-J.; Zeng, M.; Kou, H.-Z. Chiral Zn_3Ln_3 Hexanuclear Clusters of an Achiral Flexible Ligand. *Inorg. Chem.* **2023**, *62*, 12814–12821. [CrossRef]
30. Guo, Y.-N.; Xu, G.-F.; Gamez, P.; Zhao, L.; Lin, S.-Y.; Deng, R.-P.; Tang, J.-K.; Zhang, H.-J. Two-Step Relaxation in a Linear Tetranuclear Dysprosium(III) Aggregate Showing Single-Molecule Magnet Behavior. *J. Am. Chem. Soc.* **2010**, *132*, 8538–8539. [CrossRef]
31. Armelao, L.; Dell'Amico, D.B.; Bottaro, G.; Bellucci, L.; Labella, L.; Marchetti, F.; Mattei, C.A.; Mian, F.; Pineider, F.; Poneti, G.; et al. 1D hetero-bimetallic regularly alternated 4f–3d coordination polymers based on N-oxide-4,4′-bipyridine (bipyMO) as a linker: Photoluminescence and magnetic properties. *Dalton Trans.* **2018**, *47*, 8337–8345. [CrossRef]
32. Borrás-Almenar, J.J.; Clemente-Juan, J.M.; Coronado, E.; Tsukerblat, B.S. MAGPACK1 A package to calculate the energy levels, bulk magnetic properties, and inelastic neutron scattering spectra of high nuclearity spin clusters. *J. Comput. Chem.* **2001**, *22*, 985–991. [CrossRef]
33. Gschneidner, K.A.; Pecharsky, V.K. Thirty years of near room temperature magnetic cooling: Where we are today and future prospects. *Int. J. Refrig.* **2008**, *31*, 945–961. [CrossRef]
34. Evangelisti, M.; Brechin, E.K. Recipes for enhanced molecular cooling. *Dalton Trans.* **2010**, *39*, 4672–4676. [CrossRef]
35. Koskelo, E.C.; Liu, C.; Mukherjee, P.; Kelly, N.D.; Dutton, S.E. Free-Spin Dominated Magnetocaloric Effect in Dense Gd^{3+} Double Perovskites. *Chem. Mater.* **2022**, *34*, 3440–3450. [CrossRef]
36. Lorusso, G.; Sharples, J.W.; Palacios, E.; Roubeau, O.; Brechin, E.K.; Sessoli, R.; Rossin, A.; Tuna, F.; McInnes, E.J.L.; Collison, D.; et al. A Dense Metal-Organic Framework for Enhanced Magnetic Refrigeration. *Adv. Mater.* **2013**, *25*, 4653–4656. [CrossRef]
37. Yang, Y.; Zhang, Q.-C.; Pan, Y.-Y.; Long, L.-S.; Zheng, L.-S. Magnetocaloric effect and thermal conductivity of $Gd(OH)_3$ and $Gd_2O(OH)_4(H_2O)_2$. *Chem. Commun.* **2015**, *51*, 7317–7320. [CrossRef]
38. Palacios, E.; Rodríguez-Velamazán, J.A.; Evangelisti, M.; McIntyre, G.J.; Lorusso, G.; Visser, D.; de Jongh, L.J.; Boatner, L.A. Magnetic structure and magnetocalorics of $GdPO_4$. *Phys. Rev. B Condens. Mater. Phys.* **2014**, *90*, 214423. [CrossRef]
39. Chen, Y.-C.; Qin, L.; Meng, Z.-S.; Yang, D.-F.; Wu, C.; Fu, Z.; Zheng, Y.-Z.; Liu, J.-L.; Tarasenko, R.; Orendáč, M.; et al. Study of a magnetic-cooling material $Gd(OH)CO_3$. *J. Mater. Chem. A* **2014**, *2*, 9851–9858. [CrossRef]
40. Chen, Y.-C.; Prokleška, J.; Xu, W.-J.; Liu, J.-L.; Liu, J.; Zhang, W.-X.; Jia, J.-H.; Sechovský, V.; Tong, M.-L. A brilliant cryogenic magnetic coolant: Magnetic and magnetocaloric study of ferromagnetically coupled GdF_3. *J. Mater. Chem. C* **2015**, *3*, 12206–12211. [CrossRef]
41. Xu, Q.F.; Liu, B.L.; Ye, M.Y.; Zhuang, G.L.; Long, L.S.; Zheng, L.S. $Gd(OH)F_2$: A Promising Cryogenic Magnetic Refrigerant. *J. Am. Chem. Soc.* **2022**, *144*, 13787–13793. [CrossRef]
42. Chen, Y.W.; Gong, P.F.; Guo, R.X.; Fan, F.D.; Shen, J.; Zhang, G.C.; Tu, H. Improvement on Magnetocaloric Effect through Structural Evolution in Gadolinium Borate Halides $Ba_2Gd(BO_3)_2X$ (X = F, Cl). *Inorg. Chem.* **2023**, *62*, 15584–15592. [CrossRef]
43. Galván, I.F.; Vacher, M.; Alavi, A.; Angeli, C.; Aquilante, F.; Autschbach, J.; Bao, J.J.; Bokarev, S.I.; Bogdanov, N.A.; Carlson, R.K.; et al. OpenMolcas: From Source Code to Insight. *J. Chem. Theory Comput.* **2019**, *15*, 5925–5964. [CrossRef]
44. Chibotaru, L.F.; Ungur, L.; Soncini, A. The Origin of Nonmagnetic Kramers Doublets in the Ground State of Dysprosium Triangles: Evidence for a Toroidal Magnetic Moment. *Angew. Chem. Int. Ed.* **2008**, *47*, 4126–4129. [CrossRef]
45. Ungur, L.; Van den Heuvel, W.; Chibotaru, L.F. Ab initio investigation of the non-collinear magnetic structure and the lowest magnetic excitations in dysprosium triangles. *New J. Chem.* **2009**, *33*, 1224–1230. [CrossRef]
46. Chibotaru, L.F.; Ungur, L.; Aronica, C.; Elmoll, H.; Pilet, G.; Luneau, D. Structure, Magnetism, and Theoretical Study of a Mixed-Valence $Co^{II}{}_3Co^{III}{}_4$ Heptanuclear Wheel: Lack of SMM Behavior despite Negative Magnetic Anisotropy. *J. Am. Chem. Soc.* **2008**, *130*, 12445–12455. [CrossRef]
47. Lunghi, A.; Totti, F.; Sessoli, R.; Sanvito, S. The role of anharmonic phonons in under-barrier spin relaxation of single molecule magnets. *Nat. Commun.* **2017**, *8*, 14620–14626. [CrossRef]
48. Lu, F.; Ding, M.M.; Li, J.X.; Wang, B.L.; Zhang, Y.Q. Why lanthanide Er^{III} SIMs cannot possess huge energy barriers: A theoretical investigation. *Dalton Trans.* **2020**, *49*, 14576–14583. [CrossRef]
49. Lu, F.; Guo, W.X.; Zhang, Y.Q. Largely Enhancing the Blocking Energy Barrier and Temperature of a Linear Cobalt(II) Complex through the Structural Distortion: A Theoretical Exploration. *Inorg. Chem.* **2022**, *61*, 295–301. [CrossRef]

50. Ding, M.M.; Shang, T.; Hu, R.; Zhang, Y.Q. Understanding the magnetic anisotropy for linear sandwich $[Er(COT)]^+$-based compounds: A theoretical investigation. *Dalton Trans.* **2022**, *51*, 3295–3303. [CrossRef]
51. Shang, T.; Lu, F.; Tao, J.; Zhang, Y.Q. Understanding the Magnetic Relaxation Mechanism in Mixed-Valence Dilanthanide Complexes with Metal–Metal Bonding: A Theoretical Investigation. *J. Phys. Chem. A* **2023**, *127*, 3088–3095. [CrossRef]
52. Chilton, N.F.; Collison, D.; McInnes, E.J.L.; Winpenny, R.E.P.; Soncini, A. An electrostatic model for the determination of magnetic anisotropy in dysprosium. *Nat. Commun.* **2013**, *4*, 2551. [CrossRef] [PubMed]
53. Chen, Y.-C.; Tong, M.-L. Single-molecule magnets beyond a single lanthanide ion: The art of coupling. *Chem. Sci.* **2022**, *13*, 8716–8726. [CrossRef] [PubMed]
54. Malmqvist, P.Å.; Roos, B.O.; Schimmelpfennig, B. The restricted active space (RAS) state interaction approach with spin–orbit coupling. *Chem. Phys. Lett.* **2002**, *357*, 230–240. [CrossRef]
55. Heß, B.A.; Marian, C.M.; Wahlgren, U.; Gropen, O. A mean-field spin-orbit method applicable to correlated wavefunctions. *Chem. Phys. Lett.* **1996**, *251*, 365–371. [CrossRef]
56. Lines, M.E. Orbital Angular Momentum in the Theory of Paramagnetic Clusters. *J. Chem. Phys.* **1971**, *55*, 2977–2984. [CrossRef]
57. Mondal, K.C.; Sundt, A.; Lan, Y.H.; Kostakis, G.E.; Waldmann, O.; Ungur, L.; Chibotaru, L.F.; Anson, C.E.; Powell, A.K. Coexistence of Distinct Single-Ion and Exchange-Based Mechanisms for Blocking of Magnetization in a $Co^{II}{}_2Dy^{III}{}_2$ Single-Molecule Magnet. *Angew. Chem. Int. Ed.* **2012**, *51*, 7550–7554. [CrossRef]
58. Langley, S.K.; Wielechowski, D.P.; Vieru, V.; Chilton, N.F.; Moubaraki, B.; Abrahams, B.F.; Chibotaru, L.F.; Murray, K.S. A $\{Cr^{III}{}_2Dy^{III}{}_2\}$ Single-Molecule Magnet: Enhancing the Blocking Temperature through 3d Magnetic Exchange. *Angew. Chem. Int. Ed.* **2013**, *52*, 12014–12019. [CrossRef]

magnetochemistry

MDPI

Article

Enhanced Magnetocaloric Effect and Single-Molecule Magnet Behavior in a Series of Sulfur-Containing Ligand-Based Ln_9 Clusters (Ln = Gd, Tb, and Dy)

Ya-Wei Geng †, Tong Guo †, Xiao-Qin Wang and Tian Han *

School of Chemistry, Xi'an Jiaotong University, Xi'an 710049, China; gengyawei2025@163.com (Y.-W.G.); 18392876170@163.com (T.G.); 16685013752@163.com (X.-Q.W.)
* Correspondence: hantian0123@xjtu.edu.cn
† These authors contributed equally to this work.

Abstract: As an important branch of lanthanide-based complexes, clusters show unique properties in magnetocaloric effect (MCE) and single-molecule magnets (SMMs) using O/N ligands, while research on heavy p-block elements (e.g., S atom) with larger atomic radii and more diffuse p valence orbitals as coordinating atoms remains relatively scarce. Herein, using the sulfur-containing ligand of 2-pyridinethiol 1-oxide (HL), we successfully synthesized a series of hourglass-like Ln_9 clusters $[Ln_9(L)_{17}(\mu_3\text{-OH})_9(\mu_4\text{-OH})]\cdot nH_2O$ (**1**: Ln = Gd, n = 3; **2**: Ln = Tb, n = 3; **3**: Ln = Dy, n = 1). Magnetic data analysis reveals that cluster **1** shows a significant MCE, with the entropy change ($-\Delta S_m$) reaching a maximum of 34.41 J kg^{-1} K^{-1} at 2 K under ΔH = 7 T. Cluster **3**, meanwhile, exhibits distinct frequency- and temperature-dependent behavior, indicating its SMM characteristics. Interestingly, despite possessing the highest molar mass among reported Gd_9 clusters with MCE, **1** exhibits a competitive $-\Delta S_m$ value, highlighting the critical role of sulfur-containing ligand on the structure and even exchange interactions. This work offers new insights into synthesizing high-performance MCE materials and understanding magneto-structural relationships.

Keywords: lanthanide cluster; magnetocaloric effect; single-molecule magnet; sulfur-containing ligand

1. Introduction

From household refrigerators to space probes, refrigeration technology is deeply integrated into every facet of modern society [1]. While current ^{3}He gas compression technology can meet ultra-low-temperature environmental demands, it faces two long-term challenges: scarcity of ^{3}He resources [2] and the low efficiency of traditional vapor-compression refrigeration technology (5–10% of the theoretical Carnot cycle value) [3]. Therefore, developing alternatives to ^{3}He has become an urgent need to support the development of future technologies. Magnetic refrigeration [4–6] has emerged as a research focus in the cooling field due to its dual advantages of high energy efficiency and environmental friendliness. The technology's core principle relies on the magnetocaloric effect (MCE), a phenomenon where magnetic materials' temperature changes with varying magnetic fields (H), enabling cooling with 30−60% Carnot cycle efficiency while eliminating ozone-damaging emissions [7].

To achieve high magnetic entropy changes ($-\Delta S_m$), a central parameter in magnetocaloric performance, metal centers with negligible magnetic anisotropy, large spin ground state, and quenched orbital angular momentum are essential [4,8–11]. The Gd^{3+} ion, with its isotropic $^8S_{7/2}$ ground state, exemplifies these criteria and serves as a benchmark in molecular refrigerant design [4]. Efforts to enhance $-\Delta S_m$ have primarily focused on two strategies: increasing the magnetic density and engineering weak magnetic interactions [8,12]. On the one hand, the theoretical upper limit of $-\Delta S_m$ is $R\ln(2S + 1)$ (R is the gas constant and S is the spin state), typically expressed in molar units (J mol^{-1} K^{-1}) [13]. However, experimental data are often normalized to mass units (J kg^{-1} K^{-1}) or volume units (mJ cm^{-3} K^{-1}), where higher magnetic density directly correlates with larger $-\Delta S_m$. Thus, most current high-performance molecular-based magnetic refrigerants employ light ligands such as inorganic anions [14] and carboxylate ligands [15,16]. On the other hand, when magnetic interactions between metal centers are weak, $-\Delta S_m$ is a linear superposition of each ion, giving $-\Delta S_m{}^{max} = nR\ln(2S + 1)$ (n is the number of paramagnetic centers). In contrast, strong antiferromagnetic coupling reduces $-\Delta S_m$ due to the cancellation of adjacent magnetic moments, unless spin frustration [17] exists in the system; meanwhile, strong ferromagnetic coupling leads to a large ground-state spin that decreases entropy change to $R\ln(2nS + 1)$ [12,13]. The MCE performance of Gd^{3+}-based clusters depends on a delicate balance between magnetic ion density and exchange coupling. For example, **Gd_{140}** (M_W = 46730.00 g mol^{-1}, magnetic density N_{Gd}/M_W = 0.00300, Curie–Weiss constant $\theta = -2.4$ K) [15] shows $-\Delta S_m$ = 38.0 J kg^{-1} K^{-1} at 2 K for ΔH = 7 T, which is lower than **Gd_{60}**'s $-\Delta S_m$ of 48.0 J kg^{-1} K^{-1} ($\theta = -3.71$ K, N_{Gd}/M_W = 0.00302, M_W = 19871.97 g mol^{-1}) despite weaker antiferromagnetic coupling [16]. Conversely, **Gd_{24}** (M_W = 8169.48 g mol^{-1}, N_{Gd}/M_W = 0.00294, $\theta = -0.16$ K) [18], with the lowest magnetic ion density and weakest coupling, delivers $-\Delta S_m$ of 46.12 J kg^{-1} K^{-1} at 2.5 K and ΔH = 7 T, outperforming **Gd_{140}**.

Given the oxophilicity of lanthanide ions [19], employing oxygen-containing bidentate or multidentate ligands with controlled hydrolysis to generate $\mu_{2/3/4/5}$-O/OH groups promotes the aggregation of metal cores, which is one of the effective strategies for synthesizing structurally unique clusters [12,20,21]. However, synthesis involving heavier p-block elements, such as S atoms, is usually more challenging, primarily due to their reduced coordination affinity for lanthanide ions and the inherently less predictable/controllable nature of their coordination modes [22–26]. Notably, the larger atomic radii and diffuse p valence orbitals of these elements impart unique physicochemical properties to the resulting clusters [27,28], enabling structures ideal for studying magneto-structural relationships [29,30] and entropy change mechanisms. In complex cluster structures, when ligands containing heavy p-block elements coordinate in a terminal mode, they are likely to increase certain key parameters—such as intermetallic distances and metal–ligand bond lengths/angles—that directly influence magnetic coupling [31]. A substantial increase in the distance between Gd^{3+} ions, for example, may promote magnetic decoupling and reduce exchange pathways [32], ultimately leading to a significantly enhanced magnetic entropy change. Conversely, when ligands containing heavy p-block elements coordinate in a bridging mode, the diffuse p orbitals can facilitate stronger magnetic exchange coupling, thereby improving the relaxation dynamics in single-molecule magnets (SMMs) [33]. Additionally, strategic placement of soft S donors in the equatorial plane effectively weakens the transverse component of the crystal field, which contributes to enhanced SMM performance [27–30].

To this end, we employed 2-pyridinethiol 1-oxide (HL), a rarely explored ligand in lanthanide chemistry [34–40] that features both O and S donor sites, to construct lanthanide-based clusters. Using solvothermal methods, we successfully synthesized a new family of Ln_9 clusters $[Ln_9(L)_{17}(\mu_3\text{-OH})_9(\mu_4\text{-OH})]\cdot nH_2O$ (**1**: Ln = Gd, n = 3; **2**: Ln = Tb, n = 3; **3**: Ln = Dy, n = 1). Single-crystal X-ray diffraction analysis reveals that the nine metal centers adopt an hourglass geometric configuration in space. Magnetic studies of clusters

1–3 show distinct behaviors: cluster **1** features dominant antiferromagnetic interactions and a significant MCE with $-\Delta S_m$ = 34.41 J kg^{-1} K^{-1} at ΔH = 7 T and 2 K, while cluster **3** exhibits SMM behavior. This family enables comparison with previously reported MCE of Ln_9 clusters supported by O/N donor ligands [41–43] and helps to understand the influence of S coordination on cluster assembly and magnetic properties.

2. Materials and Methods

2.1. Materials and Characterization

All the starting materials were commercially available reagents of analytical grade and were used without further purification. 2-pyridinethiol 1-oxide was purchased from Aladdin, Shanghai, China; $LnCl_3{\cdot}6H_2O$ (Ln = Gd, Tb, Dy) were purchased from 3A Material, Anqing, Anhui, China; triethylamine, and all solvents were purchased from Energy Chemical, Shanghai, China. Elemental analyses (C, N, H, and S) were carried out on a EUROVECTOR EA3000 elemental analyzer (Euro Vector, Sesena Comasco, MI, Italy). The powder X-ray diffraction (PXRD) measurements were recorded on a Bruker D8 ADVANCE X-ray diffractometer (Bruker, Karlsruhe, BW, Germany). Direct-current (dc) magnetic susceptibilities were measured on a Quantum Design MPMS-squid VSM magnetometer (Quantum Design, San Diego, CA, USA). Diamagnetic corrections were made with Pascal's constants for all the constituent atoms and the sample holder. Alternating-current (ac) magnetic susceptibility data were collected on the same instrument employing a 1.5 Oe oscillating field at frequencies up to 1000 Hz.

2.2. Syntheses of Complexes 1–3

$[Gd_9(L)_{17}(\mu_3\text{-}OH)_9(\mu_4\text{-}OH)]{\cdot}3H_2O$ (**1**): A mixture of $GdCl_3{\cdot}6H_2O$ (185.85 mg, 0.5 mmol), ligand HL (127.16 mg, 1.0 mmol; HL = 2-pyridinethiol 1-oxide), 278.0 μL of triethylamine (2.0 mmol), and 8 mL of CH_3OH was sealed in a Teflon-lined autoclave, which was heated at 130 °C for 24 h. After cooling to room temperature in 48 h, light-yellow block crystals of **1** suitable for single-crystal X-ray diffraction analysis were obtained. The crystals were washed with methanol (5 × 4.0 mL) and subsequently examined under an optical microscope to separate the white flocculent impurities from the target crystals. Drying the crystals under air at room temperature afforded a desolvated and hydrated species that was analyzed as **1**. Yield: 49.6% (based on Gd). Elemental analysis calcd (%) for **1** ($C_{85}H_{84}N_{17}S_{17}Gd_9O_{30}$): C 26.98, N 6.29, H 2.24, S 14.41; found: C 26.69, N 6.30, H 2.36, S 13.92.

$[Tb_9(L)_{17}(\mu_3\text{-}OH)_9(\mu_4\text{-}OH)]{\cdot}3H_2O$ (**2**): This compound was synthesized by the same method as **1**, except that $GdCl_3{\cdot}6H_2O$ was replaced with $TbCl_3{\cdot}6H_2O$, affording light-yellow block crystals. The crystals were washed with methanol (5 × 4.0 mL) and subsequently examined under an optical microscope to separate the white flocculent impurities from the target crystals. Drying the crystals under air at room temperature afforded a desolvated and hydrated species that was analyzed as **2**. Yield: 38.7% (based on Tb). Elemental analysis calcd (%) for **2** ($C_{85}H_{84}N_{17}S_{17}Tb_9O_{30}$): C 26.87, N 6.27, H 2.23, S 14.35; found: C 26.73, N 6.26, H 2.20, S 13.92.

$[Dy_9(L)_{17}(\mu_3\text{-}OH)_9(\mu_4\text{-}OH)]{\cdot}H_2O$ (**3**): This compound was synthesized by the same method as **1**, except that $GdCl_3{\cdot}6H_2O$ was replaced with $DyCl_3{\cdot}6H_2O$, affording light-yellow block crystals. The crystals were washed with methanol (5 × 4.0 mL) and subsequently examined under an optical microscope to separate the white flocculent impurities from the target crystals. Drying the crystals under air at room temperature afforded a desolvated species that was analyzed as **3**. Yield: 42.8% (based on Dy). Elemental analysis calcd (%) for **3** ($C_{85}H_{80}N_{17}S_{17}Dy_9O_{28}$): C 26.90, N 6.27, H 2.12, S 14.36; found: C 26.54, N 6.76, H 2.24, S 14.22.

2.3. *Single-Crystal Structure Determination*

Single-crystal X-ray diffraction data were recorded on a Bruker D8 Venture diffractometer (Bruker, Karlsruhe, BW, Germany) with MoKα radiation. Diffraction data for crystals of **1**, **2**, and **3** were collected at 150 K using the APEX4 program. The structures were solved by the SHELXT-2024 software package and refined using a full-matrix least-squares method on F^2 with anisotropic thermal parameters for all nonhydrogen atoms. Hydrogen atoms were located geometrically and refined with a riding model with relative isotropic displacement parameters. Structural refinement revealed disorder and pronounced thermal motion in some lattice solvent molecules, triggering CheckCIF alerts (e.g., missing O–H acceptors). These effects are common in highly solvated crystals and do not compromise the integrity of the main framework. Details can be found in the Cambridge Crystallographic Data Centre using the CCDC numbers 2475935 (**1**), 2475936 (**2**), and 2475937 (**3**).

3. Results and Discussion

3.1. *Crystal Structure*

Single-crystal X-ray diffraction analyses reveal that complexes **1** and **2** are isostructural, both crystallizing in the triclinic *P*-1 space group, whereas complex **3** crystallizes in the monoclinic $P2_1/c$ space group (Table S1). Thus, the structures of complexes **1** and **3** are described in detail.

The asymmetric units of both complexes **1** and **3** feature nine metal ions, seventeen L^- ligands, nine μ_3-OH and one μ_4-OH groups, and some free solvent molecules. The eight Ln^{3+} (Ln1–Ln4 and Ln6–Ln9, Ln = Gd, Dy) ions occupy the vertices of a square antiprism, and the ninth (Ln5) resides at its geometric center, forming a canonical hourglass geometric configuration. This configuration consists of two quadrilateral base planes and eight triangular planes, resulting in an $[Ln_9(\mu_3\text{-OH})_9(\mu_4\text{-OH})]^{17+}$ core encapsulated by seventeen bidentate L^- ligands. To facilitate the description of ligand coordination modes, three types are identified: (1) μ_1-η^1:η^1: terminal bidentate coordination through both S and O donors to a single lanthanide center; (2) μ_2-η^1:η^2: monodentate S-coordination to one lanthanide ion coupled with μ_2-O bridging to a neighboring metal center; and (3) μ_2-η^1:η^2: monodentate O-coordination to one lanthanide ion accompanied by μ_2-S bridging to an adjacent metal ion (Figure 1).

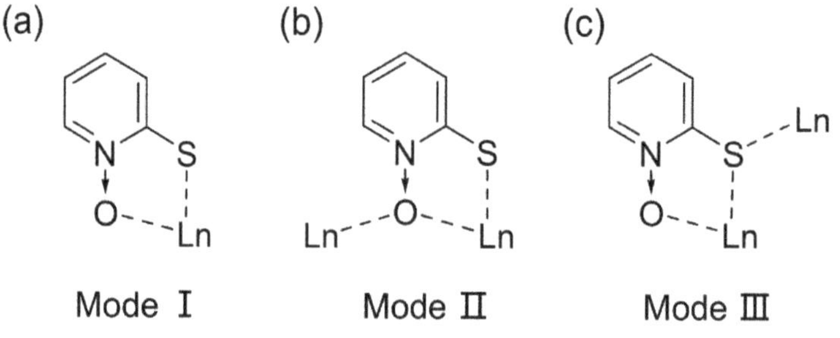

Figure 1. Ligand coordination modes: Modes I (**a**), II (**b**), and III (**c**).

In complex **1**, a μ_4-OH group bridges the four Gd^{3+} ions (Gd1–Gd4) of the upper quadrilateral base plane (O atom 0.57 Å above the plane (Figure 2a); adjacent Gd···Gd distances in the range of 3.6107(6)–3.6700(6) Å). Adjacent Gd^{3+} ions in this plane are further interconnected by μ_2-η^1:η^2 bridges from ligand L^- (Mode II: μ_2-O and monodentate S), with the exception of the Gd1–Gd4 edge, where ligand disorder results in mixed μ_2-η^1:η^2 bridging (approximately 60% Mode II (μ_2-O) and 40% Mode III (μ_2-S)) (Figure 3a). Four μ_3-OH groups connect the upper quadrilateral plane to Gd5, forming four triangular planes.

The Gd1–Gd4 ions are all eight-coordinate (Figure S1), adopting a triangular dodecahedron (D_{2d}) local geometric configuration as determined by the SHAPE 2.1 software [44] (Table S2), with each coordination sphere containing two μ_2-η^1:η^2 ligands, one μ_1-η^1:η^1 ligand, one μ_4-OH, and two μ_3-OH groups. The central Gd5 ion occupies the geometric center, coordinated to eight μ_3-OH groups that connect it to both the upper (Gd1–Gd4) and lower (Gd6–Gd9) quadrilateral planes, adopting a square antiprism (D_{4d}) local geometry. The centers of the two quadrilateral base planes are separated by 5.5476 (2) Å with an interplanar torsion angle of 47.318(105)°. The lower quadrilateral base plane (Gd6–Gd9) exhibits distinct structural features (Figure S1). A μ_3-OH group (O atom 0.53 Å below the plane) bridges Gd6, Gd7, and Gd9, while the plane displays significant distortion from ideal square geometry (adjacent Gd···Gd distances ranging from 3.5658(7) to 3.9598(7) Å and the bond lengths of Gd–μ_3-O varying between 2.413(4) and 2.540(1) Å; Tables S3 and S4). This geometric asymmetry results in varied coordination environments for Gd6–Gd9 compared to the more symmetric upper Gd1–Gd4 plane, highlighting the overall molecular asymmetry. While adjacent Gd^{3+} ions in the lower plane are similarly interconnected by μ_2-η^1:η^2 bridges from ligand L^- (Mode II: μ_2-O and monodentate S), their coordination environments diverge significantly. Gd6 and Gd7 maintain eight-coordinate triangular dodecahedron (D_{2d}) geometries, each binding to two μ_2-η^1:η^2 ligands (combing one S atom and two μ_2-O), one μ_1-η^1:η^1 L^- ligand, and three μ_3-OH groups. In contrast, Gd9 expands to nine-coordination through two μ_2-η^1:η^2 modes (two S atoms and two μ_2-O), one μ_1-η^1:η^1 L^- ligand, and three μ_3-OH groups, resulting in a spherical capped square antiprism (C_{4v}) local geometry. Gd8 remains eight-coordinate with a triangular dodecahedron (D_{2d}) local geometry but exhibits a distinct arrangement, coordinating to two μ_2-O donors (from two μ_2-η^1:η^2 L^-), two μ_1-η^1:η^1 L^- ligands, and two μ_3-OH groups.

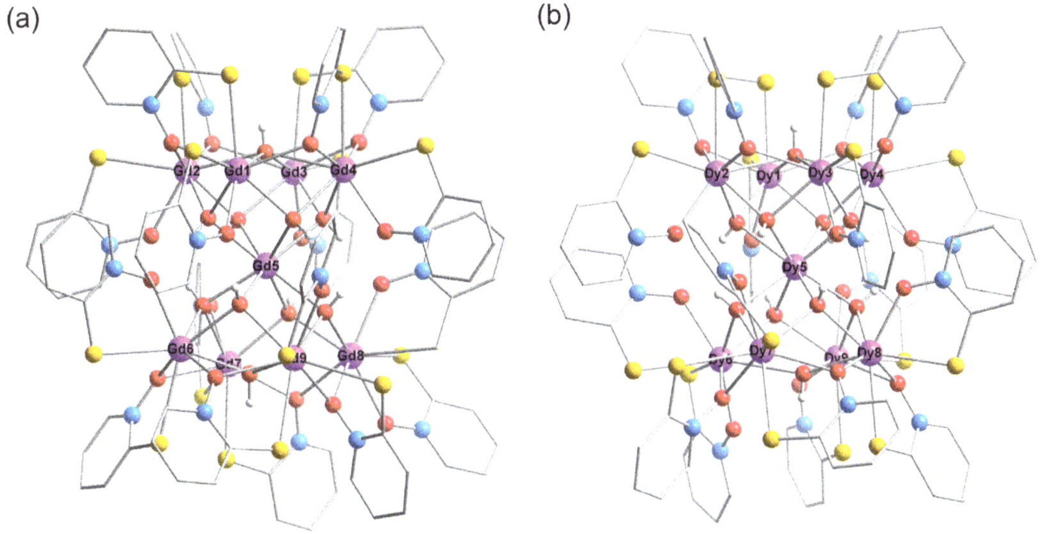

Figure 2. The crystal structure diagram of **1** (**a**) and **3** (**b**). Color code: O (red), C (gray), N (blue), Gd/Dy (purple), S (yellow), and H (white). A disordered L^- with μ_2-S (Mode III) coordination in **1,** disordered L^- ligands in **3**, and free solvent molecules and hydrogen atoms of pyridine rings in **1** and **3** are omitted for clarity.

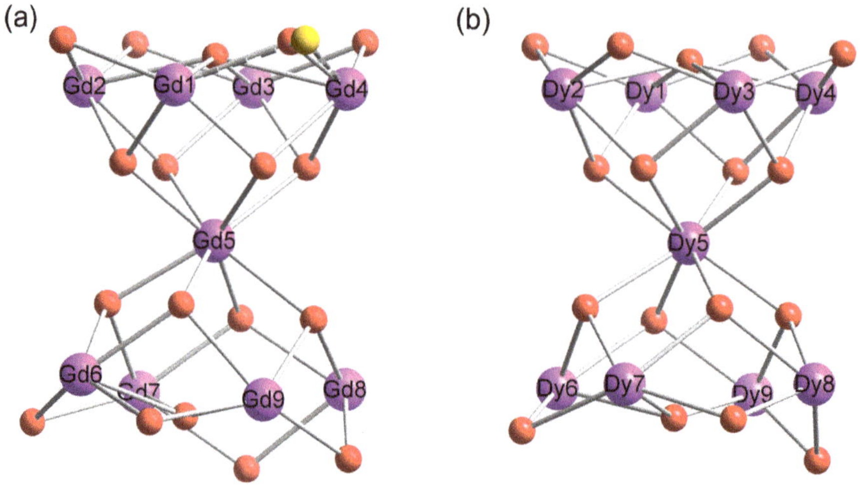

Figure 3. Representation of the core structure of **1** (**a**) and **3** (**b**). Color code: O (red), Gd/Dy (purple), and S (yellow). The gray dotted line represents the disordered components within the structure.

According to the packing diagram (Figure S2), the shortest intermolecular Gd$\cdots$Gd distance is 9.0353(7) Å. The presence of hydroxyl groups and solvent molecules within the cluster forms intra- and intermolecular hydrogen bonds (Table S5). Further analysis reveals that the pyridine rings mediate intra- and intermolecular $\pi\cdots\pi$ stacking (dihedral angle < 20°) (Table S6) and intermolecular C–H$\cdots\pi$ interactions, with distances between the hydrogen atoms and the ring centroids ranging from 2.68 to 2.98 Å (Table S7). These noncovalent interactions notably enhance the thermal and structural stability of the air-stable metal clusters.

In complex **3**, the spatial arrangement of metal ions (Dy1–Dy9) and their linkage pattern across the eight triangular planes are identical to those in complex **1**, forming a $[Dy_9(\mu_3\text{-}OH)_9(\mu_4\text{-}OH)]^{17+}$ core structure (Figures 2b and 3b). Notably, all Dy^{3+} ions are eight-coordinate. Dy1−Dy5 display coordination geometries that closely resemble those of their Gd^{3+} analogs (Gd1–Gd5), with Dy1–Dy4 adopting D_{2d} local geometries and Dy5 exhibiting a D_{4d} geometry (Table S8). Dy9 shows identical coordination to Gd6 and Gd7 (Figure S3), also featuring a D_{2d} local geometry. However, Dy6 and Dy7 share a distinct coordination sphere comprising one μ_3-OH from the lower plane, two μ_3-OH bridges to central Dy5, one μ_2-O donor (from one μ_2-η^1:η^2 L^-), and two μ_1-η^1:η^1 L^- ligands (Figure S4). Despite this similarity, their local geometries differ slightly, adopting D_{4d} and D_{2d} configurations, respectively. Dy8 adopts a unique environment, coordinated by two μ_3-OH groups, two μ_2-η^1:η^2 L^- ligands (containing two S and two μ_2-O), and one μ_1-η^1:η^1 L^- ligand, exhibiting a D_{2d} local geometry. The adjacent Dy$\cdots$Dy distances for Dy6–Dy9 range from 3.5824(7) to 3.8408(7) Å, and the Dy−μ_3-O bond lengths display a wider variation (2.470(5)–2.715(5) Å) than those found in **1** (Tables S9 and S10). The packing analysis reveals that the shortest intermolecular Dy$\cdots$Dy distance is 9.8618(6) Å (Figure S5), longer than that in **1**. The structure also features an enhanced hydrogen bonding network (Table S5). Notably, the pyridine rings are only engaged in intramolecular $\pi\cdots\pi$ stacking (Table S6) and intermolecular C–H$\cdots\pi$ interactions, with distances between the hydrogen atom and the ring centroid ranging from 2.66 to 2.85 Å (Table S7).

Overall, in contrast to previously reported symmetric hourglass-like Ln_9 clusters [24,41–43,45–69], the present Ln_9 family exhibits significant structural asymmetry: Ln1–Ln4 and Ln6–Ln9 differ entirely in coordination environments, while the upper and lower basal planes feature different bridging modes (μ_4-OH and μ_3-OH, respectively). To facilitate subsequent studies on magneto-structural relationships, we compared cluster **1** with four previously reported Gd_9 clusters that exhibit MCE [41–43]. Gd1–Gd5 share similar bridging modes: adjacent Gd1–Gd4 ions are connected by μ_2-O bridges (Gd···Gd distance ≈ 3.6 Å) with a central μ_4-OH group (diagonal Gd···Gd distance in the upper plane ≈ 5.2 Å), while Gd5 links to eight Gd^{3+} ions via eight μ_3-OH bridges ($Gd_{1-4}\cdots Gd_5$ distance ≈ 3.8 Å). Despite these similarities in connectivity and distances, cluster **1** exhibits distinct bond angles compared to the four cases: its adjacent Gd–μ_2-O–Gd angles for Gd1–Gd4 (96.1–100.6°) are in a medium range (Table S11), its diagonal Gd–μ_4-O–Gd angles in the upper plane (155.2–156.2°) are smaller, and its Gd_{1-4}–μ_3-O–Gd_5 angles (107.0–108.1°) show a larger average (Table S11). These differences become even more pronounced in Gd6–Gd9, which display greater variation in adjacent Gd···Gd distances (3.6–4.0 Å) with Gd–μ_2-O–Gd angles (93.3–107.1°, Table S11). Here, the diagonal Gd6···Gd8 and Gd7···Gd9 distances are 5.7 Å and 4.8 Å, respectively, and the Gd_{6-9}–O–Gd_5 angles range from 102.4 to 109.9°. Considering that the key difference in cluster **1** lies in the partial replacement of conventional N/O donors by S atoms from the ligands, with a disordered segment containing μ_2-S bridging Gd1 and Gd4, we propose that the asymmetric structure and variations in bond angles and Gd···Gd distances likely arise from the coordination and bridging of sulfur-containing ligands.

The powder X-ray diffraction patterns of polycrystalline samples of **1** and **2** are distinctly different from that of **3**, matching well with their respective single-crystal X-ray diffraction simulations (Figure S6). This agreement confirms the high phase purity of all samples.

3.2. Magnetic Properties

Variable-temperature magnetic susceptibility measurements of complexes **1–3** were performed on polycrystalline samples under a 1 kOe direct current (dc) magnetic field over the temperature range of 2–300 K. At 300 K, the $\chi_M T$ products for complexes **1–3** are 68.26, 104.02, and 125.64 cm^3 K mol^{-1}, respectively (Figure 4a), in close agreement with the theoretical values calculated for non-interacting magnetic centers: 70.92 $cm^3 \cdot K \cdot mol^{-1}$ for nine independent Gd^{3+} ions ($S = 7/2$, $g = 2.0$), 106.38 $cm^3 \cdot K \cdot mol^{-1}$ for nine isolated Tb^{3+} ions ($S = 3$, $g = 3/2$), and 127.53 cm^3 K mol^{-1} for nine non-coupled Dy^{3+} ions ($S = 5/2$, $g = 4/3$). Upon cooling, complex **1** exhibits nearly constant $\chi_M T$ values above 80 K, reflecting the dominant paramagnetic behavior of non-interacting Gd^{3+} ions; below 80 K, $\chi_M T$ decreases gradually before undergoing a steep decline to 34.66 cm^3 K mol^{-1} at 2 K, indicating the onset of antiferromagnetic interactions. Curie–Weiss analysis (2–300 K) yields Curie constant C = 69.35 cm^3 K mol^{-1} and Curie temperature $\theta = -2.15$ K, with the negative θ confirming antiferromagnetic interactions between metal ions (Figure 4b). Complex **2** exhibits a subtle decrease in $\chi_M T$ above 100 K, while complex **3** shows a more pronounced reduction in this regime. Both systems undergo progressive declines followed by steep reductions below 50 K, reaching minima of 39.12 and 76.30 cm^3 K mol^{-1} at 2 K for **2** and **3**, respectively. This behavior originates from thermal depopulation of Stark sublevels, with potential contributions from weak antiferromagnetic exchange.

To further explore the magnetic interactions between Gd^{3+} ions in **1**, we classify the exchange pathways into six distinct types (J_1–J_6) based on their bridging motifs (Figure 5a). The magnetic interaction denoted as J_1 stems from the coupling between Gd1 and Gd2, which are bridged by one μ_4-OH, one μ_3-OH, and a μ_2-O ligand. Notably, ligand disorder along the Gd1–Gd4 edge, which leads to a mixed μ_2-η^1:η^2 bridging mode, is not specifically considered in the magnetic model, similar to J_1. J_2 characterizes the diagonal magnetic interaction between Gd^{3+} ions in the upper Gd1–Gd4 quadrilateral base plane via a μ_4-OH bridge. J_3 describes the magnetic interaction between Gd5 and its adjacent Gd^{3+} ions through two μ_3-OH groups. J_4 accounts for the magnetic coupling between Gd^{3+} ions mediated by two μ_3-OH and one μ_2-O ligand. J_5 quantifies the interaction between adjacent Gd^{3+} ions bridged by one μ_3-OH and one μ_2-O ligand. Finally, J_6 defines the magnetic exchange interaction between Gd7 and Gd9, which are connected by a single μ_3-OH bridge. Thus, the spin Hamiltonian of the system is derived by integrating these exchange interactions.

$$\hat{H} = -2J_1(S_1S_2 + S_2S_3 + S_3S_4 + S_1S_4) - 2J_2(S_1S_3 + S_2S_4) - 2J_3(S_1S_5 + S_2S_5 + S_3S_5 + S_4S_5 + S_6S_5 + S_7S_5 + S_8S_5 + S_9S_5) - 2J_4(S_6S_7 + S_6S_9) - 2J_5(S_7S_8 + S_9S_8) - 2J_6(S_7S_9) \quad (1)$$

In this model, S_1–S_9 represent the spin of Gd1–Gd9, respectively, where positive values of J indicate ferromagnetic interactions, and negative values correspond to antiferromagnetic interactions.

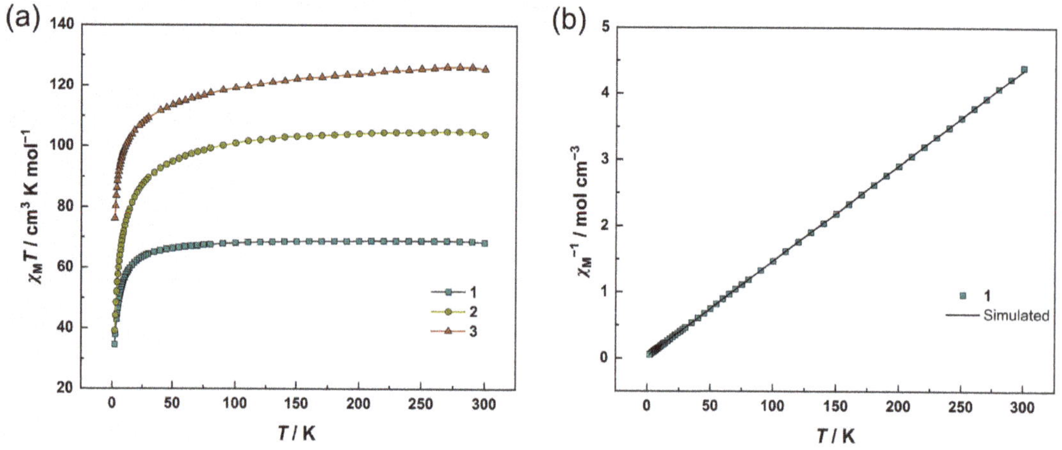

Figure 4. (**a**) Temperature-dependent susceptibilities for **1–3** under a dc magnetic field of 1000 Oe; (**b**) the χ_M^{-1} vs. T plot of **1**. The solid black line represents the best fits with the Curie–Weiss law.

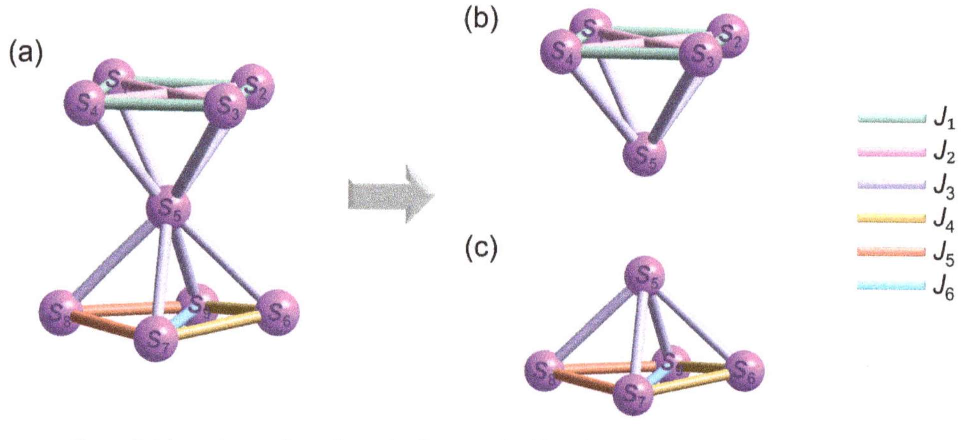

Figure 5. Magnetic coupling schematic diagram of **1** (**a**), **Gd$_5$** including Gd1–Gd5 (**b**), and **Gd$_5$** including Gd5–Gd9 (**c**) systems in **1**. Color code: Gd (purple), J_1 (cyan), J_2 (pink), J_3 (light purple), J_4 (yellow), J_5 (orange), and J_6 (blue).

To address computational challenges in modeling the magnetic properties of a nonanuclear cluster, we implement a reduced pentanuclear approximation. Given the prohibitively large exchange matrix for the full Gd_9 system, we strategically retain the core magnetic unit comprising the central Gd5 ion and four vertex ions (Gd_5: Gd1–Gd4 from the upper plane or Gd6–Gd9 from the lower plane) (Figure 5b,c). This simplification exploits the inherent S_4 symmetry of the cluster, where Gd5 occupies the inversion center and the upper/lower Gd_4 planes exhibit congruent coordination environments. The resultant Hamiltonian matrix enables computationally tractable fitting while preserving the dominant exchange pathways.

$$\hat{H} = -2J_1(S_1S_2 + S_2S_3 + S_3S_4 + S_1S_4) - 2J_2(S_1S_3 + S_2S_4) - 2J_3(S_1S_5 + S_2S_5 + S_3S_5 + S_4S_5) \quad (2)$$

$$\hat{H} = -2J_3(S_6S_5 + S_7S_5 + S_8S_5 + S_9S_5) - 2J_4(S_6S_7 + S_6S_9) - 2J_5(S_7S_8 + S_9S_8) - 2J_6(S_7S_9) \quad (3)$$

Considering the approximate treatments applied during the fitting, the $\chi_M T$ data for both Gd_5 models were derived by converting the $\chi_M T$ values of **1** (M_W = 3784.03 g mol^{-1}): $\chi_M T$ (Gd_5) = $\chi_M T$ (**1**)/9 × 5 (Figure S7). The optimal fitting of the experimental $\chi_M T$ vs. T data yields the magnetic parameters $J_1 = -0.055$ cm^{-1}, $J_2 = -0.114$ cm^{-1}, $J_3 = 0.019$ cm^{-1}, and $g = 1.979$ using Equation (2), and $J_3 = 0.015$ cm^{-1}, $J_4 = -0.097$ cm^{-1}, $J_5 = -0.082$ cm^{-1}, $J_6 = -0.072$ cm^{-1}, and $g = 1.980$ using Equation (3) [70]. These results are consistent with established magneto-structural relationships [42,71,72]. The ferromagnetic J_3 coupling originates from μ_3-OH bridges, aligning with empirical models where the Gd–O–Gd bond angles between 105 and 135° favor ferromagnetic interactions in Gd1–Gd5 (Table S4, bold font) [42,71,72]. However, in Gd5–Gd9, only 62.5% of the Gd–O–Gd bond angles fall within the 105–135° range. This may account for the smaller magnitude of the J_3 in Equation (3) compared to that in Equation (2). In contrast, the antiferromagnetic J_1, J_2, J_4, and J_6 arise from μ_4-OH/μ_3-OH/μ_2-O bridges, which is consistent with the correlation that excessively large or small bond angles result in antiferromagnetic coupling (Table S4, underlined font and wavy font, respectively). Compared with other hourglass-like Ln_9 clusters [24,41–43,45–69], the structural differences in compound **1**—beyond variations in ligands and symmetry—are mostly reflected in bond angles and Gd···Gd distances. For the upper base, the presence of μ_4-OH bridging groups restricts the ability of sulfur-containing ligands to modify the distances between central metal ions during coordination. For the lower base, however, as only three Gd^{3+} ions are connected by one μ_3-OH bridging group, the influence of sulfur-containing ligands becomes more prominent. They widen the Gd···Gd distances (Gd7···Gd8 = 3.96 Å, Gd8···Gd9 = 3.87 Å), make contributions to reducing magnetic exchange pathways [32], and also affect the distribution of Gd–O–Gd bond angles (Table S11). All these factors directly influence the magnetic coupling between metal ions.

The field-dependent magnetization of complex **1** was measured at discrete temperatures (2–10 K) under applied magnetic fields up to 7 T (Figure 6a). The magnetization (M) versus field curves exhibit a nearly linear increase at low fields (< 2 T), followed by a gradual approach to saturation, reaching a value of 62.13 Nβ at 2 K and 7 T, which is in close agreement with the theoretical saturation magnetization of 63 Nβ for nine non-interacting Gd^{3+} ions. Given the weak magnetic interactions observed in **1** and the well-known MCE in Gd^{3+}-based clusters, the $-\Delta S_m$ values were calculated by the magnetic data using the Maxwell equation:

$$\Delta Sm(T) = \int [\partial M(T, H)/\partial T]_H dH \quad (4)$$

where T is the testing temperature, H is the applied magnetic field, and M is magnetization.

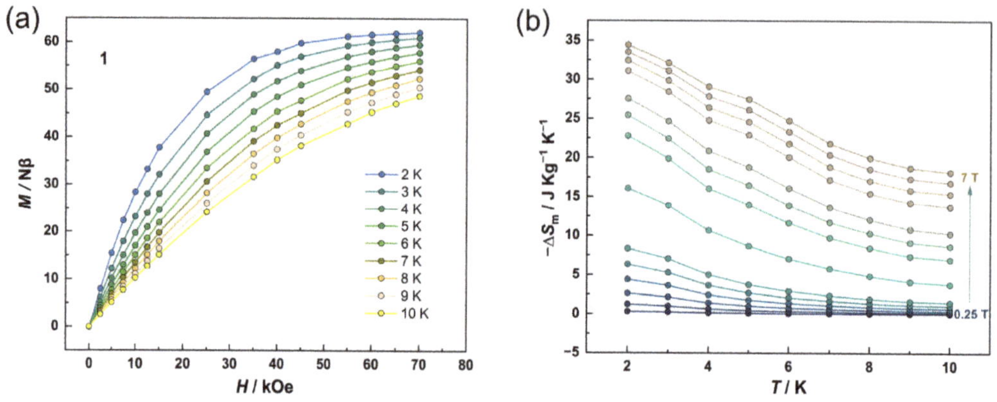

Figure 6. (**a**) Plots of M vs. H at various temperatures for **1**. (**b**) Plots of experimental magnetic entropy change versus T for **1** at 0.25–7 T from 2 to 10 K.

As shown in Figure 6b, the $-\Delta S_m$ values demonstrate a monotonic increase with decreasing temperature and increasing magnetic field change (ΔH), reaching a maximum value of $-\Delta S_m$ = 34.41 J kg^{-1} K^{-1} at 2 K under ΔH = 7 T (Figure 6b). This value is approximately 84% of the theoretical limit of 41.2 J kg^{-1} K^{-1} calculated by $-\Delta S_m{}^{max} = nR\ln(2S + 1)/M_W$, with the reduction primarily attributed to the presence of weak interactions between adjacent metal ions. Notably, while moderate among molecular magnetic coolers generally [4,8–11], this $-\Delta S_m$ value ranks as the second highest reported for Gd_9 clusters (Table S12). When the number of central metal ions is the same, magnetic density—a key parameter that determines entropy change—can be readily inferred from the molecular weight. For instance, **1** has the largest molar mass (M_W = 3784.03 g mol^{-1}), corresponding to the lowest value of N_{Gd}/M_W among these Gd_9 clusters [41,42]. In MCE materials research, a larger molar mass and lower magnetic density are generally considered unfavorable for achieving high $-\Delta S_m$ values, as a lower concentration of magnetic ions per unit mass tends to limit the degree of magnetic ordering that contributes to entropy change under an external magnetic field [4]. Despite this, **1** exhibits a $-\Delta S_m$ value comparable to that of $Gd_9(L^1)_4(\mu_4\text{-}OH)_2(\mu_3\text{-}OH)_8(\mu_2\text{-}OCH_3)_4(NO_3)_8(H_2O)_8\}(OH)\cdot 2H_2O$ (HL^1 = 2,6-dimethoxyphenol, M_W = 3015.28 g mol^{-1}). This compound, although not the lightest, displays weak antiferromagnetic and ferromagnetic interactions and holds the highest $-\Delta S_m$ value (40.60 J kg^{-1} K^{-1} at 2 K under ΔH = 7 T) reported for Gd_9 clusters to date (Table S12) [41–43]. Meanwhile, **1** has a $-\Delta S_m$ slightly larger than that of the third-ranked cluster, $[Gd_9(L^2)_4(NO_3)_{12}(\mu_3\text{-}OH)_8(\mu_4\text{-}OH)_2(CH_3CH_2OH)_4]\cdot 4.5CH_3CH_2OH\cdot 3Et_3NH$ (H_2L^2 = *N*-methyldiethanolamine, M_W = 3288.90 g mol^{-1}, $-\Delta S_m$ = 33.65 J kg^{-1} K^{-1} at 3 K under ΔH = 7 T) [41], and much larger than that of the fourth-ranked cluster, $[Gd_9(L^3)_8(NO_3)_8(\mu_3\text{-}OH)_8(\mu_4\text{-}OH)_2]\cdot NO_3$ (HL^3 = 2-(2-aminoethoxy)ethanol; acac = acetylacetonate; M_W = 2976.45 g mol^{-1}, $-\Delta S_m$ = 25.22 J kg^{-1} K^{-1} at 3 K under ΔH = 7 T) [41] with the highest magnetic density and lowest molecular weight in Gd_9 clusters with MCE (Table S12).

Research findings reveal that both magnetic density and magnetic interactions affect the $-\Delta S_m$ value. For complex **1**, magnetic density seems to play a minor role in determining the large $-\Delta S_m$. We thus propose three potential factors contributing to its enhanced magnetocaloric performance. First, the incorporation of the heavy p-block S atoms from the HL ligands alters the cluster architecture, particularly affecting the Gd^{3+} ions at the vertices of the upper and lower basal planes, which are bridged by a μ_4-OH and μ_3-OH, respectively. Next, these sulfur-containing ligands in terminal coordination mode modulate key structural parameters—specifically Gd–$\mu_2/\mu_3/\mu_4$-O–Gd bond angles and Gd···Gd distances—rendering the structure distinct from previously reported Gd_9 clusters ligated solely by N/O donors. Last, these geometric perturbations are expected to influence the nature and strength of magnetic exchange pathways, including both ferromagnetic and antiferromagnetic interactions. Collectively, these effects enhance the magnetic entropy change, despite the higher atomic mass of sulfur compared to conventional N/O-coordinating atoms.

Magnetization versus field measurements for complex **2** at 2 K reveal a gradual increase to a maximum moment of 46.31 Nβ at 7 T without a saturation plateau (Figure S8). Alternating-current (ac) magnetic susceptibility measurements under zero dc field with an oscillating field of 1.5 Oe show no detectable out-of-phase magnetic susceptibility (χ'') signal down to 2 K, suggesting the absence of magnetic relaxation (Figure S9). Consistently, hysteresis measurements at 2 K at a sweep rate of 200 Oe s^{-1} exhibit vanishingly small coercivity (Figure S10). This collective behavior precludes SMM behavior under these conditions.

Magnetization versus field measurements for complex **3** at 2 K show a gradual approach to 50.91 Nβ at 7 T (Figure S11). The absence of distinct saturation and temperature-dependent non-superposition of $M(H)$ curves possibly originates from strong intrinsic magnetic anisotropy. Frequency dependence of χ'' under 12 K confirms the slow magnetic relaxation characteristic of SMM behavior (Figures 7 and S12). However, the absence of a discernible peak in the χ'' signal prevents the determination of relaxation time and thus the relaxation energy barrier. This behavior is similar to most reported nonanuclear Dy^{3+}-based compounds [48,50,56,63]—except for one Dy_9 compound with an energy barrier of U_{eff} = 9.24 K [47]. Zero-field-cooled and field-cooled (ZFC–FC) susceptibility measurements at H = 50 Oe show no divergence, ruling out long-range magnetic ordering or superparamagnetic transitions (Figure S13). Additionally, hysteresis measurements at 2 K at a sweep rate of 200 Oe s^{-1} exhibit minimal coercivity (Figure S14).

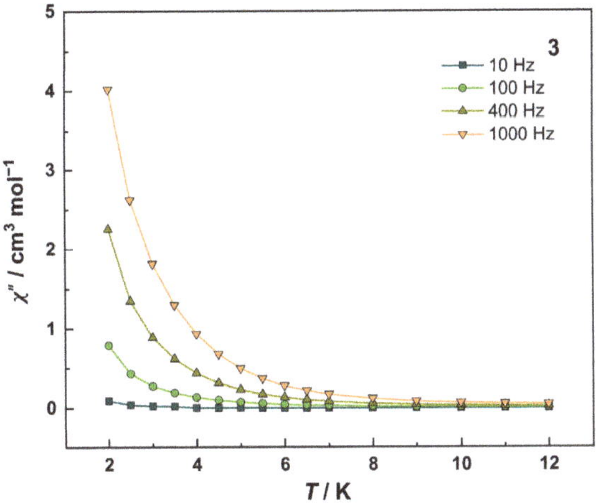

Figure 7. Temperature dependence of the χ'' ac susceptibility for **3** under zero dc field in the temperature range of 2–12 K.

4. Conclusions

In summary, three nonanuclear lanthanide-based clusters $[Ln_9(L)_{17}(\mu_3\text{-}OH)_9(\mu_4\text{-}OH)]\cdot nH_2O$ (**1**: Ln = Gd, n = 3; **2**: Ln = Tb, n = 3; **3**: Ln = Dy, n = 1) were synthesized under hydrothermal conditions with sulfur-containing bidentate ligand 2-pyridinethiol 1-oxide. Remarkably, all three clusters exhibit a unique asymmetric architecture in which the upper and lower Ln_9 cores are bridged by μ_4-OH and μ_3-OH groups, respectively. Of these three clusters, 1 exhibits an excellent cryogenic MCE with $-\Delta S_m$ = 34.41 J kg^{-1} K^{-1} at 2 K under ΔH = 7 T, and 3 shows the SMM behavior. It is worth noting that, compared with the other four reported Gd_9 clusters with MCE, 1 exhibits a magnetic entropy change value that is second only to the highest among Gd_9 systems, despite having the largest molar mass and the lowest magnetic density, indicating the dominant role of magnetic coupling in determining the magnetocaloric performance. Specifically, the sulfur-containing bidentate ligands, coordinated in a terminal mode, influence key structural parameters of the nonanuclear cluster, including increasing the distances between Gd^{3+} ions, modifying the Gd–O–Gd bond angles, and reducing the number of bridging groups. This work not only expands research on the MCE of Gd_9 clusters but also provides a new synthetic strategy for constructing high-performance molecular-based magnetic cooling materials using ligands containing heavy p-block elements.

Supplementary Materials: The following supporting information can be downloaded at https://www.mdpi.com/article/10.3390/magnetochemistry11090070/s1. Figure S1: The coordination environment diagram of compound **1**. Color code: O (red), Gd (purple), S (yellow), and H (white). A disordered L^- with μ_2-S (Mode III) coordination, free solvent molecules, and pyridine rings are omitted for clarity; Figure S2: The stacking pattern within a unit cell of **1**. Color code: O (red), C (gray), N (blue), Gd (purple), and S (yellow). Disordered L^- with μ_2-S, free solvent molecules, and H atoms are omitted for clarity; Figure S3: The part crystal structure diagram of compound **1** (**a**) and **3** (**b**) containing Ln6–Ln9. Color code: O (red), C (gray), N (blue), Gd/Dy (purple), S (yellow), and H (white). The extra carbon atoms on the pyridine rings with their hydrogen atoms and free solvent molecules are omitted for clarity; Figure S4: The coordination environment diagram of compound **3**. Color code: O (red), C (gray), N (blue), Dy (purple), S (yellow), and H (white). Disordered L^- ligands, free solvent molecules, and pyridine rings are omitted for clarity; Figure S5: The stacking pattern within a unit cell of **3**. Color code: O (red), C (gray), N (blue), Dy (purple), and S (yellow). Unordered L^- ligands, free solvent molecules, and H atoms are omitted for clarity; Figure S6: The PXRD curves of fresh samples of **1**–**3**. The two calculated curves are simulated from the single crystal data of **1** and **3**; Figure S7. Temperature dependence of χ_M and $\chi_M T$ for Gd_5: Gd1–Gd5 (**a**) and Gd5–Gd9 (**b**). The experimental data of Gd_5 were obtained by converting the data of **1**, while the solid lines represent the best fits through the PHI program [70]; Figure S8: The plot of M vs. H at 2 K for **2**; Figure S9: Temperature dependence of the in-phase (χ') (**a**) and out-of-phase (χ'') (**b**) ac susceptibility for **2** under zero dc field in the temperature range of 2–12 K; Figure S10: Hysteresis loop for **2** measured at 2 K with a sweep rate of 200 Oe/s; Figure S11: The plots of M vs. H at various temperatures for **3**; Figure S12: Temperature dependence of the in-phase (χ') (**a**) and out-of-phase (χ'') (**b**) ac susceptibility for **3** under zero dc field in the temperature range of 2–12 K; Figure S13: FC and ZFC of **3** at an applied field of 50 Oe; Figure S14: Hysteresis loop for **3** measured at 2 K with a sweep rate of 200 Oe/s. Illustration: an enlarged view of the area near the zero field. Table S1: Crystallographic data and structure refinement for **1**–**3**; Table S2: Continuous Shape Measures (CShMs) calculation for **1**; Table S3: Selected bond lengths (Å) for **1**; Table S4: Selected angles (deg) for **1**; Table S5: Hydrogen bonding parameters in clusters **1**–**3**; Table S6: $\pi\cdots\pi$ interaction parameters (Å, deg) for **1** and **3**; Table S7: C–H$\cdots\pi$ interaction parameters (Å, deg) for clusters **1** and **3**; Table S8: Continuous Shape Measures (CShMs) calculation for **3**; Table S9: Selected bond lengths (Å) for **3**; Table S10: Selected angles (deg) for **3**; Table S11: The selected Gd–O–Gd angles of the reported Gd9 clusters with MCE [41–43]; Table S12: The MCE properties of the previously reported Gd_9 clusters [41–43].

Author Contributions: Conceptualization, T.H.; methodology, Y.-W.G. and T.G.; investigation, Y.-W.G. and X.-Q.W.; data curation, T.G. and Y.-W.G.; writing—original draft preparation, Y.-W.G.; writing—review and editing, T.H.; supervision, T.H.; funding acquisition, T.H. Of the authors, Y.-W.G. and T.G. contributed equally to this work. All authors have read and agreed to the published version of the manuscript.

Funding: This research was funded by the National Natural Science Foundation of China (funding number: 22271228) and the Fundamental Research Funds for the Central Universities (funding number: xzy012023031).

Institutional Review Board Statement: Not applicable.

Informed Consent Statement: Not applicable.

Data Availability Statement: The original contributions presented in this study are included in the article and the Supplementary Materials.

Acknowledgments: We thank Chang Huang and Dan He at the Instrument Analysis Center of Xi'an Jiaotong University for assistance with single-crystal X-ray diffraction measurement and magnetic measurement, respectively.

Conflicts of Interest: The authors declare no conflicts of interest.

References

1. Lucia, U.; Grisolia, G. Magnetocaloric refrigeration in the context of sustainability: A review of thermodynamic bases, the state of the art, and future prospects. *Energies* **2024**, *17*, 3585. [CrossRef]
2. Zhao, S.; Wang, P.; Chen, W.; Wang, L.; Wang, Q.C.; Chen, W.Q. Supply and demand conflicts of critical heavy rare earth element: Lessons from gadolinium. *Resour. Conserv. Recycl.* **2023**, *199*, 1–9. [CrossRef]
3. Nehan, P.Z.Z.; Vitayaya, O.; Munazat, D.R.; Manawan, M.T.E.; Darminto, D.; Kurniawan, B. The magnetocaloric effect properties for potential applications of magnetic refrigerator technology: A review. *Phys. Chem. Chem. Phys.* **2024**, *26*, 14476–14504. [CrossRef]
4. Zheng, Y.-Z.; Zhou, G.-J.; Zheng, Z.; Winpenny, R.E.P. Molecule-based magnetic coolers. *Chem. Soc. Rev.* **2014**, *43*, 1462–1475. [CrossRef] [PubMed]
5. Pecharsky, V.K.; Gschneidner, K.A. Magnetocaloric effect and magnetic refrigeration. *J. Magn. Magn. Mater.* **1999**, *200*, 44–56. [CrossRef]
6. De Haas, W.J. Extremely low temperatures. *Nature* **1933**, *132*, 372–373. [CrossRef]
7. Phan, M.H.; Yu, S.C. Review of the magnetocaloric effect in manganite materials. *J. Magn. Magn. Mater.* **2007**, *308*, 325–340. [CrossRef]
8. Hu, J.J.; Peng, Y.; Liu, S.J.; Wen, H.R. Recent advances in lanthanide coordination polymers and clusters with magnetocaloric effect or single-molecule magnet behavior. *Dalton Trans.* **2021**, *50*, 15473–15487. [CrossRef]
9. Konieczny, P.; Sas, W.; Czernia, D.; Pacanowska, A.; Fitta, M. Magnetic cooling: A molecular perspective. *Dalton Trans.* **2022**, *51*, 12762–12780. [CrossRef] [PubMed]
10. Zhong, X.; Hu, J.-J.; Yao, S.-L.; Zhang, R.-J.; Wang, J.-J.; Cai, D.-G.; Luo, T.-K.; Peng, Y.; Liu, S.-J.; Wen, H.-R. Gd(III)-based inorganic polymers, metal-organic frameworks and coordination polymers for magnetic refrigeration. *CrystEngComm* **2022**, *24*, 2370–2382. [CrossRef]
11. Fan, C.H.; Bai, K.P.; Luo, Q.C.; Han, T.; Huang, C.; Zheng, Y.Z.; Chen, W.P. A new family of SO_4^{2-}-templated 3d-4f high-nuclearity clusters: Syntheses, structures, and magnetic properties. *Chin. J. Chem.* **2024**, *42*, 2438–2444. [CrossRef]
12. Liu, S.J.; Han, S.D.; Zhao, J.P.; Xu, J.L.; Bu, X.H. In-situ synthesis of molecular magnetorefrigerant materials. *Coord. Chem. Rev.* **2019**, *394*, 39–52. [CrossRef]
13. Liu, J.L.; Chen, Y.C.; Tong, M.L. Molecular design for cryogenic magnetic coolants. *Chem. Rec.* **2016**, *16*, 825–834. [CrossRef]
14. Han, S.D.; Miao, X.H.; Liu, S.J.; Bu, X.H. Magnetocaloric effect and slow magnetic relaxation in two dense (3,12)-connected lanthanide complexes. *Inorg. Chem. Front.* **2014**, *1*, 549–552. [CrossRef]
15. Zheng, X.Y.; Jiang, Y.H.; Zhuang, G.L.; Liu, D.P.; Liao, H.G.; Kong, X.J.; Long, L.S.; Zheng, L.S. A gigantic molecular wheel of $\{Gd_{140}\}$: A new member of the molecular wheel family. *J. Am. Chem. Soc.* **2017**, *139*, 18178–18181. [CrossRef]
16. Luo, X.M.; Hu, Z.B.; Lin, Q.F.; Cheng, W.W.; Cao, J.P.; Cui, C.H.; Mei, H.; Song, Y.; Xu, Y. Exploring the performance improvement of magnetocaloric effect based Gd-exclusive cluster Gd_{60}. *J. Am. Chem. Soc.* **2018**, *140*, 11219–11222. [CrossRef] [PubMed]
17. Zhitomirsky, M.E. Enhanced magnetocaloric effect in frustrated magnets. *Phys. Rev. B* **2003**, *67*, 1–7. [CrossRef]
18. Chang, L.X.; Xiong, G.; Wang, L.; Cheng, P.; Zhao, B. A 24-Gd nanocapsule with a large magnetocaloric effect. *Chem. Commun.* **2013**, *49*, 1055–1057. [CrossRef]
19. Ho, T.L. Hard soft acids bases (HSAB) principle and organic-chemistry. *Chem. Rev.* **1975**, *75*, 1–20. [CrossRef]
20. Zheng, X.Y.; Kong, X.J.; Zheng, Z.P.; Long, L.S.; Zheng, L.S. High-nuclearity lanthanide-containing clusters as potential molecular magnetic coolers. *Acc. Chem. Res.* **2018**, *51*, 517–525. [CrossRef]

21. Zheng, X.Y.; Xie, J.; Kong, X.J.; Long, L.S.; Zheng, L.S. Recent advances in the assembly of high-nuclearity lanthanide clusters. *Coord. Chem. Rev.* **2019**, *378*, 222–236. [CrossRef]
22. Iki, N.; Tanaka, T.; Hiro-oka, S.; Shinoda, K. Self-assembly of a trilanthanide(III) core sandwiched between two thiacalix[4] arene ligands. *Eur. J. Inorg. Chem.* **2016**, *2016*, 5020–5027. [CrossRef]
23. Tian, H.W.; Hong, J.Q.; Wang, K.; del Rosal, I.; Maron, L.; Zhou, X.G.; Zhang, L.X. Unprecedented reaction mode of phosphorus in phosphinidene rare-earth complexes: A joint experimental-theoretical study. *J. Am. Chem. Soc.* **2018**, *140*, 102–105. [CrossRef] [PubMed]
24. Li, Z.P.; Wang, D.; Zhou, Z.H.; Zhao, G.Y.; Li, Q.; Bi, Y.F.; Zheng, Z.P. Thiacalix[4]arene-sandwiched sandglass-like Ln_9 Clusters (Ln = Tb and Eu): Insights into the selective luminescence quenching properties by p-nitrobenzene derivatives. *Inorg. Chem.* **2022**, *61*, 20814–20823. [CrossRef] [PubMed]
25. Ying, J.; Gao, N.; Mou, H.C.; Tian, A.X. Rare earth mono-substituted Keggin dimmers: Structures, electrochemical, supercapacitor and magnetic properties. *J. Solid State Chem.* **2022**, *314*, 1–11. [CrossRef]
26. Ohama, K.; Komiya, M.; Yamatoya, T.; Sawamura, R.; Karashimada, R.; Gao, S.; Ozawa, Y.; Osada, K.; Aoki, I.; Nagasaki, T.; et al. Albumin nanoparticle electrostatically loaded with highly anionic Gd-thiacalixarene complex for MRI-guided neutron capture therapy. *Colloids Surf., A* **2024**, *699*, 1–13. [CrossRef]
27. Marin, R.; Gálico, D.A.; Gayfullina, R.; Moilanen, J.O.; Carlos, L.D.; Jaque, D.; Murugesu, M. A zero-field single-molecule magnet with luminescence thermometry capabilities containing soft donors. *J. Mater. Chem. C* **2022**, *10*, 13946–13953. [CrossRef]
28. Chen, S.M.; Xiong, J.; Zhang, Y.Q.; Yuan, Q.; Wang, B.W.; Gao, S. A soft phosphorus atom to "harden" an erbium(III) single-ion magnet. *Chem. Sci.* **2018**, *9*, 7540–7545. [CrossRef]
29. Liu, S.S.; Lang, K.; Zhang, Y.Q.; Yang, Q.; Wang, B.W.; Gao, S. A distinct magnetic anisotropy enhancement in mononuclear dysprosium-sulfur complexes by controlling the Dy-ligand bond length. *Dalton Trans.* **2016**, *45*, 8149–8153. [CrossRef]
30. Canaj, A.B.; Dey, S.; Céspedes, O.; Wilson, C.; Rajaraman, G.; Murrie, M. There is nothing wrong with being soft: Using sulfur ligands to increase axiality in a Dy(III) single-ion magnet. *Chem. Commun.* **2020**, *56*, 1533–1536. [CrossRef] [PubMed]
31. Tziotzi, T.G.; Gracia, D.; Dalgarno, S.J.; Schnack, J.; Evangelisti, M.; Brechin, E.K.; Milios, C.J. A $\{Gd_{12}Na_6\}$ molecular quadruple-wheel with a record magnetocaloric effect at low magnetic fields and temperatures. *J. Am. Chem. Soc.* **2023**, *145*, 7743–7747. [CrossRef]
32. Braun, J.; Seufert, D.; Anson, C.E.; Tang, J.K.; Powell, A.K. An effectively uncoupled Gd_8 cluster formed through fixation of atmospheric CO_2 showing excellent magnetocaloric properties. *Int. J. Mol. Sci* **2024**, *25*, 264. [CrossRef]
33. Wang, Y.F.; Wang, Y.X.; Yang, Q.Q.; Yin, B. Auxiliary rather than dominant. the role of direct Dy-S coordination in single-molecule magnet unveiled via ab initio study. *J. Phys. Chem. A* **2024**, *128*, 5285–5297. [CrossRef]
34. Kumari, N.; Jassal, M.; Agrawal, A.K. Effect of metal ion on UV protective and antimicrobial properties of in-situ synthesized pyrithione complexes on cellulosic textiles. *Mater. Today Commun.* **2023**, *34*, 1–13. [CrossRef]
35. Han, T.; Wang, X.Q.; Guo, T.; Cao, H.S.; Chen, W.P.; He, L. Modulation of the magnetic dynamics in two air-stable sulfur-ligated dysprosium complexes via polymerization. *Dalton Trans.* **2023**, *52*, 7551–7557. [CrossRef]
36. Yuan, F.; Li, T.; Zhang, M.; Qian, H. Synthesis and crystal structure of tri(pyrithione) cerium complex and its catalytic property for oxidation of benzoin to benzil. *Synth. React. Inorg. Met.-Org. Nano-Met. Chem.* **2013**, *43*, 1510–1513. [CrossRef]
37. Niu, D.Z.; Mu, L.L.; Lu, Z.S.; Song, B.L.; Sun, B.W. Synthesis and crystal structure of $Pr_3(DMSO)_2$ (Hmpo=2-mercaptopyridine N-oxide). *J. Chem. Crystallogr.* **2004**, *34*, 195–198. [CrossRef]
38. Xiong, R.G.; Zuo, J.L.; You, X.Z.; Huang, X.Y. Synthesis and X-ray structure of the first lanthanide 2-mercaptopyridine N-oxide complex, tris-(1-hydroxy-2(1H)-pyridinethionato-O,S) bis sulfinylbis(methane) samarium(III). *Polyhedron* **1996**, *15*, 3321–3324. [CrossRef]
39. Hu, Y.H.; Weng, L.H.; Huang, L.G.; Chen, X.T.; Wu, D.X.; Kang, B.S. Structure of tris(2-mercaptopyridine 1-oxido)cobalt(III) acetonitrile solvate. *Acta Crystallogr. Sect. C Cryst. Struct. Commun.* **1991**, *47*, 2655–2656. [CrossRef]
40. West, D.X.; Frank, C.A. 2 series of rare-earth complexes of the 2-thiopyridine N-oxide anion. *J. Inorg. Nucl. Chem.* **1979**, *41*, 49–53. [CrossRef]
41. Wang, T.; Wang, H.; Sun, W. Sandglass-like Ln_9 nanoclusters with magnetocaloric effect and lanthanide-centered luminescence. *Cryst. Growth Des.* **2024**, *24*, 8076–8084. [CrossRef]
42. Pandey, P.; Chauhan, D.; Walawalkar, M.G.; Gupta, S.K.; Meyer, F.; Rajaraman, G.; Murugavel, R. Hourglass-shaped homo- and heteronuclear nonanuclear lanthanide clusters: Structures, magnetism, photoluminescence, and theoretical analysis. *Inorg. Chem.* **2024**, *63*, 11963–11976. [CrossRef]
43. Hu, P.; Li, S.H.; Cao, L.H.; Liu, A.G.; Zhuang, G.L.; Ji, L.D.; Li, B. Construction of a high nuclear gadolinium cluster with enhanced magnetocaloric effect through structural transition. *Acs Omega* **2022**, *7*, 38782–38788. [CrossRef]
44. Casanova, D.; Cirera, J.; Llunell, M.; Alemany, P.; Avnir, D.; Alvarez, S. Minimal distortion pathways in polyhedral rearrangements. *J. Am. Chem. Soc.* **2004**, *126*, 1755–1763. [CrossRef]

45. Panguluri, S.P.K.; Jourdain, E.; Chakraborty, P.; Klyatskaya, S.; Kappes, M.M.; Nonat, A.M.; Charbonnière, L.J.; Ruben, M. Yb-to-Eu cooperative sensitization upconversion in a multifunctional molecular nonanuclear lanthanide cluster in solution. *J. Am. Chem. Soc.* **2024**, *146*, 13083–13092. [CrossRef] [PubMed]
46. Knighton, R.C.; Soro, L.K.; Lecointre, A.; Pilet, G.; Fateeva, A.; Pontille, L.; Francés-Soriano, L.; Hildebrandt, N.; Charbonnière, L.J. Upconversion in molecular hetero-nonanuclear lanthanide complexes in solution. *Chem. Commun.* **2021**, *57*, 53–56. [CrossRef] [PubMed]
47. Hu, P.; Cao, L.H.; Liu, A.G.; Zhang, Y.Q.; Zhang, T.L.; Li, B. Modulating the relaxation dynamics via structural transition from a dinuclear dysprosium cluster to a nonanuclear cluster. *Dalton Trans.* **2021**, *50*, 12814–12820. [CrossRef] [PubMed]
48. Huang, Y.; Qin, Y.R.; Ge, Y.; Cui, Y.F.; Zhang, X.M.; Li, Y.H.; Yao, J.L. Rationally assembled nonanuclear lanthanide clusters: Dy_9 displays slow relaxation of magnetization and Tb_9 serves as luminescent sensor for Fe^{3+}, $CrO_4{}^{2-}$ and $Cr_2O_7{}^{2-}$. *New J. Chem.* **2019**, *43*, 19344–19354. [CrossRef]
49. Griffiths, K.; Kühne, I.A.; Tizzard, G.J.; Coles, S.J.; Kostakis, G.E.; Powell, A.K. Twists to the spin structure of the Ln_9-diabolo motif exemplified in two $\{Zn_2Ln_2\}Ln_9\{Zn_2\}$ coordination clusters. *Inorg. Chem.* **2019**, *58*, 2483–2490. [CrossRef]
50. Wang, H.S.; Long, Q.Q.; Yin, C.L.; Xu, Z.W.; Pan, Z.Q. Syntheses, crystal structures and magnetic properties of sandglass $Dy_9{}^{III}$ and irregular tetrahedron $Dy_4{}^{III}$ complexes. *Polyhedron* **2018**, *141*, 69–76. [CrossRef]
51. Chen, H.F.; Yang, X.P.; Jiang, D.M.; Shi, D.L.; Zhang, L.J. Construction of NIR luminescent polynuclear lanthanide-based nanoclusters with sensing properties towards metal ions. *Dalton Trans.* **2018**, *47*, 13880–13886. [CrossRef]
52. Zheng, X.Y.; Peng, J.B.; Livera, M.; Luo, Y.; Wang, Y.Y.; Kong, X.J.; Long, L.S.; Zheng, Z.P.; Zheng, L.S. Selective formation of chromogen I from N-acetyl-D-glucosamine upon lanthanide coordination. *Inorg. Chem.* **2017**, *56*, 110–113. [CrossRef]
53. Mylonas-Margaritis, I.; Kitos, A.A.; Panteli, C.C.; Skordi, K.; Tasiopoulos, A.J.; Bekiari, V.; Escuer, A.; Perlepes, S.P. 2-hydroxybenzophenone-controlled self-assembly of enneanuclear lanthanide(III) hydroxo coordination clusters with an "hourglass"-like topology. *Inorg. Chem. Commun.* **2017**, *83*, 118–122. [CrossRef]
54. Omagari, S.; Nakanishi, T.; Kitagawa, Y.; Seki, T.; Fushimi, K.; Ito, H.; Meijerink, A.; Hasegawa, Y. Critical role of energy transfer between terbium ions for suppression of back energy transfer in nonanuclear terbium clusters. *Sci. Rep.* **2016**, *6*, 1–11. [CrossRef] [PubMed]
55. Greisch, J.F.; Chmela, J.; Harding, M.E.; Klopper, W.; Kappes, M.M.; Schooss, D. Gas-phase photoluminescence characterization of stoichiometrically pure nonanuclear lanthanoid hydroxo complexes comprising europium or gadolinium. *Inorg. Chem.* **2016**, *55*, 3316–3323. [CrossRef] [PubMed]
56. Zou, H.H.; Sheng, L.B.; Chen, Z.L.; Liang, F.P. Lanthanide nonanuclear clusters with sandglass-like topology and the SMM behavior of dysprosium analogue. *Polyhedron* **2015**, *88*, 110–115. [CrossRef]
57. Omagari, S.; Nakanishi, T.; Seki, T.; Kitagawa, Y.; Takahata, Y.; Fushimi, K.; Ito, H.; Hasegawa, Y. Effective photosensitized energy transfer of nonanuclear terbium clusters using methyl salicylate derivatives. *J. Phys. Chem. A* **2015**, *119*, 1943–1947. [CrossRef]
58. Nakanishi, T.; Suzuki, Y.; Doi, Y.; Seki, T.; Koizumi, H.; Fushimi, K.; Fujita, K.; Hinatsu, Y.; Ito, H.; Tanaka, K.; et al. Enhancement of optical faraday effect of nonanuclear Tb(III) complexes. *Inorg. Chem.* **2014**, *53*, 7635–7641. [CrossRef]
59. Greisch, J.F.; Harding, M.E.; Schäfer, B.; Ruben, M.; Klopper, W.; Kappes, M.M.; Schooss, D. Characterization of nonanuclear europium and gadolinium complexes by gas-phase luminescence spectroscopy. *J. Phys. Chem. Lett.* **2014**, *5*, 1727–1731. [CrossRef]
60. Zhang, B.; Xiao, T.; Liu, C.M.; Li, Q.; Zhu, Y.Y.; Tang, M.S.; Du, C.X.; Song, M.P. Systematic study of the luminescent europium-based nonanuclear clusters with modified 2-hydroxybenzophenone ligands. *Inorg. Chem.* **2013**, *52*, 13332–13340. [CrossRef]
61. Singh-Wilmot, M.A.; Sinclair, R.A.; Andrews, M.; Rowland, C.; Cahill, C.L.; Murugesu, M. Nonanuclear lanthanide(III) nanoclusters: Structure, luminescence and magnetic properties. *Polyhedron* **2013**, *53*, 187–192. [CrossRef]
62. Xu, X.B.; Zhao, L.; Xu, G.F.; Guo, Y.N.; Tang, J.K.; Liu, Z.L. A diabolo-shaped Dy_9 cluster: Synthesis, crystal structure and magnetic properties. *Dalton Trans.* **2011**, *40*, 6440–6444. [CrossRef]
63. Alexandropoulos, D.I.; Mukherjee, S.; Papatriantafyllopoulou, C.; Raptopoulou, C.P.; Psycharis, V.; Bekiari, V.; Christou, G.; Stamatatos, T.C. A new family of nonanuclear lanthanide clusters displaying magnetic and optical properties. *Inorg. Chem.* **2011**, *50*, 11276–11278. [CrossRef] [PubMed]
64. Baril-Robert, F.; Petit, S.; Pilet, G.; Chastanet, G.; Reber, C.; Luneau, D. Site-selective lanthanide doping in a nonanuclear yttrium(III) cluster revealed by crystal structures and luminescence spectra. *Inorg. Chem.* **2010**, *49*, 10970–10976. [CrossRef]
65. Petit, S.; Baril-Robert, F.; Pilet, G.; Reber, C.; Luneau, D. Luminescence spectroscopy of europium(III) and terbium(III) penta-, octa- and nonanuclear clusters with β-diketonate ligands. *Dalton Trans.* **2009**, 6809–6815. [CrossRef] [PubMed]
66. Volpe, M.; Bombieri, G.; Clemente, D.A.; Foresti, E.; Grillone, M.D. Synthesis and physico-chemical characterization of a new series of hydroxide ion acetylacetonate lanthanide(III)-ditungsten decacarbonyl hydride complexes. *J. Alloys Compd.* **2004**, *374*, 382–386. [CrossRef]
67. Addamo, M.; Bombieri, G.; Foresti, E.; Grillone, M.D.; Volpe, M. Assembling process of charged nonanuclear cationic lanthanide(III) clusters assisted by dichromium decacarbonyl hydride. *Inorg. Chem.* **2004**, *43*, 1603–1605. [CrossRef]

68. Xu, G.; Wang, Z.M.; He, Z.; Lu, Z.; Liao, C.S.; Yan, C.H. Synthesis and structural characterization of nonanuclear lanthanide complexes. *Inorg. Chem.* **2002**, *41*, 6802–6807. [CrossRef]
69. Hubert-Pfalzgraf, L.G.; Miele-Pajot, N.; Papiernik, R.; Vaissermann, J. A novel example of self-assembly in lanthanide chemistry: Synthesis and molecular structure of $Na(EtOH)_6Y_9(\mu_4\text{-}O)_2(\mu_3\text{-}OH)_8\{\mu\text{-}\eta^2\text{-}MeC(O)CHC(O)OEt\}_8\{\eta^2\text{-}MeC(O)CHC(O)OEt\}_8$. *J. Chem. Soc. Dalton Trans.* **1999**, 4127–4130. [CrossRef]
70. Chilton, N.F.; Anderson, R.P.; Turner, L.D.; Soncini, A.; Murray, K.S. PHI: A powerful new program for the analysis of anisotropic monomeric and exchange-coupled polynuclear d- and f-block complexes. *J. Comput. Chem.* **2013**, *34*, 1164–1175. [CrossRef]
71. Singh, M.K.; Rajeshkumar, T.; Kumar, R.; Singh, S.K.; Rajaraman, G. Role of (1,3) {Cu-Cu} interaction on the magneto-caloric effect of trinuclear $\{Cu^{II}\text{-}Gd^{III}\text{-}Cu^{II}\}$ complexes: Combined DFT and experimental studies. *Inorg. Chem.* **2018**, *57*, 1846–1858. [CrossRef] [PubMed]
72. Rajeshkumar, T.; Singh, S.K.; Rajaraman, G. A computational perspective on magnetic coupling, magneto-structural correlations and magneto-caloric effect of a ferromagnetically coupled $\{Gd^{III}\text{-}Gd^{III}\}$ pair. *Polyhedron* **2013**, *52*, 1299–1305. [CrossRef]

magnetochemistry

MDPI

Article

A Dinuclear Dysprosium(III) Single Molecule Magnet of Benzo[*h*]quinolin-10-ol

Limin Zhou [1,†], Hongling Lv [1,†], Yuning Liang [1,*], Dongcheng Liu [1], Zaiheng Yao [1], Shuchang Luo [2,*] and Zilu Chen [1,*]

1 Key Laboratory for Chemistry and Molecular Engineering of Medicinal Resources (Ministry of Education of China), Guangxi Key Laboratory of Chemistry and Molecular Engineering of Medicinal Resources, University Engineering Research Center for Chemistry of Characteristic Medicinal Resources (Guangxi), School of Chemistry and Pharmaceutical Sciences, Guangxi Normal University, Guilin 541004, China; zhoulm@gxnu.edu.cn (L.Z.); 17357113719@163.com (H.L.); ldcheng@mailbox.gxnu.edu.cn (D.L.); 3120503957@stu.gxnu.edu.cn (Z.Y.)

2 College of Chemical Engineering, Guizhou University of Engineering Science, Bijie 551700, China

* Correspondence: liangyuning008@163.com (Y.L.); luosc@gues.edu.cn (S.L.); zlchen@mailbox.gxnu.edu.cn (Z.C.)

† These authors contribute equally to this work.

Abstract: To develop single molecule magnets, a dinuclear complex $[Dy_2(HOBQ)_4Cl_6]$ (**1**) was prepared from the reaction of $DyCl_3$ with benzo[*h*]quinolin-10-ol (HOBQ). Each Dy(III) ion shows a compressed octahedral geometry and the two Dy(III) ions in **1** are bridged by two Cl^- ligands to construct a dinuclear structure with the four HOBQ ligands on the axial positions and six Cl^- ligands in the equatorial plane. Magnetic measurements showed that complex **1** is a field-induced single molecule magnet having an obvious magnetic hysteresis loop with an energy barrier of 71(2) K. These experimental results are corroborated by the ab initio complete active space self-consistent field (CASSCF) calculations which also interpret the magneto-structural correlation. It is a typical example to achieve Dy(III) SMM through regulating coordination geometry, i.e., lengthening equatorial coordination bonds and shortening axial ones to form a compressed octahedral geometry.

Keywords: dysprosium(III) complex; benzo[*h*]quinolin-10-ol; single molecule magnet; coordination bond; compressed octahedral geometry

1. Introduction

In recent years, single-molecule magnet (SMM) has attracted a significant interest in many fields such as chemistry, physics, and material science [1–7]. It has undergone a considerable development, acquiring some surprising achievements such as large barrier energy, high blocking temperature and long relaxation time especially in mononuclear Dy(III) SMMs [8–11]. However, these achievements are still far from the requirement for their practical applications.

These studies revealed some rules on the synthetic strategy and SMM performance tuning. There are many factors influencing SMM performances, which include ligand design, symmetry control, counter ions and even coordinated or lattice solvent molecules [12–16]. In a word, any factors which can influence the anisotropy of the whole molecule will ultimately tune the SMM performance. For an example, the group of Sanping Chen found that lattice water molecules can tune SMM performance in Dy_2-SMMs through adjusting magnetic interactions [17]. In the latest several years, high SMM performances were also achieved in low-coordinate Dy(III) single-molecule magnets [18]. The group Eduard B.

Yagubskii discussed the competition between the apical and equatorial crystal fields in tuning SMM performances in seven-coordinate pentagonal bipyramidal Dy(III) complexes [19]. They pointed out that their SMM performances are presumably improved by enhancing the apical crystal field through the use of more negatively charged ligands to obtain more short axial bonds, as well as by weakening equatorial crystal field through longer equatorial bonds and less equatorial negative charge.

Compared with transition metal ions, lanthanide ions, especially Dy(III) ions, have large magnetic moments and magnetic anisotropy due to their large unquenched orbital angular momentum. Thus, lanthanide ions are usually the optimum selection for developing SMMs with high performance. In addition to the anisotropy of individual lanthanide ion, intermetallic interactions can also influence the SMM performance in some extent. However, the 4f-electrons of lanthanide ions are shielded by their outer 5s and 5p electron shells, which leads to a weak inter-lanthanide magnetic interaction. So the high SMM performance of many lanthanide clusters and polymers are mainly from individual lanthanide ions [20–23]. In this sense, the design of ligands and the symmetry control of lanthanide ions in some special coordination geometries are of significant importance for optimizing SMM performances. In addition, a rational arrangement of lanthanide ions is also crucial for maximizing the SMM performances of complexes bearing two or more lanthanide ions. A significant success has been achieved in dilanthanide complexes, such as $\{Dy[15\text{-}MC_{Cu}\text{-}5]\}_2$ [24], $[Er^{III}{}_2(COT'')_3]$ [25], phosphino-supported Er_2 SMMs [26] and $(Cp^{iPr5})_2Dy_2I_3$ [27], highlighting the importance of intramolecular anisotropy axis alignment for enhancing SMM performance.

Based on the above-mentioned consideration, we aimed to use chloride ions to coordinate to Dy(III) ion in the equatorial plane and to use HOBQ to coordinate to Dy(III) at axial sites, attempting to achieve short axial coordination bonds and long equatorial coordination bonds. As expected, $[Dy_2(HOBQ)_4Cl_6]$ (**1**) was finally prepared from the reaction of $DyCl_3$ with HOBQ. Its magnetic properties were investigated in detail, which demonstrated a field-induced single molecule magnet for complex **1**.

2. Materials and Methods

2.1. Materials

All materials for this work were used as commercially received (benzo[*h*]quinolin-10-ol, Bide Pharmatech Ltd., Shanghai, China; $DyCl_3{\cdot}6H_2O$, Macklin Biochemical Co., Ltd., Shanghai, China; ethanol and acetonitrile, Asia Chemical Co., Ltd., Chengdu, China). Their IR, powder and single crystal X-ray diffraction (PXRD and SCXRD) analyses were carried out on Spectrum Two FTIR equipment (PerKin-Elmer, Waltham, MA, USA), Rigaku D/max 2500v/pc diffractometer (Rigaku, Tokyo, Japan) and XtaLAB Synergy-DS single crystal X-ray diffractometer (Rigaku, Tokyo, Japan), respectively. Elemental analyses (C, H and N) were made on an Elementar Micro cube CHN elemental analyzer (Elementar, Langenselbold, Germany). Magnetic measurements were performed on a MPMS 3 SQUID magnetometer (Quantum Design, San Diego, CA, USA) furnished with a 7 T magnet. Diamagnetic corrections were applied for dealing with their magnetic susceptibilities.

2.2. Synthesis of $[Dy_2(HOBQ)_4Cl_6]$ (1)

After one end of a 20 cm long glass tube was sealed, $DyCl_3{\cdot}6H_2O$ (0.1 mmol, 0.0368 g), benzo[*h*]quinolin-10-ol (0.1 mmol, 0.0195 g), ethanol (0.5 mL) and acetonitrile (1.5 mL) were added into this tube. It was then evacuated and completely sealed before being heated in an oven (80 °C) for 10 h. Finally the sealed tube was cooled overnight, yielding pale yellow crystals with a yield of 42% (calculated from HOBQ). Elemental analysis calculated for $C_{52}H_{36}Cl_6Dy_2N_4O_4$: C, 47.37; H, 2.75; N, 4.25%. Found: C, 47.51; H, 2.83; N, 4.32.

IR (Figure S1, KBr pellet, cm^{-1}): 3377(s), 3067(w), 2885(w), 1630(m), 1576(m), 1543(m), 1462(m), 1435(s), 1355(m), 1327(m), 1263(S), 1227(m), 1189(w), 1145(m), 1085(w), 1060(w), 1028(m), 993(w), 983(w), 920(w), 881(w), 861(w), 836(S), 808(m), 784(w), 761(w), 714(m), 681(w), 635(w), 585(m), 556(m), 522(w), 509(w).

2.3. X-Ray Data Collection and Structure Refinement

The structure of **1** was determined from single crystal X-ray diffraction analysis on a XtaLAB Synergy diffractometer with graphite-monochromate Cu K_α radiation (λ = 1.54184 Å). After being solved with SHELXS, its structure was then refined using the SHELXL program. Hydrogen atoms binding with C were geometrically added using a riding model and hydrogen atoms on N atoms were located from difference Fourier map. The detailed crystallographic data are listed in Table S1 with the selected bond distances and angles given in Table S2.

3. Results

3.1. Synthesis and Structural Characterization of 1

The crystalline sample of **1** was obtained from the solvothermal reaction of $DyCl_3 \cdot 6H_2O$ with HOBQ in a mixed solvent of ethanol and acetonitrile in a glass tube. Its purity was confirmed through the high agreement of experimental powder X-ray diffraction patterns with its simulated one derived from its single crystal structural data (Figure S2).

Complex **1** crystallizes in the $P\bar{1}$ space group of triclinic system and shows a dinuclear structure (Figure 1a) with six Cl^- ligands in the equatorial plane and four HOBQ ligands on the axial positions. Two of the six Cl^- ligands bridge the two Dy(III) ions with the other four Cl^- ligands terminally coordinating to the two Dy(III) ions. There exist intramolecular H-bonds of N–H···O (Table S3) in the HOBQ ligands which terminally coordinate to Dy(III) ions (Figure 1b). Between the adjacent two HOBQ ligands, $\pi \cdots \pi$ stacking interactions were found as indicated by the dihedral angle of 3.806(2)° between the two benzo[*h*]quinoline planes and a shortest centroid-to-centroid distance of 3.5866(2) Å. Thus, each Dy(III) ion is six coordinated by four equatorial Cl^- ligands and two axial HOBQ ligands. The Dy–O and Dy–Cl bond lengths are in the range of 2.186(4)–2.209(4) and 2.5844(16)–2.7880(14) Å, respectively. This means that the equatorial coordination bonds are much longer than the axial ones. The axial O–Dy–O bond angle is 173.74(15)°, near to 180°. The Cl–Dy–Cl bond angles in the equatorial plane are in the range of 80.20(4)–101.41(6) and 167.64(5)–170.98(6)°. The O–Dy–Cl bond angles between axial and equatorial coordination atoms are in the range of 82.54(12)–93.56(12)°. All of these bond lengths and bond angles support that the Dy(III) ions have distorted compressed octahedral geometries (Figure 1c and Table S4).

The neighboring coordination molecules of **1** interact with each other through hydrogen bonds (C10–H10···Cl1B, Table S3), leading to the construction of a hydrogen-bonded chain along axis a (Figure S3). Each coordination molecule further interacts with another four ones in the (1$\bar{1}$0) plane through $\pi \cdots \pi$ stacking interactions as demonstrated by their parallel benzo[*h*]quinoline planes with the nearest centroid-to-centroid distance of 3.4924(2) and 3.6525(2) Å, building a $\pi \cdots \pi$ stacking 2D framework in the (1$\bar{1}$0) plane (Figure S4). These hydrogen bonds and $\pi \cdots \pi$ stacking interactions finally connect the coordination molecules of **1** into a 3D supramolecular framework (Figure 2).

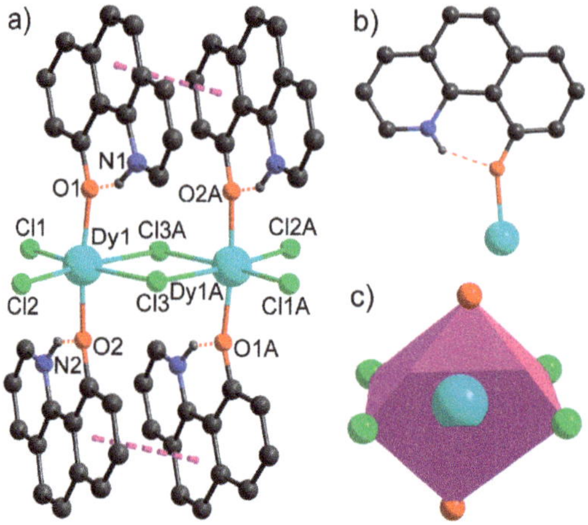

Figure 1. (**a**) Molecular structure of **1** with C-bounded H atoms omitted and only selected atoms labeledfor clarity, (**b**) coordination mode of HOBQ ligand and (**c**) coordination geometry of Dy(III) ion. Symmetry code: (A) $1 - x, 1 - y, 1 - z$. Thick dashed pink lines and thin broken red lines represent $\pi \cdots \pi$ stacking interactions and hydrogen bonds, respectively.

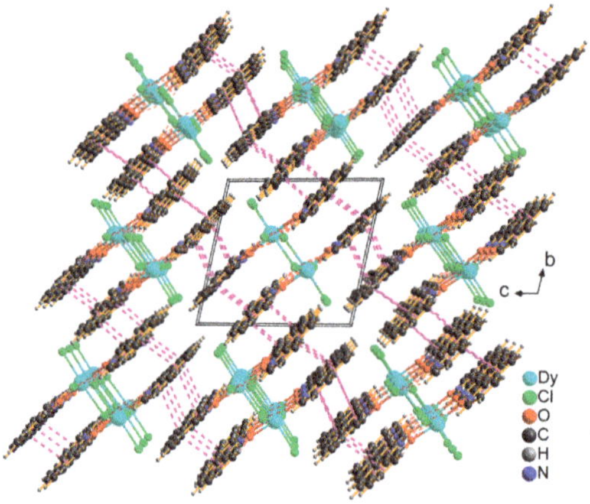

Figure 2. Packing diagram of **1** through hydrogen bonds (thin broken red lines) and $\pi \cdots \pi$ stacking interactions (thick dashed pink lines).

3.2. Magnetic Properties

The magnetic susceptibilities of complex **1** from 2 to 300 K were measured under 1000 Oe, which are plotted into Figure 3a in the form of $\chi_M T$ versus T. The 300 K $\chi_M T$ value of **1** is 27.12 cm^3 K mol^{-1}, a little lower than the theoretical spin-only values (28.34 cm^3 K mol^{-1}) for two independent Dy(III) ions ($g = 4/3$, $J = 15/2$, $L = 5$, $S = 5/2$, $^6H_{15/2}$) [28,29]. The $\chi_M T$ values of complex **1** decrease with lowering temperature to 15.04 cm^3 K mol^{-1} at 2 K. The $\chi_M T$-T profiles might result from the thermal depopulation of the Stark sublevels of Dy(III) ions and/or the crystal field splitting, and also possible weak antiferromagnetic interactions [30–35]. The linear fitting of $\chi_M{}^{-1}$ versus T plot (Figure S5) for **1** in the temperature range of 2.0–300 K based on Curie-Weiss law gave C = 27.38 cm^3 K mol^{-1} and $\theta = -3.84$ K. The negative Weiss constant supports the presence of dominant weak antiferromagnetic interaction.

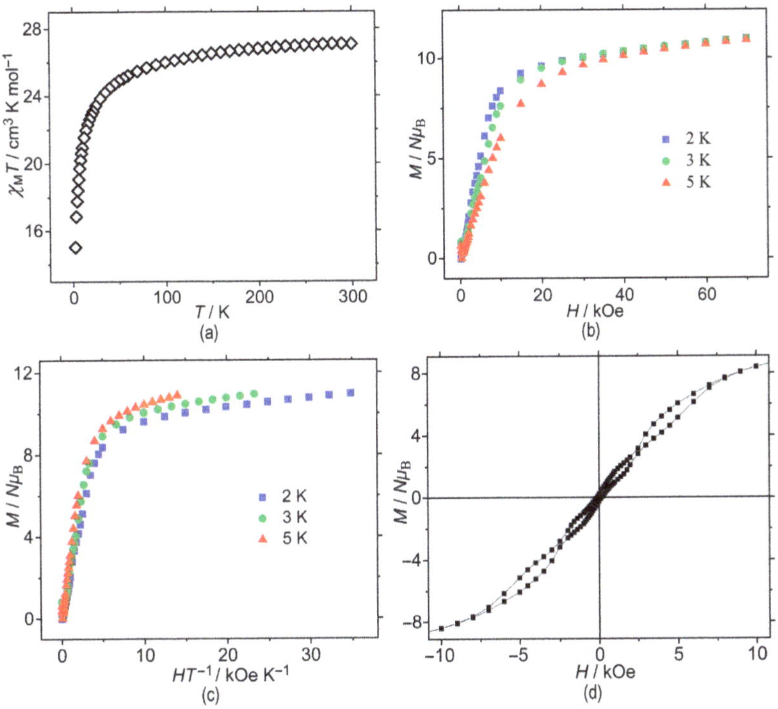

Figure 3. Plots of $\chi_M T$ versus T under an applied field of 1000 Oe (**a**), M versus H (**b**), M versus HT^{-1} (**c**) and magnetic hysteresis loop (**d**) for **1**.

Field-related magnetizations of **1** at 2, 3 and 5 K were measured and drawn as M–H and M–HT^{-1} plots (Figure 3b,c). Its magnetization increases slowly in the high field region after a rapid increase in the low field region, reaching a maximum of 10.99 Nμ_β at an applied field up to 70 kOe measured at 2 K, which is consistent with those reported in the literature [36–38], but far from saturation values. The difference in M values at different temperatures become larger with the increasing HT^{-1} values as revealed in Figure 3c, leading to the non-superposition of M–HT^{-1} plots at different temperatures. All of the above-mentioned features suggest a large magnetic anisotropy of Dy(III) ions in **1**. Interestingly, an obvious small hysteresis loop (Figure 3d) was observed for **1** at 2 K [39].

In order to study the slow magnetic relaxation behavior of complex **1** in detail, its dynamic magnetic properties were investigated. The optimal field for measuring its in-phase (χ') and out-of-phase (χ'') alternating current (ac) magnetic susceptibilities (Figure 4) was first explored at a temperature of 2 K under a series of external fields (0–1800 Oe). At zero dc field, the χ'' values of **1** are very small with no maximum in its χ''–ν curve. When a dc field was applied, the χ'' values of **1** become larger and the χ''–ν curves show maximums which become more significant before 1200 Oe and then less obvious and move to a little higher frequency upon increasing the applied dc fields. After an overall consideration of the above-mentioned information, 1200 Oe was chosen as the optimal dc field for the further measurement of ac magnetic susceptibility.

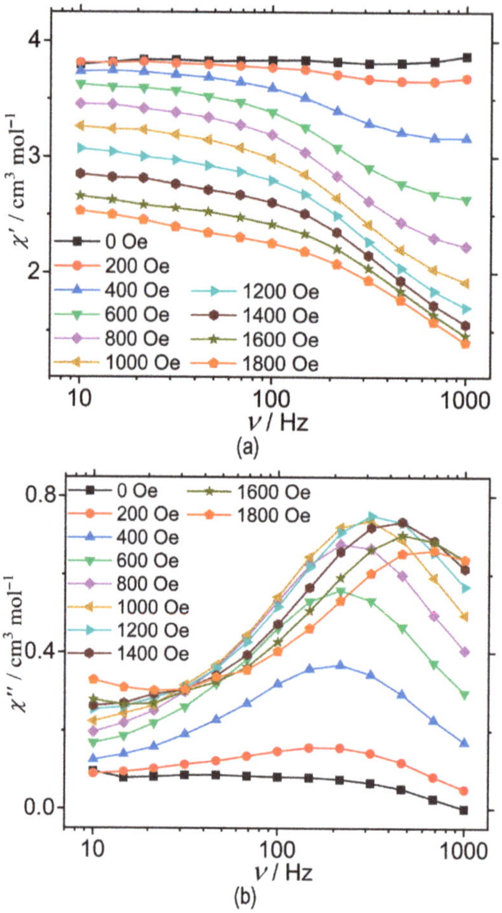

Figure 4. Frequency-dependent χ' (**a**) and χ'' (**b**) for **1** under a series of different dc fields with solid lines guiding for clarity.

Thus the frequency-dependent χ' and χ'' of **1** were collected under 1200 Oe with its data and the derived data plotted into Figure 5 [40,41]. As shown in Figure 5b, **1** present clearly non-zero χ'' values under the optimal dc field and the χ''-ν curves show maximums which became less important upon increasing the measuring temperature and disappeared after 3.5 K, with the maximum at 2 K appearing at 284.71 Hz. These experimental facts told us that **1** exhibits slow magnetic relaxation behavior and is a field-induced SMM [42–45]. Then, the Cole-Cole diagram (Figure 5c) of **1** was drawn based on these frequency-dependent χ' and χ'' data collected under the optimal dc field. The best fitting of these χ'' versus χ' curves on Debye model gave the distribution coefficients $\alpha = 0.12 - 0.17$ and relaxation time $\tau = 3.7 \times 10^{-5} - 4.4 \times 10^{-4}$ s. The derived relaxation times at different temperatures were plotted into a curve of $\ln(\tau/s)$ versus T^{-1} (Figure 5d) for analyzing its spin-lattice relaxation. Considering that the ac magnetic susceptibilities were measured under the optimal field, only direct relaxation process ($AH^{m}T$), Raman process (CT^{n}) and Orbach process ($\tau_0{}^{-1}\exp(-U_{eff}/k_BT)$) were taken into the fitting of $\ln(\tau/s)$ versus T^{-1} curve with the quantum tunneling process (QTM) omitted [46–48]. The best simulation based the equation of $\tau^{-1} = \tau_0{}^{-1}\exp(U_{eff}/k_BT) + CT^{n} + AH^{m}T$ gave $\tau_0 = 1.3(7) \times 10^{-13}$ s, $U_{eff}/k_B = 71(2)$ K, $C = 112(3)$ s^{-1} K^{-n}, $n = 3.0(1)$, $AH^{m} = 1356.1(4)$. The low Raman exponent (n = 3.0(1)) might be resulted from an acoustic-optic two-phonon Raman process as suggested in documents [49,50], which is presumably originated from the dinuclear skeleton of **1** bearing six equatorial Cl^- ligands and four axial ligands. A nearly linear slope can be found in the ln(τ/s) versus T^{-1} curve above 3.2 K, suggesting a dominant Orbach relaxation mechanism in this temperature region. Raman process accounts for the curvature of the ln(τ/s) versus T^{-1} curve at intermediate temperature region.

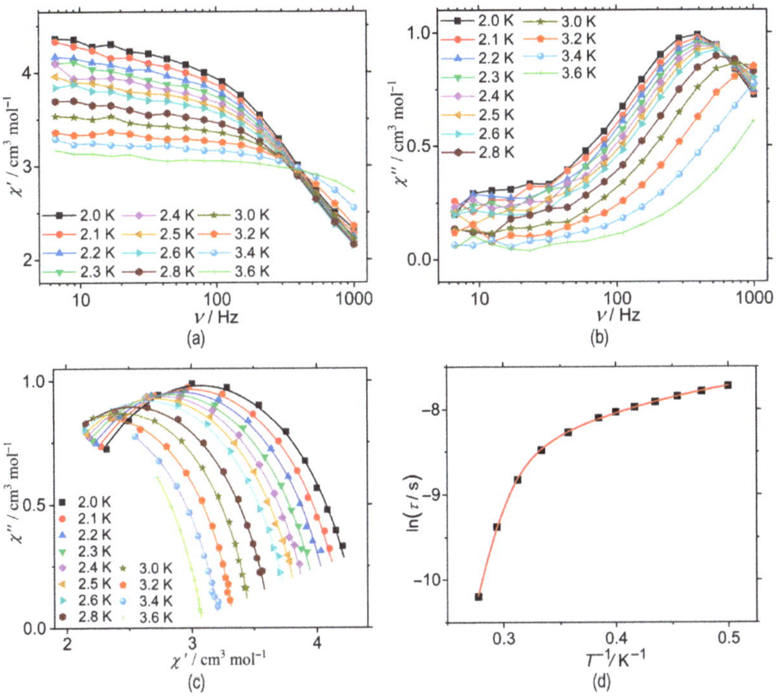

Figure 5. Frequency-dependent χ' (**a**) and χ'' (**b**) of **1** measured under 1200 Oe at a series of different temperatures with solid lines for guiding the eyes, Cole–Cole diagram (**c**) with solid lines representing the best fit to Debye model for **1**, and ln(τ/s) versus T^{-1} (**d**) for **1** with the solid line representing the best fitting on Arrhenius equation.

All of these investigations confirmed that we succeeded in obtaining a field-induced single molecule magnet. It might be associated with the structure which has six chloride ions coordinating to Dy(III) ions in the equatorial plane and four HOBQ ligands coordinating to Dy(III) ions in the axial positions, with the equatorial coordination bonds (Dy–Cl) much longer than the axial coordination bonds (Dy–O). This kind of structure is facile for improving the anisotropy of Dy(III) ion, achieving to obtain SMMs. However, an optimal dc field is necessary for suppressing the QTM and the energy barrier of **1** is not high as expected. It is presumably caused by two deficiencies in the structure. One is that the HOBQ ligands coordinate to Dy(III) ions in the neutral form with the H atoms still between O and N atoms, which leads to the absence of necessary negative charge on the axial coordinated O atoms for enhancing the axial anisotropy of Dy(III) ion. The second one is that the equatorial negative charge is superfluous although the equatorial bond lengths are long enough, which is not facile for improving the axial anisotropy of Dy(III) ion. This also gave us some hints in the performance improvement of this kind of SMMs. We can remove the protons on the axial ligands to increase axial negative charges and replace the four terminal Cl^- ligands using neutral chelating ligands to reduce the equatorial negative charge, achieving high axial anisotropy of Dy(III) through their cooperation. The further work is presumably to construct another perfect story.

3.3. Theoretical Calculation

To further understand the magnetic properties and magnetic exchange interactions of compound **1**, we performed electronic structure calculations based on single-crystal diffraction data without optimizing the compound's structure. Here, we used version 5.0.3 of the ORCA software [51,52] to perform ab initio complete active space self-consistent field (CASSCF) calculations [53]. To further calculate the magnetic interactions between the two dysprosium ions, our calculations were conducted in two steps: First, replacing one of the dysprosium ions with an antiferromagnetic Lu^{3+} ion, we constructed two fragments, **1a** and **1b** (Figure S6). Using the CASSCF method and the SINGLE_ANISO module [54,55], we calculated the anisotropic g-factor, the eight lowest-energy Kramers doublet energy levels, and the magnetic axis for **1a** and **1b**. Second, we calculated the magnetic interaction between the two dysprosium ions based on the results from the first step using the Lines model and the POLY_ANISO module [56–59]. The equations used for calculation are shown below:

$$\hat{H} = -J_{total}\tilde{\hat{S}}_{1z}\tilde{\hat{S}}_{2z} = -(J_{exch}\tilde{\hat{S}}_{1z}\tilde{\hat{S}}_{2z} + J_{dip}\tilde{\hat{S}}_{1z}\tilde{\hat{S}}_{2z}) \quad (1)$$

$$J_{dip} = \frac{\mu_{Bohr}^2}{|r|^3}[\vec{g}_{1z} \cdot \vec{g}_{2z} - 3(\vec{g}_{1z} \cdot \vec{r})(\vec{r} \cdot \vec{g}_{2z})] \quad (2)$$

$$J_{total} = J_{exch} + J_{dip} \quad (3)$$

In the calculations, the SARC2-DKH-QZVP basis set for Dy and Lu atoms [60], DKH-def2-TZVP for O atoms, and DKH-def2-SVP [61] for other atoms. The "Autoaux" feature was used to automatically generate auxiliary basis sets for RIJCOSX approximations. In CASSCF calculations, the active space included 9 electrons in 7 orbitals for the Dy^{3+} ion (CASSCF(9,7)). In RASSI calculations, all 21 sextets were considered, and selecting all sextets to describe Dy-SMMs is reliable [62–64]. The relativistic effects were considered using the second-order Douglas-Kroll-Hess (DKH2) method [65]. Additionally, the effective energy barrier was calculated using the method proposed by Yin et al. [20,66], and the magnetic relaxation mechanism of **1a** and **1b** was analyzed, which involves Equations (4)–(8).

$$U_{\mathrm{eff}}(T) = \sum_{i=1}^{n} \frac{K_i(T)}{\sum_i K_i(T)} E_i \quad (4)$$

$$K_i(T) = \frac{\exp(-E_i/k_B T)}{\sum_i \exp(-E_i/k_B T)} k_{\mathrm{QT},i} \quad (5)$$

$$k_{\mathrm{QT},i} = \frac{g_{\mathrm{XY},i}^2}{2\sqrt{(g_{\mathrm{XY},i}^2 + g_{\mathrm{Z},i}^2)}}, \; g_{XY} = \sqrt{(g_{\mathrm{X}}^2 + g_{\mathrm{Y}}^2)} \quad (6)$$

$$N = \sum_i K_i(T) \quad (7)$$

$$\tau_{\mathrm{QTM},i}^{-1} = \frac{\beta B_{ave}}{h} k_{\mathrm{QT},i} \quad (8)$$

In **1**, the two Dy^{3+} ions have similar coordination environments, so the calculated saturated effective barriers for fragments **1a** and **1b** are close, with values of **1a** (U_{eff}(300 K) = 601.5 cm^{-1}) and 1b (U_{eff}(300 K) = 601.4 cm^{-1}) (Figure S7). From the energy gaps of the two fragments, **1a** (856.6 cm^{-1}) and **1b** (856.5 cm^{-1}) are also similar (Tables S5 and S6), further indicating that the two Dy^{3+} ions have similar coordination environments. From the perspective of magnetic relaxation mechanisms and wave functions, the lowest-energy Kramers doublets (KDs) of **1a** and **1b** are both pure $| \pm 15/2>$, with wave function compositions of 99.8%. In the ground state, the probabilities of magnetic quantum tunneling between the ground state KDs of the two fragments are 2.63×10^{-4} μ_B for **1a** and 2.64×10^{-4} μ_B for **1b** (Figure S8), indicating that the QTM process is almost quenched in the ground state. The magnetic relaxation process occurs via the thermally activated third excited state (KD_2). At the same time, the calculated g_{xy} values are 1.151×10^{-3} and 1.156×10^{-3} for **1a** and **1b**, respectively, both of which are less than 0.015 [67]. This also indicates that **1** can be considered a candidate for zero-field single molecule magnets.

To further investigate the magnetic interaction between two Dy^{3+} ions, the magnetic coupling constant was calculated using the POLY_ANISO module [54,55] based on the Lines model. The J_{total} of -0.752 cm^{-1} ($J_{exch} = -0.187$ cm^{-1} and $J_{dip} = -0.565$ cm^{-1}) indicates that there is a weak antiferromagnetic coupling between the two Dy^{3+} ions, which is consistent with experimental measurements. From the perspective of the main magnetic axis direction of the ground state of compound **1**, its main magnetic axis in the ground state is along the O–Dy–O axis and does not lie in the Dy_2Cl_6 plane (Figure S9).

4. Conclusions

In summary, we have successfully synthesized a field-induced dinuclear Dy(III) SMM. Each Dy(III) ion has a compressed octahedral geometry with four Cl^- ligands on the four equatorial positions and two oxygen atoms from two benzo[*h*]quinolin-10-ol ligands on the two axial positions. The two Dy(III) ions are bridged by two Cl^- ligands in the equatorial plane to construct the dinuclear structure with the four HOBQ ligands on the axial positions and six Cl^- ligands in the equatorial plane, which leads to much shorter axial coordination bonds than the equatorial ones. Its static and dynamic magnetic properties were investigated in detail. It revealed a field-induced single molecule magnet for **1** with an obvious magnetic hysteresis loop and an energy barrier of 71(2) K. The ab initio complete active space self-consistent field (CASSCF) calculation results support the experimental results. It is a typical example for the construction of Dy(III) SMM through regulating coordination geometry which was achieved by controlling axial coordination bonds shorter than the equatorial ones. Although the performance of this SMM is not as high as expected, it can give us some guiding for the construction of SMMs, as well as some hints for further modification to achieve high-performance SMMs.

Supplementary Materials: The following supporting information can be downloaded at https://www.mdpi.com/article/10.3390/magnetochemistry11090073/s1, Figure S1: FT-IR spectrum of **1**; Figure S2: PXRD patterns of **1**; Figure S3: H-bonded chain (broken red lines) of **1** along axis a; Figure S4: 2D supramolecular framework of **1** in the $(1\bar{1}0)$ plane formed through $\pi\cdots\pi$ stacking interactions (dashed pink lines); Figure S5: The χ_M^{-1} vs. T plot of for **1** in the temperature range of 2.0–300 K with the solid line representing the linear fitting based on Curie-Weiss law; Figure S6: The molecular structure of **1a** (left) and **1b** (right) for calculation; Figure S7: The calculated U_{eff} and relative contribution of each Kramer's doublet to the relaxation calculated for **1a** (top) and **1b** (bottom); Figure S8: Possible relaxation mechanism based on transition magnetic moment for **1a** (top) and **1b** (bottom); Figure S9: The orientation of ground state main magnetic axis in **1**; Table S1: Crystallographic data for complexes **1**; Table S2: Selected bond lengths/Å and bond angles/° for **1**; Table S3: Hydrogen bonds for **1** (Å and °); Table S4: Computational analysis of the coordination configuration of Dy(III) in **1**; Table S5: The energy level, *g*-factor, the angle (θ) between the ground state and the excited state of the g_z magnetic axis, magnetic relaxation time (τ_{QTM}), and wave function for **1a**; Table S6: The energy level, *g*-factor, the angle (θ) between the ground state and the excited state of the g_z magnetic axis, magnetic relaxation time (τ_{QTM}), and wave function for **1b**.

Author Contributions: The manuscript was prepared with the contributions of all authors. They specifically contributed as follows: L.Z.—Investigation, writing—original draft; H.L.—Investigation; Y.L.—Formal analysis and supervision; D.L.—Review and editing, supervision; Z.Y.—Formal analysis; S.L.—Review and editing, theoretical calculation; Z.C.—Project administration, supervision. All authors have read and agreed to the published version of the manuscript.

Funding: This work was supported by National Natural Science Foundation of China (Nos. 12064002 and 22361005) and Guangxi Technology Base and Talent Subject, China (grant no. GUIKE AD23026067).

Institutional Review Board Statement: Not applicable.

Informed Consent Statement: Not applicable.

Data Availability Statement: CCDC 2469036 contains the supplementary crystallographic data for complex **1**. These data can be obtained free of charge via https://www.ccdc.cam.ac.uk/structures/ (accessed on 5 February 2025) or from Cambridge Crystallographic Data Centre, 12 Union Road, Cambridge CB2 1EZ, UK; Fax: (+44)-1223-336-033; or e-mail: deposit@ccdc.cam.ac.uk. The original contributions presented in this study are included in the article and Supplementary Materials; further inquiries can be directed to the corresponding authors.

Acknowledgments: We gratefully acknowledge Bing Yin (Lab of Theoretical Molecular Magnetism, College of Chemistry and Materials Science, Northwest University, Xi'an 710127, China) for his constructive suggestions on theoretical calculations.

Conflicts of Interest: The authors declare no conflicts of interest.

References

1. Benner, F.; Jena, R.; Odom, A.L.; Demir, S. Magnetic Hysteresis in a Dysprosium Bis(amide) Complex. *J. Am. Chem. Soc.* **2025**, *147*, 8156–8167. [CrossRef]
2. Kong, H.; Wang, J.-Y.; Liu, J.-C.; Zhang, L.; Liao, P.-Y.; Qi, Y.-Q.; Liu, Z.; Wu, S.-G.; Tong, M.-L. Photochromic Dysprosium Single-Molecule Magnet Featuring Reversible Redox Transformation of Polyoxomolybdate Moiety. *Angew. Chem. Int. Ed.* **2025**, *64*, e202422557. [CrossRef]
3. Ling, B.-K.; Chang, M.; Zhai, Y.-Q.; Deng, J.; Kofu, M.; Guo, H.; Zhao, J.; Fu, Z.; Zheng, Y.-Z. Quantum Spin Wave Excited from a Cr–Dy Single-Molecule Magnet. *J. Am. Chem. Soc.* **2025**, *147*, 10935–10942. [CrossRef] [PubMed]
4. Sessoli, R.; Gatteschi, D.; Caneschi, A.; Novak, M.A. Magnetic bistability in a metal-ion cluster. *Nature* **1993**, *365*, 141–143. [CrossRef]
5. Qu, Y.-X.; Liao, P.-Y.; Chen, Y.-C.; Tong, M.-L. Functional lanthanide complexes with N,N′-bis(2-hydroxybenzyl)-N,N′-bis(pyridin-2-ylmethyl)ethylenediamine (H_2bbpen) derivatives: Coordination chemistry, single-molecule magnetism and optical properties. *Coord. Chem. Rev.* **2023**, *475*, 214880. [CrossRef]
6. Vieru, V.; Gómez-Coca, S.; Ruiz, E.; Chibotaru, L.F. Increasing the Magnetic Blocking Temperature of Single-Molecule Magnets. *Angew. Chem. Int. Ed.* **2023**, *63*, e202303146. [CrossRef] [PubMed]
7. Hu, Z.-B.; Gui, L.-A.; Li, L.-H.; Xiao, T.-T.; Hand, A.T.; Tin, P.; Ozerov, M.; Peng, Y.; Ouyang, Z.; Wang, Z.; et al. Co^{II} single-ion magnet and its multi-dimensional aggregations: Influence of the structural rigidity on magnetic relaxation process. *Chin. Chem. Lett.* **2025**, *36*, 109600. [CrossRef]
8. Zhu, Z.; Zhao, C.; Feng, T.; Liu, X.; Ying, X.; Li, X.-L.; Zhang, Y.-Q.; Tang, J. Air-Stable Chiral Single-Molecule Magnets with Record Anisotropy Barrier Exceeding 1800 K. *J. Am. Chem. Soc.* **2021**, *143*, 10077–10082. [CrossRef]
9. Xu, W.-J.; Luo, Q.-C.; Li, Z.-H.; Zhai, Y.-Q.; Zheng, Y.-Z. Bis-Alkoxide Dysprosium(III) Crown Ether Complexes Exhibit Tunable Air Stability and Record Energy Barrier. *Adv. Sci.* **2024**, *11*, 2308548. [CrossRef]
10. Goodwin, C.A.P.; Ortu, F.; Reta, D.; Chilton, N.F.; Mills, D.P. Molecular magnetic hysteresis at 60 kelvin in dysprosocenium. *Nature* **2017**, *548*, 439–442. [CrossRef]
11. Ding, Y.-S.; Chilton, N.F.; Winpenny, R.E.P.; Zheng, Y.-Z. On Approaching the Limit of Molecular Magnetic Anisotropy: A Near-Perfect Pentagonal Bipyramidal Dysprosium(III) Single-Molecule Magnet. *Angew. Chem. Int. Ed.* **2016**, *55*, 16071–16074. [CrossRef] [PubMed]
12. Peng, S.; Ouyang, X.; Wang, Y.; Teng, Q.; Li, Y.; Zhang, X.; Hu, Z.; Wang, K.; Liang, F. In situ synthesized homochiral dysprosium-oxo clusters with threonine Schiff bases. *Chin. Chem. Lett.* **2023**, *34*, 108044. [CrossRef]

3. Armenis, A.S.; Bakali, G.P.; Paul, S.; Pantelis, K.N.; Cunha-Silva, L.; Tang, J.; Alexandropoulos, D.I.; Wernsdorfer, W.; Moreno-Pineda, E.; Stamatatos, T.C. Deciphering the Magnetostructural Criteria in Dy^{III} and Ho^{III} Macrocycle-Based Single-Molecule Magnets with Pseudo-D_{5h} Symmetry: A Combined Single-Crystal Hysteresis and Theoretical Study. *Inorg. Chem.* **2025**, *64*, 11476–11489. [CrossRef]
4. Baka, K.H.; Liu, D.; Paul, S.; Wernsdorfer, W.; Tang, J.; Chibotaru, L.F.; Stamatatos, T.C. Unveiling the Werner-Type Cluster Chemistry of Heterometallic 4f/Post-Transition Metals: A $\{Dy_3Bi_8\}$ Complex Exhibiting Quantum Tunneling Steps in the Hysteresis Loops and its 1-D Congener. *Inorg. Chem.* **2025**, *64*, 1962–1974. [CrossRef] [PubMed]
5. Armenis, A.S.; Mondal, A.; Giblin, S.R.; Alexandropoulos, D.I.; Tang, J.; Layfield, R.A.; Stamatatos, T.C. 'Kick-in the head': High-performance and air-stable mononuclear Dy^{III} single-molecule magnets with pseudo-D_{6h} symmetry from a [1 + 1] Schiff-base macrocycle approach. *Inorg. Chem. Front.* **2025**, *12*, 1214–1224. [CrossRef]
6. Zhu, Z.; Paul, S.; Zhao, C.; Wu, J.; Ying, X.; Ungur, L.; Wernsdorfer, W.; Meyer, F.; Tang, J. Record Quantum Tunneling Time in an Air-Stable Exchange-Bias Dysprosium Macrocycle. *J. Am. Chem. Soc.* **2024**, *146*, 18899–18904. [CrossRef]
7. Du, J.; Wei, S.; Jiang, Z.; Ke, H.; Sun, L.; Zhang, Y.; Chen, S. Influence of lattice water molecules on the magnetization dynamics of binuclear dysprosium(III) compounds: Insights from magnetic and ab initio calculations. *Phys. Chem. Chem. Phys.* **2023**, *25*, 11717–11724. [CrossRef]
8. Zhu, Z.; Ying, X.; Zhao, C.; Zhang, Y.-Q.; Tang, J. A new breakthrough in low-coordinate Dy(III) single-molecule magnets. *Inorg. Chem. Front.* **2022**, *9*, 6061–6066. [CrossRef]
9. Bazhenova, T.A.; Kopotkov, V.A.; Korchagin, D.V.; Yureva, E.A.; Zhidkov, M.V.; Dmitriev, A.I.; Yakushev, I.A.; Efimov, N.N.; Babeshkin, K.A.; Mironov, V.S.; et al. Pentagonal-bipyramidal dysprosium(III) complexes with two apical phosphine oxide ligands and equatorial pentadentate N_3O_2 Schiff-base ligands: Breakdown of the apical magnetic axiality by a strong equatorial crystal field. *Inorg. Chem. Front.* **2025**, *12*, 231–252. [CrossRef]
10. Wu, X.; Li, J.-F.; Yin, B. The interpretation and prediction of lanthanide single-ion magnets from ab initio electronic structure calculation: The capability and limit. *Dalton Trans.* **2022**, *51*, 14793–14816. [CrossRef]
11. Long, B.-F.; Li, Y.-L.; Zhu, Z.-H.; Wang, H.-L.; Liang, F.-P.; Zou, H.-H. Assembly of pinwheel/twist-shaped chiral lanthanide clusters with rotor structures by an annular/linear growth mechanism and their magnetic properties. *Dalton Trans.* **2022**, *51*, 17040–17049. [CrossRef]
12. Chen, Y.-C.; Tong, M.-L. Single-molecule magnets beyond a single lanthanide ion: The art of coupling. *Chem. Sci.* **2022**, *13*, 8716–8726. [CrossRef]
13. Liu, J.-L.; Chen, Y.-C.; Tong, M.-L. Symmetry strategies for high performance lanthanide-based single-molecule magnets. *Chem. Soc. Rev.* **2018**, *47*, 2431–2453. [CrossRef]
14. Wang, J.; Li, Q.-W.; Wu, S.-G.; Chen, Y.-C.; Wan, R.-C.; Huang, G.-Z.; Liu, Y.; Liu, J.-L.; Reta, D.; Giansiracusa, M.J.; et al. Opening Magnetic Hysteresis by Axial Ferromagnetic Coupling: From Mono-Decker to Double-Decker Metallacrown. *Angew. Chem. Int. Ed.* **2021**, *60*, 5299–5306. [CrossRef]
15. Le Roy, J.J.; Ungur, L.; Korobkov, I.; Chibotaru, L.F.; Murugesu, M. Coupling Strategies to Enhance Single-Molecule Magnet Properties of Erbium–Cyclooctatetraenyl Complexes. *J. Am. Chem. Soc.* **2014**, *136*, 8003–8010. [CrossRef] [PubMed]
16. Hilgar, J.D.; Bernbeck, M.G.; Rinehart, J.D. Million-fold Relaxation Time Enhancement across a Series of Phosphino-Supported Erbium Single-Molecule Magnets. *J. Am. Chem. Soc.* **2019**, *141*, 1913–1917. [CrossRef]
17. Gould, C.A.; McClain, K.R.; Reta, D.; Kragskow, J.G.C.; Marchiori, D.A.; Lachman, E.; Choi, E.-S.; Analytis, J.G.; Britt, R.D.; Chilton, N.F.; et al. Ultrahard magnetism from mixed-valence dilanthanide complexes with metal-metal bonding. *Science* **2022**, *375*, 198–202. [CrossRef] [PubMed]
18. Liu, W.; Qiu, Y.; Xu, W.; Cao, M.; Deng, X.; Lin, W. Slow magnetic relaxation behaviors and diamagnetic dilution studies of a carboxyl bridged dinuclear dysprosium(III) complex. *J. Rare Earths* **2024**, *42*, 2088–2096. [CrossRef]
19. Long, J.; Habib, F.; Lin, P.-H.; Korobkov, I.; Enright, G.; Ungur, L.; Wernsdorfer, W.; Chibotaru, L.F.; Murugesu, M. Single-Molecule Magnet Behavior for an Antiferromagnetically Superexchange-Coupled Dinuclear Dysprosium(III) Complex. *J. Am. Chem. Soc.* **2011**, *133*, 5319–5328. [CrossRef]
20. Long, J.; Tolpygin, A.O.; Cherkasov, A.V.; Lyssenko, K.A.; Guari, Y.; Larionova, J.; Trifonov, A.A. Single-molecule magnet behavior in heterolopetic Dy^{3+}-chloro-diazabutadiene complexes: Influence of the nuclearity and ligand redox state. *Dalton Trans.* **2020**, *49*, 11890–11901. [CrossRef] [PubMed]
21. Acharya, J.; Ahmed, N.; Flores Gonzalez, J.; Kumar, P.; Cador, O.; Singh, S.K.; Pointillart, F.; Chandrasekhar, V. Slow magnetic relaxation in a homo dinuclear Dy(III) complex in a pentagonal bipyramidal geometry. *Dalton Trans.* **2020**, *49*, 13110–13122. [CrossRef]
22. Kalita, P.; Ahmed, N.; Bar, A.K.; Dey, S.; Jana, A.; Rajaraman, G.; Sutter, J.-P.; Chandrasekhar, V. Pentagonal Bipyramidal Ln(III) Complexes Containing an Axial Phosphine Oxide Ligand: Field-induced Single-ion Magnetism Behavior of the Dy(III) Analogues. *Inorg. Chem.* **2020**, *59*, 6603–6612. [CrossRef]

33. Zhu, W.-H.; Xiong, X.; Gao, C.; Li, S.; Zhang, Y.; Wang, J.; Zhang, C.; Powell, A.K.; Gao, S. A family of one-dimensional lanthanide complexes bridged by two distinct carboxylate ligands with the Dy analogue displaying magnetic relaxation behaviour. *Dalton Trans.* **2017**, *46*, 14114–14121. [CrossRef] [PubMed]
34. Wang, Y.-H.; Gao, Z.-N.; Liang, S.; Wei, W.-J.; Han, S.-D.; Zhang, Y.-Q.; Hu, J.-X.; Wang, G.-M. Synergism of Light-Induced [4 + 4] Cycloaddition and Electron Transfer Toward Switchable Photoluminescence and Single-Molecule Magnet Behavior in a Dy_4 Cubane. *Research* **2024**, *7*, 0411. [CrossRef]
35. Zhang, Q.; Han, S.-D.; Li, Q.; Hu, J.-X.; Wang, G.-M. Electron transfer photochromism of Ln-based (Ln = Dy, Tb) coordinated polymers for reversibly switching off/on single-molecule magnetic behavior. *Sci. China Mater.* **2022**, *65*, 788–794. [CrossRef]
36. Chen, L.; Niu, B.; Mansikkamäki, A.; Liu, K.; Sun, L.; Wang, W.; Xie, G.; Chen, S.; Ke, H. Alcohol-bridged dinuclear dysprosium single-molecule magnets. *J. Mol. Struct.* **2025**, *1320*, 139673. [CrossRef]
37. Zhang, K.; Guo, F.-S.; Wang, Y.-Y. Synthesis, crystal structure and magnetic properties of a dinuclear dysprosium single-molecule magnet. *Inorg. Chem. Commun.* **2017**, *76*, 95–99. [CrossRef]
38. Ke, H.; Yang, Y.; Wei, W.; Jiang, Y.; Zhang, Y.-Q.; Xie, G.; Chen, S. Synergistic effect of mixed ligands on the anisotropy axis of two dinuclear dysprosium complexes. *Dalton Trans.* **2020**, *49*, 10594–10602. [CrossRef] [PubMed]
39. Huang, Z.-N.; Sun, R.; Ma, F.; Yang, J.; Huang, X.-X.; Sun, H.-L.; Gao, S. Coordination-driven assembly of dysprosium layers into a framework featuring enhanced energy barriers and quantum tunneling of magnetization. *Inorg. Chem. Front.* **2024**, *11*, 4686–4694. [CrossRef]
40. Yao, X.; Yan, P.; An, G.; Li, Y.; Li, W.; Li, G. Investigation of magneto-structural correlation based on a series of seven-coordinated β-diketone Dy(III) single-ion magnets with C_{2v} and C_{3v} local symmetry. *Dalton Trans.* **2018**, *47*, 3976–3984. [CrossRef]
41. Ge, Y.; Huang, Y.; Montenegro, J.L.B.; Cui, Y.; Liu, W.; Li, Y.; Wang, B.-L. Synthesis, Structures, and Single-Molecule Magnetic Properties of Three Dy_2 Complexes. *Chem. Asian J.* **2019**, *14*, 986–994. [CrossRef]
42. Wang, T.-T.; Che, Z.-W.; Chen, J.-T.; Yan, H.; Zhou, T.-D.; Zhang, Y.-Q.; Sun, W.-B. Tetranuclear lanthanide complexes display significant slow magnetic relaxation with an open hysteresis loop and the magnetocaloric effect. *Inorg. Chem. Front.* **2023**, *10*, 1501–1510. [CrossRef]
43. Klementyeva, S.V.; Gamer, M.T.; Schulze, M.; Suryadevara, N.; Bogomyakov, A.S.; Abramov, P.A.; Konchenko, S.N.; Ruben, M.; Wernsdorfer, W.; Moreno-Pineda, E. Dinuclear Rare-Earth β-Diketiminates with Bridging 3,5-Ditert-butyl-catecholates: Synthesis, Structure, and Single-Molecule Magnet Properties. *Inorg. Chem.* **2025**, *64*, 701–715. [CrossRef] [PubMed]
44. Bartolomé, E.; Arauzo, A.; Luzón, J.; Melnic, S.; Shova, S.; Prodius, D.; Nlebedim, I.C.; Bartolomé, F.; Bartolomé, J. High relaxation barrier in neodymium furoate-based field-induced SMMs. *Dalton Trans.* **2019**, *48*, 15386–15396. [CrossRef] [PubMed]
45. Xémard, M.; Cordier, M.; Molton, F.; Duboc, C.; Le Guennic, B.; Maury, O.; Cador, O.; Nocton, G. Divalent Thulium Crown Ether Complexes with Field-Induced Slow Magnetic Relaxation. *Inorg. Chem.* **2019**, *58*, 2872–2880. [CrossRef] [PubMed]
46. Zhong, X.; Rana, R.; Luo, T.-K.; Fang Wu, Y.; Peng, Y.; Liu, S.-J.; Rajaraman, G.; Wen, H.-R. Systematically study of ligands field effect on a pseudo-D_{6h} Dy(III) single-ion magnet with axially asymmetric ligands. *J. Mol. Struct.* **2025**, *1338*, 142270. [CrossRef]
47. Luo, T.-K.; Luo, Q.-C.; Cao, C.; Hu, Z.-B.; Zhong, X.; Peng, Y.; Niu, Z.-G.; Zheng, Y.-Z.; Liu, S.-J.; Wen, H.-R. Remote substituents effect on the valence and magnetic behavior in mononuclear cobalt complexes. *J. Mol. Struct.* **2025**, *1344*, 142997. [CrossRef]
48. Athanasopoulou, A.A.; Baldoví, J.J.; Carrella, L.M.; Rentschler, E. Field-induced slow magnetic relaxation in the first Dy(III)-centered 12-metallacrown-4 double-decker. *Dalton Trans.* **2019**, *48*, 15381–15385. [CrossRef]
49. Goodwin, C.A.P.; Reta, D.; Ortu, F.; Chilton, N.F.; Mills, D.P. Synthesis and Electronic Structures of Heavy Lanthanide Metallocenium Cations. *J. Am. Chem. Soc.* **2017**, *139*, 18714–18724. [CrossRef]
50. Shrivastava, K.N. Theory of Spin-Lattice Relaxation. *Phys. Status Solidi B* **1983**, *117*, 437–458. [CrossRef]
51. Neese, F.; Wennmohs, F.; Becker, U.; Riplinger, C. The ORCA quantum chemistry program package. *J. Chem. Phys.* **2020**, *152*, 224108. [CrossRef] [PubMed]
52. Neese, F. Software update: The ORCA program system—Version 5.0. *WIREs Comput. Mol. Sci.* **2022**, *12*, e1606. [CrossRef]
53. Malmqvist, P.-Å.; Roos, B.O. The CASSCF state interaction method. *Chem. Phys. Lett.* **1989**, *155*, 189–194. [CrossRef]
54. Chibotaru, L.F.; Ungur, L. Ab initio calculation of anisotropic magnetic properties of complexes. I. Unique definition of pseudospin Hamiltonians and their derivation. *J. Chem. Phys.* **2012**, *137*, 064112. [CrossRef]
55. Ungur, L.; Chibotaru, L.F. Ab Initio Crystal Field for Lanthanides. *Chem. Eur. J.* **2017**, *23*, 3708–3718. [CrossRef] [PubMed]
56. Lines, M.E. Orbital Angular Momentum in the Theory of Paramagnetic Clusters. *J. Chem. Phys.* **1971**, *55*, 2977–2984. [CrossRef]
57. Chibotaru, L.F.; Ungur, L.; Aronica, C.; Elmoll, H.; Pilet, G.; Luneau, D. Structure, Magnetism, and Theoretical Study of a Mixed-Valence $Co^{II}_3Co^{III}_4$ Heptanuclear Wheel: Lack of SMM Behavior despite Negative Magnetic Anisotropy. *J. Am. Chem. Soc.* **2008**, *130*, 12445–12455. [CrossRef] [PubMed]
58. Chibotaru, L.F.; Ungur, L.; Soncini, A. The Origin of Nonmagnetic Kramers Doublets in the Ground State of Dysprosium Triangles: Evidence for a Toroidal Magnetic Moment. *Angew. Chem. Int. Ed.* **2008**, *47*, 4126–4129. [CrossRef]
59. Luo, Q.-C.; Yu, K.-X.; Jin, P.-B.; Liu, Y.-Y.; Zhai, Y.-Q.; Zheng, Y.-Z. Strong Antiferromagnetic Exchange-Coupling Observed in Hydride-Bridged Dimeric Dysprosium(III) Single-Molecule Magnet. *Chin. J. Chem.* **2024**, *42*, 391–396. [CrossRef]

60. Aravena, D.; Neese, F.; Pantazis, D.A. Improved Segmented All-Electron Relativistically Contracted Basis Sets for the Lanthanides. *J. Chem. Theory Comput.* **2016**, *12*, 1148–1156. [CrossRef]
61. Weigend, F.; Ahlrichs, R. Balanced basis sets of split valence, triple zeta valence and quadruple zeta valence quality for H to Rn: Design and assessment of accuracy. *Phys. Chem. Chem. Phys.* **2005**, *7*, 3297–3305. [CrossRef]
62. Briganti, M.; Garcia, G.F.; Jung, J.; Sessoli, R.; Le Guennic, B.; Totti, F. Covalency and magnetic anisotropy in lanthanide single molecule magnets: The DyDOTA archetype. *Chem. Sci.* **2019**, *10*, 7233–7245. [CrossRef] [PubMed]
63. Shi, Y.; Zhang, C.; Shi, M.; Zhou, Y.; Li, H.; Liu, R.; Zheng, S.; Yuan, A.; Chen, L.; Chen, S.-Y.; et al. A high-performance dysprosium single-ion magnet with local pseudo cubic geometry. *Inorg. Chem. Front.* **2025**, *12*, 1090–1101. [CrossRef]
64. Wang, Y.-F.; Wang, Y.-X.; Yang, Q.-Q.; Yin, B. Auxiliary Rather Than Dominant. The Role of Direct Dy–S Coordination in Single-Molecule Magnet Unveiled via ab initio Study. *J. Phys. Chem. A* **2024**, *128*, 5285–5297. [CrossRef]
65. Nakajima, T.; Hirao, K. The Douglas–Kroll–Hess Approach. *Chem. Rev.* **2012**, *112*, 385–402. [CrossRef]
66. Yin, B.; Li, C.-C. A method to predict both the relaxation time of quantum tunneling of magnetization and the effective barrier of magnetic reversal for a Kramers single-ion magnet. *Phys. Chem. Chem. Phys.* **2020**, *22*, 9923–9933. [CrossRef] [PubMed]
67. Aravena, D.; Ruiz, E. Shedding Light on the Single-Molecule Magnet Behavior of Mononuclear Dy^{III} Complexes. *Inorg. Chem.* **2013**, *52*, 13770–13778. [CrossRef] [PubMed]

magnetochemistry

MDPI

Article

A Dy_2 Complex Constructed by $TCNQ^{\cdot-}$ Radical Anions with Slow Magnetic Relaxation Behavior †

Xirong Wang [1], Shijia Qin [1], Xiulan Li [1], Wenjing Zuo [1], Qinglun Wang [1], Licun Li [1,*], Yue Ma [1,*], Jinkui Tang [2] and Bin Zhao [1]

[1] Frontiers Science Center for New Organic Matter, Key Laboratory of Advanced Energy Materials Chemistry (MOE), College of Chemistry, Nankai University, Tianjin 300071, China; 2120231054@mail.nankai.edu.cn (X.W.); 2120231050@mail.nankai.edu.cn (S.Q.); 2120231061@mail.nankai.edu.cn (X.L.); 2120241094@mail.nankai.edu.cn (W.Z.); wangql@nankai.edu.cn (Q.W.); zhaobin@nankai.edu.cn (B.Z.)

[2] State Key Laboratory of Rare Earth Resource Utilization, Changchun Institute of Applied Chemistry, Chinese Academy of Sciences, Renmin Street 5625, Changchun 130022, China; tang@ciac.ac.cn

* Correspondence: llicun@nankai.edu.cn (L.L.); maynk@nankai.edu.cn (Y.M.)

† Dedicated to Prof. Daizheng Liao on the occasion of his 85th birthday.

Abstract: A centrosymmetric dinuclear complex, $[Dy_2(H_2dapp)_2(\mu\text{-}OH)_2(H_2O)_2]\cdot 4TCNQ\cdot 2CH_3OH$, was synthesized using the $TCNQ^{\cdot-}$ radical anion (TCNQ = 7,7,8,8-tetracyanoquinodimethane) and pentadentate nitrogen-containing Schiff base ligand (H_2dapp = 2,6-diacetylpyridine)-bis(2-pyridylhydrazone). In the Dy_2 dimer, the two Dy^{III} ions adopt eight-coordinated geometries intermediate between D_{4d} and D_{2d} symmetries, linked by two OH^- groups, with ferromagnetic Dy-Dy interactions. The $TCNQ^{\cdot-}$ radical anions are uncoordinated, and they pack tightly into antiparamagnetic dimers to balance the system charge. Under zero field, weak magnetic relaxation was observed, with an approximate Δ_{eff} = 2.82 K and $\tau_0 = 6.88 \times 10^{-6}$ s. This might be attributed to the short intermolecular Dy···Dy distance of 7.97 Å, which could enhance intermolecular dipolar interactions and quantum tunneling of magnetization (QTM).

Keywords: Dy_2 complex; $TCNQ^{\cdot-}$ radical anion; slow magnetic relaxation

1. Introduction

Single-molecule magnets (SMMs), a class of nanoscale magnets that exhibit magnetic bistability at the molecular level [1], offer unprecedented opportunities for miniaturizing data storage devices and advancing quantum technologies due to their ability to retain magnetization at the single-molecule level [2–4]. Among the high-performance SMMs, rare-earth ions (especially Dy^{III}) have emerged as leading candidates due to their strong spin-orbit coupling and large magnetic anisotropy, which are critical for achieving high energy barriers (U_{eff}) and blocking temperatures (T_B) [5–7]. In particular, significant progress in dysprosium metallocene SMMs has been demonstrated by $[Dy(Cp^{ttt})_2][B(C_6F_5)_4]$ achieving a blocking temperature of 60 K [8], followed by its improved heteroligand derivative $[(Cp^{iPr5})Dy(Cp^*)][B(C_6F_5)_4]$ reaching 80 K [9]. More recently, a dysprosium bis(amide)–alkene complex of $[Dy\{N(Si^iPr_3)[Si(^iPr)_2C(CH_3)=CHCH_3]\}\{N(Si^iPr_3)(Si^iPr_2Et)\}][Al\{OC(CF_3)_3\}_4]$ has been reported to exhibit an effective energy barrier to magnetic reversal of 1843(11) cm^{-1} and show slow closing of soft magnetic hysteresis loops up to 100 K, owing to the charge-dense amide ligands and pendant alkenes that structurally enforce a large N–Dy–N angle leading to weak equatorial interactions [10]. These discoveries offer new avenues for achieving magnetic memory functionality at higher temperatures.

While for most Dy^{III}-based SMMs, a limitation of their U_{eff} and T_B lies in the existence of QTM, which can induce rapid magnetization relaxation and loss of bistability under zero external field [11]. To quench QTM, symmetry control—by placing Dy^{III} ions in crystal ligands with high symmetry (e.g., D_{4d}, D_{5h}, D_{6d}, $C_{\infty v}$)—can effectively suppress transverse crystal field components which facilitate QTM [12–14]. Alternatively, strong intramolecular magnetic interactions provided by 3d/4f metals or 2p organic radicals can also quench the QTM of Dy^{III} systems to some extent [15–17]. Radicals prove to be particularly effective due to their diffuse orbitals penetrating 4f electron clouds, creating a "giant spin" effect [18,19]. This approach has spurred the development of diverse 2p–4f heterospin systems, incorporating radicals such as nitronyl nitroxide [20,21], $TCNQ^{\cdot -}$ [22,23] and N_2^{3-} [24], etc. Among these, $TCNQ^{\cdot -}$ radical anions and its derivatives stand out due to their multiple coordination sites, strong coordination ability, and extended π-conjugation [22,25,26]. They can not only effectively modulate the ligand field environment around Ln^{III} ions but also facilitate strong magnetic coupling by coordinating to Ln^{III} ions without other bridges [27]. This unique electronic characteristic endows $TCNQ^{\cdot -}$-based systems with intrinsic semiconducting properties, effectively bridging molecular magnetism and electronic conductivity—a particularly valuable feature in the design of SMMs that offers significant multifunctional potential [28]. The Dunabr group has synthesized a hetero-tri-spin complex $[(valpn)CuTb(TCNQF)_2(H_2O)_4][TCNQF]\cdot CH_3OH\cdot 6H_2O$, where the coordinated $TCNQF^-$ radical anions link the SMMs to single-chain magnetism via supramolecular π-stacking [23]. Subsequently, their further work of $Dy(TPMA)(\mu\text{-}TCNQ)(\mu\text{-}OH)_2\cdot CH_3CN$ with partially charged $TCNQ^{\cdot -}$ radical anions was prepared, which exhibited bifunctionality of SMMs performance (5.0–8.0 K) and semiconducting performance (180–350 K), advancing the development of radical-based multifunctional materials [29].

Auxiliary ligands critically influence SMM behavior by controlling metal ion coordination symmetry and magnetic anisotropy. The recently reported zero-dimensional dinuclear complex $Dy_2(B(OMe)_4)(Hdapp)_2(tBu\text{-}DDTP)_2\cdot 3MeOH$ [30], with Dy^{III} centers in a distorted triangular dodecahedral (D_{2d}) symmetry, exhibits slow magnetic relaxation. Subsequently, the dinuclear complexes $[Dy(HDAPP)(^{Me}DDTP)_2]\cdot EtOH$ and $Dy(H_2DAPP)(^{Me}DDTP)_2\cdot 2EtOH\cdot H_2O$ [31] adopt distorted pentagonal bipyramidal (D_{5h}-like) geometries, where $H_2dapp/Hdapp^-$ acts as an N_5 equatorial pocket, exhibiting SMMs behavior. Guided by these findings and in pursuit of more $TCNQ^{\cdot -}$-Dy^{III} SMMs with interesting properties, in this study, the pentadentate nitrogen-rich Schiff base ligand 2,6-diacetylpyridine-bis(2-pyridylhydrazone) (H_2dapp) was utilized as the ancillary ligand, which might provide a weak equatorial coordination environment and leave the axial coordination sites available for $TCNQ^{\cdot -}$ radical anions (TCNQ = 7,7,8,8-tetracyanoquinodimethane). The structure of the ligands are illustrated in Scheme 1. We aimed to establish a high-symmetry D_{5h} configuration and develop bifunctional materials with both SMM properties and conductivity. The dinuclear Dy^{III} complex $[Dy_2(H_2dapp)_2(\mu\text{-}OH)_2(H_2O)_2]\cdot 4TCNQ\cdot 2CH_3OH$ (**1**) was successfully obtained. Although the $TCNQ^{\cdot -}$ radical anions do not coordinate with Dy^{III} ions, they form antiferromagnetic dimers via strong π–π interactions. These dimers are further connected into an extended chain structure through additional intermolecular π–π interactions, and the complex still exhibits slow magnetic relaxation behavior under a zero dc field.

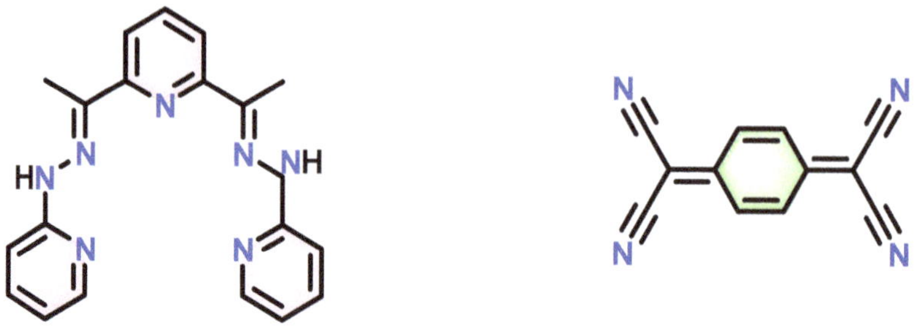

H_2dapp TCNQ$^{\cdot-}$

Scheme 1. Ligands used of **complex 1** in this work.

2. Results

2.1. Crystal Structure Description

Single-crystal X-ray diffraction analysis reveals that **complex 1** is a centrosymmetric dinuclear Dy^{III} complex, which crystallizes in the triclinic crystal system with space group *P*-1. The molecular structure consists of a $[Dy_2(H_2dapp)_2(\mu\text{-}OH)_2]^{4+}$ group, four uncoordinated TCNQ$^{\cdot-}$ radical anions, and two free methanol molecules (Figure 1). In the $[Dy_2(H_2dapp)_2(\mu\text{-}OH)_2]^{4+}$ unit, two Dy^{III} ions are bridged by oxygen atoms (O1 and O1a) from hydroxido groups. Each Dy^{III} ion is eight-coordinated by five nitrogen atoms from the H_2dapp ligand, two bridging oxygen atoms from hydroxido groups and one oxygen atom from a water molecule. The CShMs analysis (listed in Table S1) using SHAPE 2.1 software [32] reveals that the coordination geometry of Dy^{III} ions lies between square antiprismatic (D_{4d}) and triangular-dodecahedral (D_{2d}) symmetries. The Dy–N bond lengths range from 2.497(3) Å to 2.561(3) Å. The Dy–O bonds vary from 2.211(3) Å to 2.359(3) Å. The Dy···Dy distance bridged by μ-OH in Dy_2 dimer is 3.6603(3) Å, and the Dy–O–Dy bond angle is 110.34(11)°. The O–Dy–O and O–Dy–N angles are within 69.66(11)–154.73(10)° and 73.62(10)–142.17(10)°, respectively.

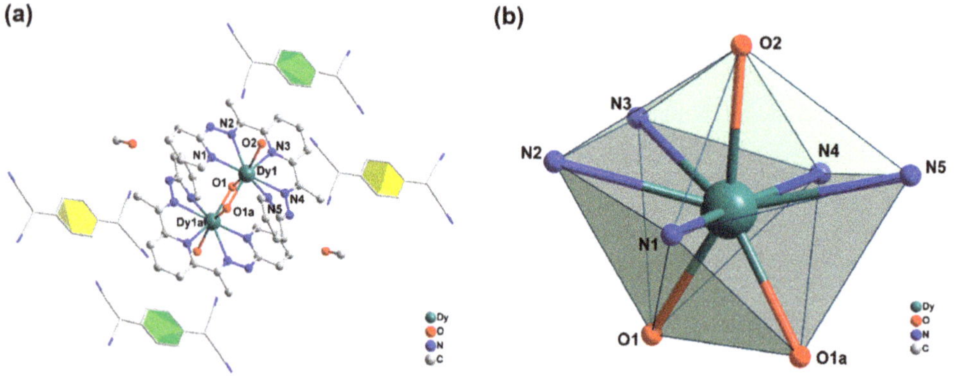

Figure 1. (**a**) Crystal structure and (**b**) coordination polyhedron configuration of Dy^{III} ion in **complex 1** (H atoms have been omitted).

Based on the electrical neutrality of **complex 1**, the approximate charge on each TCNQ$^{\cdot-}$ radical anions were calculated using Kistenmacher's empirical [22,33] (Equation (1)), where b, c, and d represent the relative average bond lengths of TCNQ$^{\cdot-}$ radical anions. Different TCNQ$^{\cdot-}$ radical anions were marked as A and B. The results indicate that all TCNQ$^{\cdot-}$ radical anions in the system carry a -1.0 charge (Table S3). The TCNQ$^{\cdot-}$ radical anions in **complex 1** are all uncoordinated, serving only to balance the charge of the system. They form strongly packed antiparamagnetic dimers [29,34] (but do not form a continuous chain structure) through strong π–π interactions, with an interplanar centroid distance of 3.2241 Å between benzene rings (As in Figure 2c). Although these TCNQ$^{\cdot-}$ dimers themselves do not form a continuous infinite one-dimensional stack, intermolecular π–π interactions exist between the TCNQ$^{\cdot-}$ radical anions and the pyridine ring of adjacent $[Dy_2(H_2dapp)_2(\mu\text{-}OH)_2]^{4+}$ units (as in Figure 2b) with distances of 3.4049 Å (C26···C10) and 3.4430 Å (C24···C9). In addition, a strong π–π interaction is present between the cyano group of TCNQ (C45 and N15) and the pyridine ring of the H_2dapp ligand involving N3, with a distance of approximately 3.1567 Å. These interactions collectively contribute to the formation of one-dimensional supramolecular chains. Hydrogen bonds exist between TCNQ$^{\cdot-}$ radical anions and coordinated water molecules (Table S2). Along the a-axis, the $[Dy_2(H_2dapp)_2(\mu\text{-}OH)_2]^{4+}$ units are effectively isolated through staggered lateral stacking of TCNQ$^{\cdot-}$ radical anions with an intermolecular Dy···Dy separation of 13.8636(3) Å (Figure S1). However, along the b-axis (Figure 2a), the TCNQ$^{\cdot-}$ dimers fail to isolate intermolecular Dy^{III} ions effectively, with a relatively shorter Dy···Dy distance of 7.9667(4) Å.

$$\delta = -41.67[c/(b+d)] + 19.83 \quad (1)$$

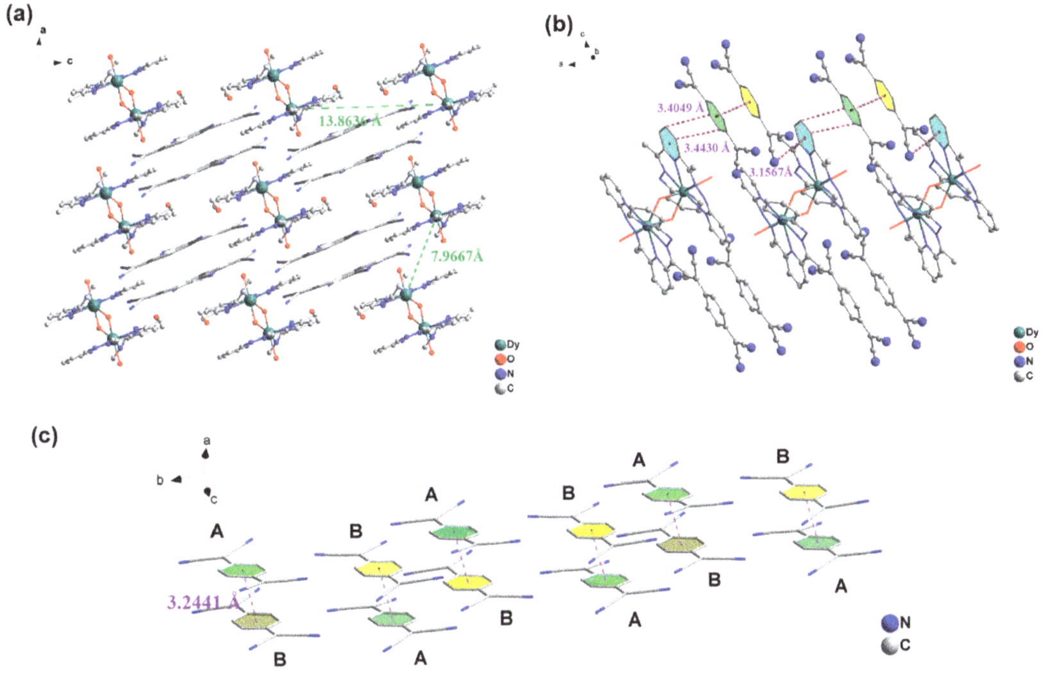

Figure 2. (**a**) Crystal stacking diagram of **complex 1** along the b-axis. (**b**) TCNQ$^{\cdot-}$ dimers form an infinite one-dimensional stacked structure through π–π interactions with $[Dy_2(H_2dapp)_2(\mu\text{-}OH)_2]^{4+}$ units. (**c**) Arrangement of TCNQ dimers.

2.2. Magnetic Properties

Variable-temperature magnetic susceptibility measurements were performed on **complex 1** at a dc (direct current) field of 1000 Oe and in the temperature range of 300–2 K. The phase purity of the synthesized crystals was confirmed by the consistency between the experimental PXRD data of **complex 1** and the simulated pattern as in Figure S2. At 300 K, the $\chi_M T$ value is 27.87 $cm^3 \cdot K \cdot mol^{-1}$ (Figure 3), which is close to the calculated value of 28.34 $cm^3 \cdot K \cdot mol^{-1}$ for two independent Dy^{III} ions ($^6H_{15/2}$, g = 4/3), with no contribution from the spin of the $TCNQ^{\cdot -}$ radical anions. As evident from the structure, the strong π–π interactions between the $TCNQ^{\cdot -}$ radical anions lead to the formation of dimers with orbital overlap, causing strong antiferromagnetic coupling between spins and resulting in diamagnetic dimers that do not contribute to magnetic properties of the system [34]. Upon cooling, the $\chi_M T$ value gradually increases to a maximum value of 29.51 $cm^3 \cdot K \cdot mol^{-1}$ at 70 K, suggesting an intramolecular ferromagnetic interaction between Dy^{III} ions [35–38]. The magnetic susceptibility data of $1/\chi_M$ versus T above 50 K (inset of Figure 3) were analyzed using the Curie–Weiss law [39,40] (Equation (2)). The fitting results show that the Curie constant C and the Weiss constant θ are 27.88 $cm^3 \cdot K \cdot mol^{-1}$ and 5.01 K, respectively, and $\theta > 0$ clearly proves the presence of ferromagnetic interactions between Dy^{III} ions [36]. However, below 70 K, the $\chi_M T$ value decreases rapidly and reaches a minimum value of 20.16 $cm^3 \cdot K \cdot mol^{-1}$ at 2 K, which might be due to the depopulation of Dy^{III} ions Stark sublevels and the intermolecular antiferromagnetic interactions between Dy^{III} ions [41,42].

$$1/\chi_M = \frac{T - \theta}{C} \tag{2}$$

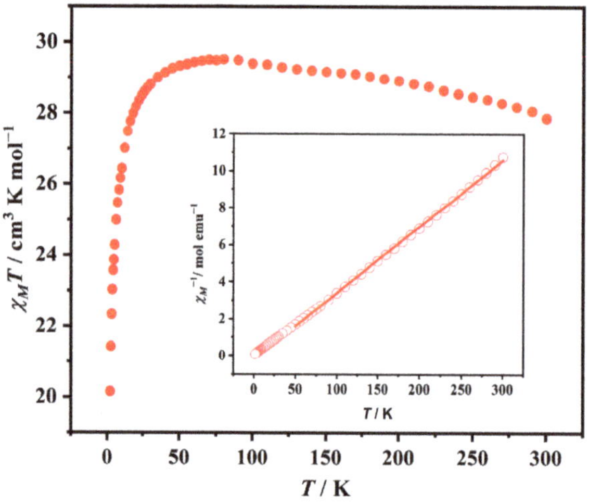

Figure 3. $\chi_M T$ versus T for **complex 1** at a 1000 Oe dc field. The inset represents temperature dependence of $1/\chi_M$ and the solid red indicates the result of a linear fit using the Curie–Weiss law.

The isothermal field-dependent magnetization (M vs. H plots in Figure 4) was measured at different temperatures from 0 to 70 kOe. It can be observed that the M values show a sharp rise as the magnetic field increases, reaching a maximum value of 13.80 μ_B under 70 kOe at 1.9 K. This maximum value is significantly lower than the expected value of 20 μ_B for two independent Dy^{III} ions ($g_J \times J = 4/3 \times 15/2 = 10$ μ_B), and the M vs. H/T curves do not overlap at various temperatures, indicating the presence of magnetic anisotropy and/or low-lying excited states.

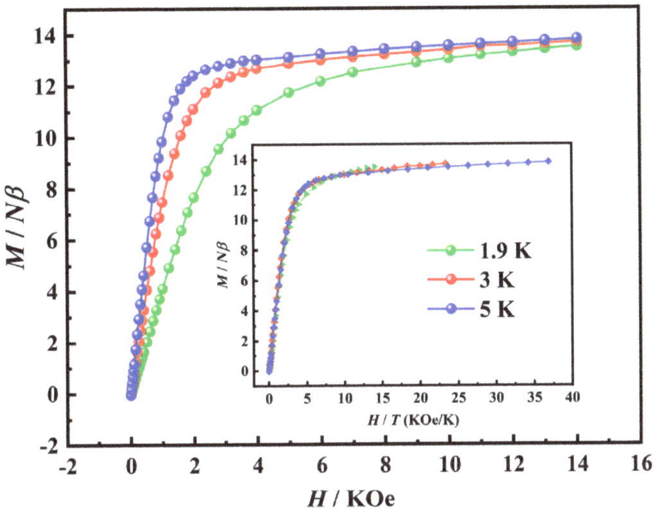

Figure 4. M vs. H plots of **complex 1** at different temperatures. The inset represents the plots of M vs. H/T.

To study the dynamic magnetic properties of **complex 1**, ac (alternating current) magnetic susceptibility measurements were conducted without an applied dc field. As shown in Figure 5, the χ'' signals exhibit clear frequency dependence in ac susceptibility measurements below 10 K, demonstrating slow magnetic relaxation behavior of **complex 1**. Unfortunately, no distinct peaks are observed, likely due to QTM effects. Then, the energy barrier (Δ_{eff}) and relaxation time (τ_0) cannot be determined through standard Arrhenius fitting. Assume that there is only one magnetization relaxation process, the ac susceptibility data can be analyzed using Equation (3) to approximate the Δ_{eff} and τ_0 [43,44]. The analysis yields an effective energy barrier Δ_{eff} = 2.82 K with a corresponding pre-exponential factor $\tau_0 = 6.88 \times 10^{-6}$ s. The observed τ_0 values fall within the characteristic range for SMMs [45]. To suppress QTM, field-dependent χ'' measurements at 1.9 K and 997 Hz revealed an optimal dc field of 400 Oe for **complex 1** (Figure S3a). Although χ'' signals were detected under this field, their minimal intensity variation and absence of distinct peaks (Figure S3b) indicate that quantum tunneling was not effectively quenched despite the applied field for **complex 1** [46–48].

$$\ln(\chi''/\chi') = \ln(\omega\tau_0) + \Delta_{\text{eff}}/k_B T \quad (3)$$

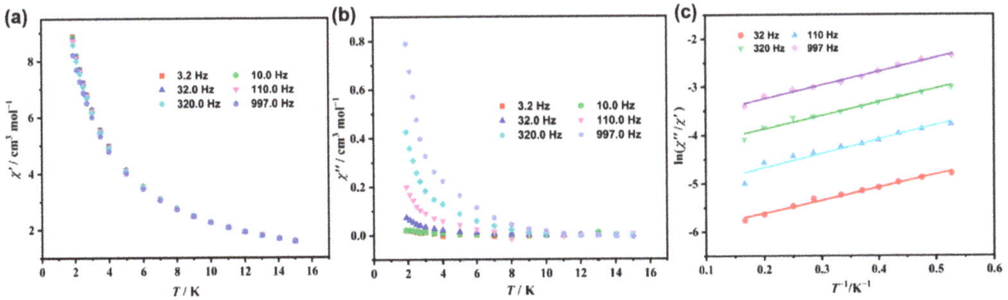

Figure 5. (**a**,**b**) Temperature-dependent ac susceptibility of **complex 1** under 0 Oe dc field. (**c**) Plots of $\ln(\chi''/\chi')$ versus $1/T$ for **complex 1**.

The magnetic anisotropy axes of the Dy^{III} ions in **complex 1** was calculated using the electrostatic model by the Magellan program [49] (Figure 6). The magnetic axis is oriented toward the μ-OH oxygen donor (O1). The shortest Dy1–O1 bond exhibits only a minor deviation of 8.504(139)° from the anisotropy axis. This near-parallel alignment likely enhances the axial electron density. The possible presence of D_{2d} symmetric geometries, combined with insufficient magnetic isolation due to the relatively shorter intermolecular Dy⋯Dy distance of 7.9667(4) Å (which enhances dipolar coupling), results in QTM and ultimately leads to poor SMM behavior. The comparison of the SMMs' behavior with that of other reported dinuclear Dy^{III} complexes is shown in Table 1 [50–54].

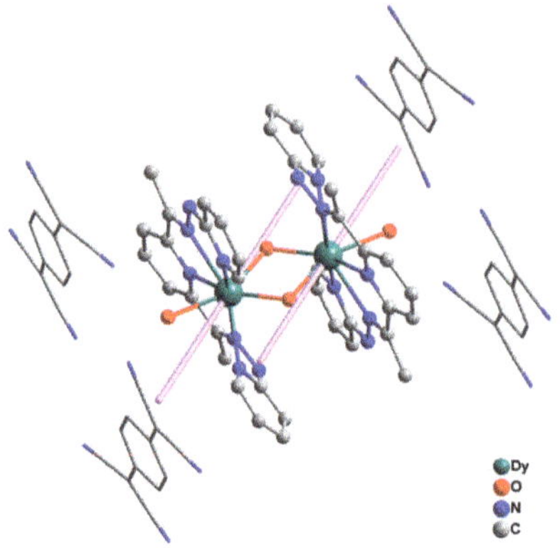

Figure 6. Calculated the magnetic anisotropy axes of Dy^{III} ions in **complex 1** using the MAGELLAN program.

Table 1. Some reported Dy_2 complexes exhibiting SMM behavior.

Dy_2 Complex	Point Group	Shortest Intermolecular Dy⋯Dy Distance	Magnetic Coupling	The Slow Relaxation Process		
				Field (Oe)	Ueff (K)	τ_0 (s)
$[Dy_2(dmpd)_6(bmp)]$	D_{4d}	10.487 Å	AF ($J = -0.524\ cm^{-1}$)	0	87.29	5.48×10^{-7}
$[Dy_2(dbm)_2(LH_3)_2]\cdot H_2O$	D_{4d}	9.423 Å	F ($J = 15.00\ cm^{-1}$)	0	76.18	2.54×10^{-8}
$[Dy(hfac)_3]_2(\mu\text{-}HMq)_2$	D_{4d}	10.028 Å	F ($J = 1.00\ cm^{-1}$)	0	9.2	1.7×10^{-5}
$[Dy_2(L_5)(tfac)_4]$	D_{4d}	9.214 Å	F ($J = 2.72\ cm^{-1}$)	0	79.0	3.3×10^{-8}
$[Dy_2(hfac)_4(L_3)_2]$	D_{2d}	10.172 Å	F (/)	0	6.8	9.12×10^{-6}
$[Dy_2(bfac)_4(L_4)_2]\cdot C_7H_{16}$	D_{2d}	9.502 Å	AF (/)	0	25.7	1.64×10^{-6}
$[Dy_2(H_2dapp)_2(\mu\text{-}OH)_2(H_2O)_2]\cdot 4TCNQ\cdot 2CH_3OH$ (this work)	D_{4d}	7.967 Å	F (/)	0	2.82	6.88×10^{-6}

The Dy-Dy magnetic interactions within the Dy_2 dimer can also be proved by the simplified Equation (4) [55,56], where θ represents the angle between the magnetic axes and the line connecting the two spin centers. Ferromagnetic coupling occurs when θ is below the threshold angle of 54.75°, whereas antiferromagnetic coupling prevails at larger angles. The θ value of 43.59° for 1 proves again the presence of ferromagnetic Dy-Dy coupling in the dimer.

$$E_{dip} = -\frac{\mu_0}{4\pi}\frac{\mu^2}{r^3}[3\cos^2\theta - 1] \quad (4)$$

3. Materials and Methods

3.1. Materials and Characterizations

The starting materials H_2dapp [57] and LiTCNQ [58] were prepared following reported literature methods. All chemical reagents were purchased from commercial suppliers and utilized directly without additional purification steps, except for distilled water and methanol, which were deoxygenated prior to use. Elemental analyses were conducted on a

Vario EL cube (Langenselbold, Germany). IR spectra data was on a Nicolet iS10 ATR-FTIR instrument (Thermo, Waltham, MA, USA), and the measured wavenumber range was 4000–400 cm^{-1}. Magnetic moment measurements were performed using a Quantum Design MPMS-7 SQUID magnetometer (Santa Clara, CA, USA). Powder X-ray diffraction (PXRD) analysis was conducted using a Rigaku Ultima IV diffractometer (Rigaku Corporation, Tokyo, Japan) equipped with a Cu Kα radiation. Diamagnetic contributions in the dc susceptibility measurements were corrected using Pascal's constants [59–61].

Single-crystal X-ray diffraction data for **complex 1** were collected at 120.01(10) K using an XtaLAB Synergy R diffractometer (Rigaku, Tokyo, Japan) equipped with a HyPix detector, with CuKα radiation (λ = 1.54184 Å) employed. The crystal structure was solved via the direct method, with analytical refinement carried out using the software package including SHELXS-2014 [62], SHELXL-2014 [63], and Olex2–1.5. Anisotropic displacement parameters were used for refining all non-hydrogen atoms. Some important bond lengths and angles for **complex 1** are listed in Table S5. The detailed crystallographic parameters of **complex 1** are summarized in Table 2. Crystallographic data have been deposited at the Cambridge Crystallographic Data Center (CCDC: 2465080).

Table 2. Crystallographic data of **complex 1**.

	Complex 1
Empirical formula	$C_{88}H_{68}Dy_2N_{30}O_6$
Formula weight	1966.70
Temperature/K	120.01(10)
Crystal system	triclinic
Space group	$P\bar{1}$
a [Å]	10.41860(10)
b [Å]	13.5613(2)
c [Å]	15.3557(2)
a [deg]	75.1470(10)
b [deg]	81.0730(10)
g [deg]	85.4510(10)
Volume [Å^3]	2069.93(5)
Z	1
ρ_{calc} [g/cm^{-3}]	1.578
μ [mm^{-1}]	10.163
F(000)	986
Reflections collected	25526
Unique/parameters	8352/577
R(int)	0.0584
Goodness-of-fit on F^2	1.161
R_1, wR_2 [$I > 2\sigma(I)$]	0.0419, 0.1132
R_1, wR_2 (all data)	0.0431, 0.1140

3.2. *Synthesis of [$Dy_2(H_2dapp)_2(\mu\text{-}OH)_2(H_2O)_2$] 4TCNQ 2$CH_3OH$*

A quantity of 0.05 mmol (22.8 mg) of $Dy(NO_3)_3\cdot 6H_2O$ and 0.05 mmol (17.0 mg) of H_2dapp were dissolved in 2 mL of a MeOH/H_2O mixture (v:v = 1:1) and stirred for 30 min at 70 °C. After stirring, the solution was allowed to stand at room temperature. A solution of 0.05 mmol (10.5 mg) LiTCNQ in 2 mL of MeOH/H_2O (v:v = 1:1) was carefully layered on top of the above solution. Dark blue crystals formed after a few days. Yield: 4 mg (16%, based on LiTCNQ). Anal. Calcd. for $C_{88}H_{66}Dy_2N_{30}O_6$: C, 31.53; H, 1.16; N, 2.94%. Found: C, 31.50; H, 1.19; N, 2.89%. IR (KBr cm^{-1}): 2175(vs, br), 2154(vs, br), 2036(s), 1614(s), 1574(vs), 1543(m), 1524(m), 1487(vs), 1088(s), 1004(s), 979(m), 797(m), 768(m), 717(m), 647(w), 572(w).

4. Conclusions

In pursuit of more radical-Ln^{III} complexes, a dinuclear Dy^{III} complex was synthesized using $TCNQ^{\cdot -}$ radical anions and a pentadentate nitrogen-rich Schiff base ligand. The Dy^{III} centers exhibit a geometry between D_{4d} and D_{2d} symmetries and demonstrate intramolecular ferromagnetic coupling. Magnetic studies reveal weak slow magnetic relaxation behavior at zero field with an approximate Δ_{eff} = 2.82 K and τ_0 = 6.88 × 10^{-6} s. This might be due to the fact that antiferromagnetic $TCNQ^{\cdot -}$ dimers do not effectively isolate the molecules (with the shortest Dy···Dy distances < 8 Å), causing dipolar Dy-Dy interactions that favor QTM and further limit magnetic performance. Further studies focus on designing more suitable co-ligands to develop $TCNQ^{\cdot -}$-coordinated Dy^{III} complexes that effectively suppress QTM while optimizing $TCNQ^{\cdot -}$ packing, potentially enabling multifunctional materials combining conductivity and magnetic properties.

Supplementary Materials: The following supporting information can be downloaded at: https://www.mdpi.com/article/10.3390/magnetochemistry11090075/s1, Figure S1. Crystal packing diagram of **complex 1** (a) along the a-axis and (b) along the c-axis (H atoms are omitted for clarity).; Figure S2. The PXRD pattern corresponding to the experimental synthesis and simulation of **complex 1**.; Figure S3 (a) The magnetic susceptibility (χ'') of **complex 1** versus magnetic field (H) at 1.9 K and 997 Hz. (b) Temperature dependence of in-phase(χ') and out-of-phase (χ'') ac magnetic susceptibilities under 400 Oe dc fields and 997 Hz for **complex 1**.; Table S1. Dy^{III} geometry analysis of **complex 1** by using SHAPE 2.1 software.; Table S2. The hydrogen bond lengths (Å) and angles (°) for **complex 1**.; Table S3. The charges (δ) of different $TCNQ^{\cdot -}$ anions in **complex 1**.; Table S4. Selected bond lengths (Å) and-angles (°) for **complex 1**.

Author Contributions: Conceptualization, X.W.; methodology, X.W. and S.Q.; software, X.W. and J.T.; validation, W.Z. and X.L.; formal analysis, X.W. and S.Q.; investigation, Q.W. and X.L.; resources, B.Z. and Y.M.; data curation, X.W. and S.Q.; writing—original draft preparation, X.W.; writing—review and editing, X.W.; visualization, Y.M. and L.L.; supervision, Y.M. and L.L.; project administration, Y.M. and B.Z.; funding acquisition, Y.M. and B.Z. All authors have read and agreed to the published version of the manuscript.

Funding: This work was supported by the National Natural Science Foundation of China, Nos. 22371138, 92161202, 22206088, 92261103.

Institutional Review Board Statement: Not applicable.

Informed Consent Statement: Not applicable.

Data Availability Statement: The original contributions presented in this study are included in the article/Supplementary Materials. Further inquiries can be directed to the corresponding authors.

Conflicts of Interest: The authors declare no conflicts of interest.

References

1. Zhu, Z.; Tang, J. Lanthanide single-molecule magnets with high anisotropy barrier: Where to from here? *Natl. Sci. Rev.* **2022**, *9*, nwac194. [CrossRef]
2. Mannini, M.; Pineider, F.; Sainctavit, P.; Danieli, C.; Otero, E.; Sciancalepore, C.; Talarico, A.M.; Arrio, M.-A.; Cornia, A.; Gatteschi, D.; et al. Magnetic memory of a single-molecule quantum magnet wired to a gold surface. *Nat. Mater.* **2009**, *8*, 194–197. [CrossRef]
3. Troiani, F.; Affronte, M. Molecular spins for quantum information technologies. *Chem. Soc. Rev.* **2011**, *40*, 3119–3129. [CrossRef]
4. Aromí, G.; Aguilà, D.; Gamez, P.; Luis, F.; Roubeau, O. Design of magnetic coordination complexes for quantum computing. *Chem. Soc. Rev.* **2012**, *41*, 537–546. [CrossRef]
5. Emerson-King, J.; Gransbury, G.K.; Whitehead, G.F.S.; Vitorica-Yrezabal, I.J.; Rouzières, M.; Clérac, R.; Chilton, N.F.; Mills, D.P. Isolation of a Bent Dysprosium Bis(amide) Single-Molecule Magnet. *J. Am. Chem. Soc.* **2024**, *146*, 3331–3342. [CrossRef]
6. Liu, M.; Chen, Y.-C.; Wang, H.; Shang, T.; Tong, M.-L.; Layfield, R.A.; Mansikkamäki, A.; Guo, F.-S. A Linear Dysprosium(II) Metallocene with a High Effective Energy Barrier and Magnetic Hysteresis up to 70 Kelvin. *J. Am. Chem. Soc.* **2025**, *147*, 18307–18316. [CrossRef]

7. Borah, A.; Murugavel, R. Magnetic relaxation in single-ion magnets formed by less-studied lanthanide ions Ce(III), Nd(III), Gd(III), Ho(III), Tm(II/III) and Yb(III). *Coord. Chem. Rev.* **2022**, *453*, 214288. [CrossRef]
8. Goodwin, C.A.P.; Ortu, F.; Reta, D.; Chilton, N.F.; Mills, D.P. Molecular magnetic hysteresis at 60 kelvin in dysprosocenium. *Nature* **2017**, *548*, 439–442. [CrossRef]
9. Guo, F.-S.; Day, B.M.; Chen, Y.-C.; Tong, M.-L.; Mansikkamäki, A.; Layfield, R.A. Magnetic hysteresis up to 80 kelvin in a dysprosium metallocene single-molecule magnet. *Science* **2018**, *362*, 1400–1403. [CrossRef]
10. Emerson-King, J.; Gransbury, G.K.; Atkinson, B.E.; Blackmore, W.J.A.; Whitehead, G.F.S.; Chilton, N.F.; Mills, D.P. Soft magnetic hysteresis in a dysprosium amide–alkene complex up to 100 kelvin. *Nature* **2025**, *643*, 125–129. [CrossRef]
11. Li, H.-D.; Wu, S.-G.; Tong, M.-L. Lanthanide–radical single-molecule magnets: Current status and future challenges. *Chem. Commun.* **2023**, *59*, 6159–6170. [CrossRef]
12. Zhu, Z.; Paul, S.; Zhao, C.; Wu, J.; Ying, X.; Ungur, L.; Wernsdorfer, W.; Meyer, F.; Tang, J. Record Quantum Tunneling Time in an Air-Stable Exchange-Bias Dysprosium Macrocycle. *J. Am. Chem. Soc.* **2024**, *146*, 18899–18904. [CrossRef]
13. Canaj, A.B.; Dey, S.; Martí, E.R.; Wilson, C.; Rajaraman, G.; Murrie, M. Insight into D6h Symmetry: Targeting Strong Axiality in Stable Dysprosium(III) Hexagonal Bipyramidal Single-Ion Magnets. *Angew. Chem. Int. Ed.* **2019**, *58*, 14146–14151. [CrossRef]
14. Liu, J.-L.; Chen, Y.-C.; Tong, M.-L. Symmetry strategies for high performance lanthanide-based single-molecule magnets. *Chem. Soc. Rev.* **2018**, *47*, 2431–2453. [CrossRef]
15. Peng, Y.; Powell, A.K. What do 3d-4f butterflies tell us? *Coord. Chem. Rev.* **2021**, *426*, 213490. [CrossRef]
16. Liu, K.; Shi, W.; Cheng, P. Toward heterometallic single-molecule magnets: Synthetic strategy, structures and properties of 3d–4f discrete complexes. *Coord. Chem. Rev.* **2015**, *289–290*, 74–122. [CrossRef]
17. Vaz, M.G.F.; Andruh, M. Molecule-based magnetic materials constructed from paramagnetic organic ligands and two different metal ions. *Coord. Chem. Rev.* **2021**, *427*, 213611. [CrossRef]
18. Demir, S.; Jeon, I.-R.; Long, J.R.; Harris, T.D. Radical ligand-containing single-molecule magnets. *Coord. Chem. Rev.* **2015**, *289–290*, 149–176. [CrossRef]
19. Liu, F.; Spree, L.; Krylov, D.S.; Velkos, G.; Avdoshenko, S.M.; Popov, A.A. Single-Electron Lanthanide-Lanthanide Bonds Inside Fullerenes toward Robust Redox-Active Molecular Magnets. *Acc. Chem. Res.* **2019**, *52*, 2981–2993. [CrossRef]
20. Jin, C.-Y.; Mei, X.-L.; Zhou, Y.; Li, L.; Sutter, J.-P. Single-Molecule Magnet Behavior in a Tb-Nitronyl Nitroxide Radical Network with [$Tb_3(NIT)_2$] Nodes. *Inorg. Chem.* **2024**, *63*, 23829–23836. [CrossRef]
21. Li, H.; Sun, Z.; Sun, J.; Xi, L.; Guo, J.; Sun, G.; Xie, J.; Ma, Y.; Li, L. Single-molecule magnet behavior in a Cu^{II}-decorated {Dy^{III}_2} complex with nitronyl nitroxide biradicals. *J. Mater. Chem. C* **2018**, *6*, 2060–2068. [CrossRef]
22. Ballesteros-Rivas, M.; Ota, A.; Reinheimer, E.; Prosvirin, A.; Valdés-Martinez, J.; Dunbar, K.R. Highly Conducting Coordination Polymers Based on Infinite M(4,4′-bpy) Chains Flanked by Regular Stacks of Non-Integer TCNQ Radicals. *Angew. Chem. Int. Ed.* **2011**, *50*, 9703–9707. [CrossRef]
23. Wang, Z.-X.; Zhang, X.; Zhang, Y.-Z.; Li, M.-X.; Zhao, H.; Andruh, M.; Dunbar, K.R. Single-Chain Magnetic Behavior in a Hetero-Tri-Spin Complex Mediated by Supramolecular Interactions with $TCNQF^{\cdot-}$ Radicals. *Angew. Chem. Int. Ed.* **2014**, *53*, 11567–11570. [CrossRef]
24. Benner, F.; La Droitte, L.; Cador, O.; Le Guennic, B.; Demir, S. Magnetic hysteresis and large coercivity in bisbenzimidazole radical-bridged dilanthanide complexes. *Chem. Sci.* **2023**, *14*, 5577–5592. [CrossRef]
25. Kaim, W.; Moscherosch, M. The coordination chemistry of TCNE, TCNQ and related polynitrileπ acceptors. *Coord. Chem. Rev.* **1994**, *129*, 157–193. [CrossRef]
26. Zhang, X.; Wang, Z.-X.; Xie, H.; Li, M.-X.; Woods, T.J.; Dunbar, K.R. A cobalt(II) spin-crossover compound with partially charged TCNQ radicals and an anomalous conducting behavior. *Chem. Sci.* **2016**, *7*, 1569–1574. [CrossRef]
27. Üngör, Ö.; Shatruk, M. Transition metal complexes with fractionally charged TCNQ radical anions as structural templates for multifunctional molecular conductors. *Polyhedron* **2020**, *177*, 114254. [CrossRef]
28. Taponen, A.I.; Ayadi, A.; Lahtinen, M.K.; Oyarzabal, I.; Bonhommeau, S.; Rouzières, M.; Mathonière, C.; Tuononen, H.M.; Clérac, R.; Mailman, A. Room-Temperature Magnetic Bistability in a Salt of Organic Radical Ions. *J. Am. Chem. Soc.* **2021**, *143*, 15912–15917. [CrossRef]
29. Zhang, X.; Xie, H.; Ballesteros-Rivas, M.; Woods, T.J.; Dunbar, K.R. Conducting Molecular Nanomagnet of Dy^{III} with Partially Charged TCNQ Radicals. *Chem.—A Eur. J.* **2017**, *23*, 7448–7452. [CrossRef]
30. Zhou, Y.-Q.; Deng, W.; Du, S.-N.; Chen, Y.-C.; Ruan, Z.-Y.; Wu, S.-G.; Liu, J.-L.; Tong, M.-L. An alkoxyborate-bridging Dy_2 single-molecule magnet with ferromagnetic coupling. *Inorg. Chem. Front.* **2024**, *11*, 1061–1069. [CrossRef]
31. Zhou, Y.-Q.; Yao, C.-Y.; Ruan, Z.-Y.; Lyu, B.-H.; Du, S.-N.; Wu, S.-G.; Chen, Y.-C.; Deng, W.; Liu, J.-L.; Tong, M.-L. Proton control of Raman relaxation in bis-hydrazone single-molecule magnets. *Chin. Chem. Lett.* **2025**, *36*, 111179. [CrossRef]
32. Llunell, M.; Casanova, D.; Cirera, J.; Alemany, P.; Alvarez, S. *SHAPE 2.1*; University of Barcelona: Barcelona, Spain, 2013.

33. Tiunova, A.V.; Korchagin, D.V.; Kazakova, A.V.; Shilov, G.V.; Buravov, L.I.; Dmitriev, A.I.; Zhidkov, M.V.; Zverev, V.N.; Yagubskii, E.B.; Aldoshin, S.M. Solvent and Dihalide Substituent Effects on Crystal Structure, Spin-Crossover Behavior and Conductivity of the Cationic Mn(III) Complexes with Electroactive TCNQ Counteranions. *Eur. J. Inorg. Chem.* **2024**, *27*, e202300749. [CrossRef]
34. Zhang, J.; Yan, P.; Li, G.; Hou, G.; Suda, M.; Einaga, Y. Systematic investigation of an array of TCNQ lanthanide complexes: Synthesis, structure and magnetic properties. *Dalton Trans.* **2009**, 10466–10473. [CrossRef]
35. Wang, J.; Li, Q.-W.; Wu, S.-G.; Chen, Y.-C.; Wan, R.-C.; Huang, G.-Z.; Liu, Y.; Liu, J.-L.; Reta, D.; Giansiracusa, M.J.; et al. Opening Magnetic Hysteresis by Axial Ferromagnetic Coupling: From Mono-Decker to Double-Decker Metallacrown. *Angew. Chem. Int. Ed.* **2021**, *60*, 5299–5306. [CrossRef]
36. Lu, Y.-B.; Jian, F.-M.; Jin, S.; Zhao, J.-W.; Xie, Y.-R.; Luo, G.-T. Three-Dimensional Extended Frameworks Constructed from Dinuclear Lanthanide(III) 1,4-Naphthalenedicarboxylate Units with Bis(2,2′-biimidazole) Templates: Syntheses, Structures, and Magnetic and Luminescent Properties. *Cryst. Growth Des.* **2014**, *14*, 1684–1694. [CrossRef]
37. Kahn, M.L.; Sutter, J.-P.; Golhen, S.; Guionneau, P.; Ouahab, L.; Kahn, O.; Chasseau, D. Systematic Investigation of the Nature of The Coupling between a Ln(III) Ion (Ln = Ce(III) to Dy(III)) and Its Aminoxyl Radical Ligands. Structural and Magnetic Characteristics of a Series of {Ln(organic radical)$_2$} Compounds and the Related {Ln(Nitrone)$_2$} Derivatives. *J. Am. Chem. Soc.* **2000**, *122*, 3413–3421. [CrossRef]
38. Xi, L.; Li, H.; Sun, J.; Ma, Y.; Tang, J.; Li, L. Designing Multicoordinating Nitronyl Nitroxide Radical Toward Multinuclear Lanthanide Aggregates. *Inorg. Chem.* **2020**, *59*, 443–451. [CrossRef]
39. Lv, X.-H.; Yang, S.-L.; Li, Y.-X.; Zhang, C.-X.; Wang, Q.-L. A family of lanthanide compounds based on nitronyl nitroxide radicals: Synthesis, structure, magnetic and fluorescence properties. *RSC Adv.* **2017**, *7*, 38179–38186. [CrossRef]
40. Ruan, L.; Tong, J.; Luo, F.; Qin, G.; Tian, F.; Jiao, X.; Zhang, X. Magnetic Improvement and Relaxation Mechanism of the Tb-Phthalocyanine Single-Molecule Magnet by Absorbing CH_2Cl_2 Molecules. *J. Phys. Chem. C* **2021**, *125*, 10165–10172. [CrossRef]
41. Tian, H.; Liu, R.; Wang, X.; Yang, P.; Li, Z.; Li, L.; Liao, D. Magnetic Slow Relaxation in Cyclic Tb^{III}-Nitronyl Nitroxide Radical Complexes. *Eur. J. Inorg. Chem.* **2009**, *2009*, 4498–4502. [CrossRef]
42. Zhao, J.; Luo, J.; Chen, L.; Yuan, J.; Li, H.; Ma, P.; Wang, J.; Niu, J. Novel 1-D double-chain organic–inorganic hybrid polyoxotungstates constructed from dimeric copper–lanthanide heterometallic silicotungstate units. *CrystEngComm* **2012**, *14*, 7981–7993. [CrossRef]
43. Luis, F.; Bartolomé, J.; Fernández, J.F.; Tejada, J.; Hernández, J.M.; Zhang, X.X.; Ziolo, R. Thermally activated and field-tuned tunneling in $Mn_{12}Ac$ studied by ac magnetic susceptibility. *Phys. Rev. B* **1997**, *55*, 11448–11456. [CrossRef]
44. Bartolomé, J.; Filoti, G.; Kuncser, V.; Schinteie, G.; Mereacre, V.; Anson, C.E.; Powell, A.K.; Prodius, D.; Turta, C. Magnetostructural correlations in the tetranuclear series of {Fe_3LnO_2} butterfly core clusters: Magnetic and Mössbauer spectroscopic study. *Phys. Rev. B* **2009**, *80*, 014430. [CrossRef]
45. Gatteschi, D.; Sessoli, R. Quantum Tunneling of Magnetization and Related Phenomena in Molecular Materials. *Angew. Chem. Int. Ed.* **2003**, *42*, 268–297. [CrossRef]
46. Xi, L.; Han, J.; Huang, X.; Li, L. Nitronyl Nitroxide Biradical-Based Binuclear Lanthanide Complexes: Structure and Magnetic Properties. *Magnetochemistry* **2020**, *6*, 48. [CrossRef]
47. Jin, C.-Y.; Zhang, Q.; Tang, J.; Cheng, P.; Li, L. Unusual zig-zag-shaped Ln-radical coordination networks derived from a nitronyl nitroxide with two imidazole groups: Single-crystal-to-single-crystal transformations and magnetic properties. *Dalton Trans.* **2025**, *54*, 7718–7725. [CrossRef]
48. Wang, X.; Han, J.; Huang, X.; Li, L. Ln^{III}-Ni^{II} heterometallic compounds linked by nitronyl nitroxides: Structure and magnetism. *Inorg. Chem. Commun.* **2021**, *134*, 108983. [CrossRef]
49. Chilton, N.F.; Collison, D.; McInnes, E.J.L.; Winpenny, R.E.P.; Soncini, A. An electrostatic model for the determination of magnetic anisotropy in dysprosium complexes. *Nat. Commun.* **2013**, *4*, 2551. [CrossRef]
50. Wang, Y.; Zeng, Z.; Luo, S.; Guo, Y.; Liu, X. Enhancing the single-molecule magnetic performance of β-diketonate Dy(III) complexes by modulating the coordination microenvironment and magnetic interaction: From a mononuclear to a dinuclear structure. *Dalton Trans.* **2024**, *53*, 17272–17280. [CrossRef]
51. Zhang, W.; Xu, S.-M.; Zhu, Z.-X.; Ru, J.; Zhang, Y.-Q.; Yao, M.-X. Strong intramolecular Dy^{III}–Dy^{III} magnetic couplings up to 15.00 cm^{-1} in phenoxyl-bridged dinuclear 4f complexes. *New J. Chem.* **2020**, *44*, 2083–2090. [CrossRef]
52. Zhang, W.-Y.; Tian, Y.-M.; Li, H.-F.; Chen, P.; Sun, W.-B.; Zhang, Y.-Q.; Yan, P.-F. A series of dinuclear Dy(III) complexes bridged by 2-methyl-8-hydroxylquinoline: Replacement on the periphery coordinated β-diketonate terminal leads to different single-molecule magnetic properties. *Dalton Trans.* **2016**, *45*, 3863–3873. [CrossRef]
53. Shen, F.-X.; Pramanik, K.; Brandão, P.; Zhang, Y.-Q.; Jana, N.C.; Wang, X.-Y.; Panja, A. Macrocycle supported dimetallic lanthanide complexes with slow magnetic relaxation in Dy2 analogues. *Dalton Trans.* **2020**, *49*, 14169–14179. [CrossRef]
54. Wang, W.-M.; Zhang, H.-X.; Wang, S.-Y.; Shen, H.-Y.; Gao, H.-L.; Cui, J.-Z.; Zhao, B. Ligand Field Affected Single-Molecule Magnet Behavior of Lanthanide(III) Dinuclear Complexes with an 8-Hydroxyquinoline Schiff Base Derivative as Bridging Ligand. *Inorg. Chem.* **2015**, *54*, 10610–10622. [CrossRef]

55. Guo, M.; Zhang, Y.-Q.; Zhu, Z.; Tang, J. Dysprosium Compounds with Hula-Hoop-like Geometries: The Influence of Magnetic Anisotropy and Magnetic Interactions on Magnetic Relaxation. *Inorg. Chem.* **2018**, *57*, 12213–12221. [CrossRef]
56. Leng, J.-D.; Liu, J.-L.; Lin, W.-Q.; Gómez-Coca, S.; Aravena, D.; Ruiz, E.; Tong, M.-L. Unprecedented ferromagnetic dipolar interaction in a dinuclear holmium(III) complex: A combined experimental and theoretical study. *Chem. Commun.* **2013**, *49*, 9341–9343. [CrossRef]
57. Curry, J.D.; Robinson, M.A.; Busch, D.H. Metal complexes derived from substituted hydrazones of 2,6-diacetylpyridine. *Inorg. Chem.* **1967**, *6*, 1570–1574. [CrossRef]
58. Melby, L.R.; Harder, R.J.; Hertler, W.R.; Mahler, W.; Benson, R.E.; Mochel, W.E. Substituted Quinodimethans. II. Anion-radical Derivatives and Complexes of 7,7,8,8-Tetracyanoquinodimethan. *J. Am. Chem. Soc.* **1962**, *84*, 3374–3387. [CrossRef]
59. Zhong, L.; Chen, W.-B.; OuYang, Z.-J.; Yang, M.; Zhang, Y.-Q.; Gao, S.; Schulze, M.; Wernsdorfer, W.; Dong, W. Unprecedented one-dimensional chain and two-dimensional network dysprosium(III) single-molecule toroics with white-light emission. *Chem. Commun.* **2020**, *56*, 2590–2593. [CrossRef]
60. Wang, H.-S.; Chen, Y.; Hu, Z.-B.; Yin, C.-L.; Zhang, Z.; Pan, Z.-Q. Modulation of the directions of the anisotropic axes of Dy^{III} ions through utilizing two kinds of organic ligands or replacing Dy^{III} ions by Fe^{III} ions. *CrystEngComm* **2019**, *21*, 5429–5439. [CrossRef]
61. Wang, H.-S.; Zhang, K.; Song, Y.; Pan, Z.-Q. Recent advances in 3d-4f magnetic complexes with several types of non-carboxylate organic ligands. *Inorganica Chim. Acta* **2021**, *521*, 120318. [CrossRef]
62. Sheldrick, G.M. *SHELXS-2014, Program for Solution of Crystal Structures;* University of Göttingen: Göttingen, Germany, 2014.
63. Sheldrick, G.M. *SHELXL-2014, Program for Refinement of Crystal Structures;* University of Göttingen: Göttingen, Germany, 2014.

magnetochemistry

MDPI

Review

Molecular Nanomagnets with Photomagnetic Properties: Design Strategies and Recent Advances

Xiaoshuang Gou [1,2], Xinyu Sun [1], Peng Cheng [1,*] and Wei Shi [1,*]

[1] Frontiers Science Center for New Organic Matter, State Key Laboratory of Advanced Chemical Power Sources and Department of Chemistry, College of Chemistry, Nankai University, Tianjin 300071, China; xsgou2023@sinano.ac.cn (X.G.); 2120241072@mail.nankai.edu.cn (X.S.)

[2] i-Lab, Suzhou Institute of Nano-Tech and Nano-Bionics (SINANO), Chinese Academy of Sciences (CAS), Suzhou 215123, China

* Correspondence: pcheng@nankai.edu.cn (P.C.); shiwei@nankai.edu.cn (W.S.)

Abstract: The magnetic properties of molecular nanomagnets can be finely modulated by light, which provides great potential in optical switches, smart sensors, and data storage devices. Light-induced spin transition, structure changes, and radical formation could tune the static and dynamic magnetic properties of molecular nanomagnets with high spatial and temporal resolutions. Herein, we summarize the design strategies of photoresponsive molecular nanomagnets and review the recent advances in transition metal/lanthanide molecular nanomagnets with photomagnetic properties. The photoresponsive mechanism based on spin transition, photocyclization, and photogenerated radicals is discussed in detail, providing insights into the photomagnetic properties of molecular nanomagnets for advanced photoresponsive materials.

Keywords: molecular nanomagnets; photomagnetic properties; photocyclization; photogenerated radical

1. Introduction

Stimuli-responsive materials have recently garnered great attention due to their physical and chemical properties being modulated by external stimuli, such as temperature, electric field, and light [1–4]. Among the diverse stimuli-responsive materials, photoresponsive molecular nanomagnets are particularly intriguing candidates, showing great potential toward applications such as high-density information storage, optical switching devices, smart sensors, and spintronic devices [5–8].

Molecular nanomagnets, including single-molecule magnets (SMMs) and single-chain magnets (SCMs), are usually constructed by magnetically anisotropic transition or lanthanide metal centers and organic ligands [9–11]. The electronic state of metal ions that is defined by the coordination sphere is crucial for the rational design of advanced molecular nanomagnets possessing tailored photomagnetic properties [12–14]. For example, photomagnetic properties were observed in cyanide-bridged spin-crossover (SCO) complexes with the light-induced excited spin state trapping (LIESST) effect. The LIESST effect was also applied to modulate the bistable states of molecular nanomagnets, such as $\{[W(CN)_8][(Fe^{II})(bib)_2](bibH)\}\cdot 2CH_3OH$ (bib = 1,4-bis(1H-imidazol-1-yl)benzene), representing an example of reversible switching between SCM and SMM behaviors via light [15]. On the other hand, lanthanide SMMs have been extensively studied, achieving groundbreaking progress, especially in effective energy barrier (U_{eff}) and blocking temperature (T_B) [16–23]. Early studies have demonstrated that light irradiation can effectively modulate the static magnetic susceptibility of lanthanide SMMs; however, regulating their dynamic magnetic susceptibility remains challenging [24–26]. Selected examples of photoresponsive lanthanide SMMs generally have photoresponsive cyclic units, such as 1,2-bis(4-pyridyl)ethylene [27], 9-diethylphosphorylmethylanthracene [28], and dipyridyldithienylethene [29], as well as photoresponsive radical precursors [30]. To gain insights into the mechanism of photomagnetic properties of molecular nanomagnets, we summarize the advancements of transition metal and lanthanide molecular nanomagnets with static and dynamic magnetic properties modulated by light (Figure 1). The roles of selecting organic ligands that can undergo light-induced ring-opening/ring-closing transitions or generate stable radicals under illumination are discussed, providing key structural factors for the rational design and synthesis of high-performance lanthanide molecular nanomagnets with photoinduced magnetic prop-erties.

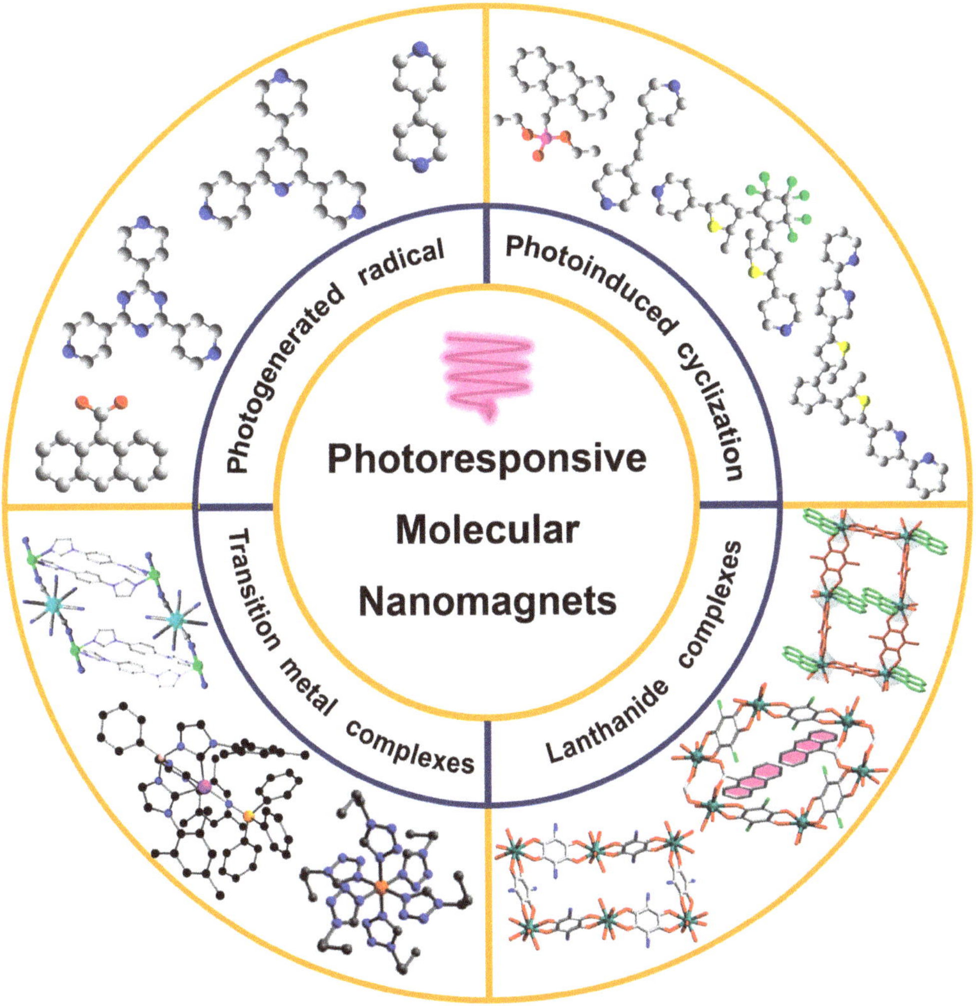

Figure 1. Molecular nanomagnets with photomagnetic properties.

2. Design Strategies for Molecular Nanomagnets with Photomagnetic Properties

Molecular nanomagnets with photomagnetic properties are primarily synthesized by introducing photoresponsive components, notably SCO units, photo-isomerizing organics, or radical-forming moieties. In transition metal molecular nanomagnets, incorporation of SCO units could facilitate optical switching of spin states to modulate the magnetic properties of SMMs and SCMs [31–34]. For lanthanide molecular nanomagnets, light-induced ligand transformations such as photocyclization or radical formation could modify the ligand field of the metal center and introduce additional spin carriers, thereby tuning room-temperature magnetic moments and/or dynamic magnetic susceptibilities [27–30].

2.1. *Photoresponsive Spin-Crossover Units*

The photoinduced electron transfer is fundamental in photophysics and photochemistry, with extensive applications in artificial light sources, solar energy conversion, and photocatalysis [35–49], enabling the regulation of properties at a molecular level [50–52]. SCO complexes undergo spin transition at the orbitals of metal sites upon illumination, leading to a change in the spin state [53–64]. However, assembling SCO units with coordination bonds to form photoresponsive compounds remains highly challenging because minor changes in the coordination environment of metal sites may influence the spin transition to some extent. Through the rational design of molecular structures based on the SCO unit, it is desired to achieve multifunctional molecular materials with controllable photomagnetic properties.

2.2. *Photoinduced Structural Change in Organic Molecules*

Photochromism refers to the reversible transformation of a chemical substance between two or more forms that exhibit distinct absorption to light [65,66]. Photochromic systems can undergo reversible transformations between distinct states with varying properties, making them widely applicable in data storage, sensors, bioimaging, and nanomachines [67–75].

Typical photochromic ligands, for example, are shown in Figure 2. Upon light illumination, the structural changes most commonly observed are trans-to-cis isomerization and ring-opening/ring-closing transitions [75–77]. Under ultraviolet light, azobenzene can undergo trans-to-cis isomerization, and the cis isomer can be reverted to the trans form through exposure to visible light or heat [76,77]. For diarylethene, a [2 + 2] photocycloaddition can occur when the distance between parallelly arranged alkene bonds is appropriate [78–82]. Similarly, 9-anthrylmethylphosphonic acid can undergo a [4 + 4] cycloaddition under ultraviolet light when the spatial requirements are satisfied [83,84]. Upon ultraviolet irradiation, dithienylethene can transform from a colorless open-ring state to a colored closed-ring state and can be switched back to its original state via visible light irradiation [85,86]. In the case of spiropyran, ultraviolet irradiation can result in the photochemical cleavage of the C–O bond, resulting in its conversion from the colorless spiropyran state to the colored merocyanine state. Upon turning off the ultraviolet light, it can revert to its initial colorless state [87,88]. Fulgides can transform from a colorless open-ring state to a colored closed-ring state under ultraviolet light and revert to the initial state through a ring-opening reaction when exposed to visible light [89,90].

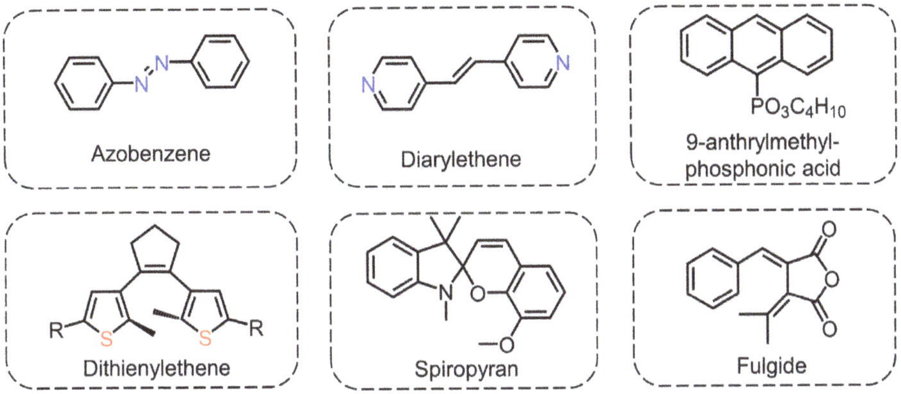

Figure 2. Selected examples of organic molecules with photoinduced structural change groups.

2.3. Photogenerated Organic Radicals

In addition to photochromic ligands, another important category of photoresponsive ligands is photogenerated organic radicals. These compounds have been used for various applications, such as smart windows, erasable and inkless printing, catalysis, sensing, artificial photosynthesis, and solar cells [91–95].

As shown in Figure 3, pyridine derivatives, viologens, triazoles, 9-anthracene carboxylic acids, and naphthalenediimides can be converted to radicals upon light illumination through electronic transition [96–98]. In recent years, there has been a growing interest in the study of not only molecular nanomagnets but also multifunctional metal–organic frameworks with photogenerated radicals [99–101].

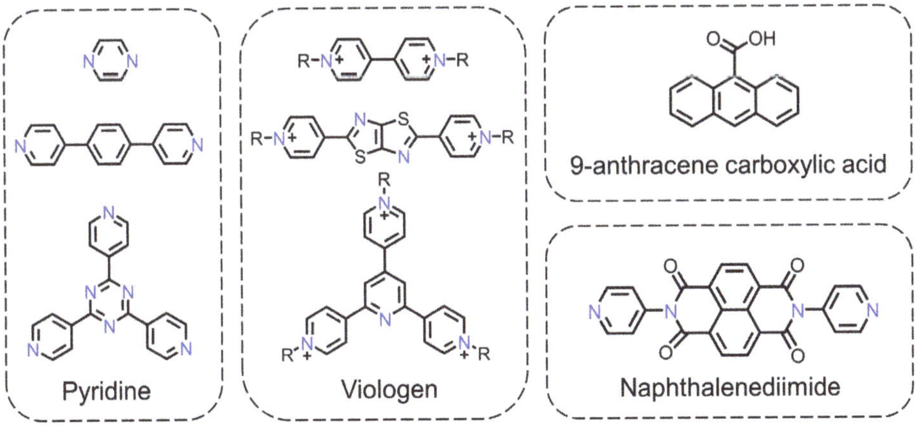

Figure 3. Selected examples of photoactive organic molecules that can generate stable radicals after illumination.

3. Photomagnetic Properties of Transition Metal Molecular Nanomagnets

The photomagnetic properties of transition metal molecular nanomagnets primarily arise from the alteration of their electronic structures and/or spin states. Light-induced spin state changes could serve as the fundamental mechanism behind photomagnetic properties. For example, in the six-coordination Fe(II) complexes, the transition from a low-spin state to a high-spin state could be induced by light illumination, thereby influencing the magnetic properties of the complexes. The SCO units have been assembled with cyanide to form chains, wherein the spin state, anisotropy, and magnetic interaction were photoswitched to tune the magnetic properties [15,31–34,102,103]. Additionally, transition metal molecular nanomagnets with photoinduced metal-to-metal electron transfer (MMET) and photoinduced structural changes in the ligands were also reported [104,105]. Reliable experimental characterization is required for understanding the photoinduced mechanism, which is typically accomplished through advanced characterization techniques. The superconducting quantum interference device (SQUID), physical property measurement system (PPMS), and pulsed electron paramagnetic resonance (EPR) spectrometer are commonly employed to determine the photomagnetic properties of molecular nanomagnets [106–109].

3.1. Photomagnetic Properties of SCO-Type Transition Metal Molecular Nanomagnets

In 2013, Long's group reported an iron(II) molecular nanomagnet with photomagnetic properties, $[Fe(1\text{-propyltetrazole})_6](BF_4)_2$ (Figure 4a) [31]. Under an external magnetic field, this complex exhibited fully reversible switching between the $S = 0$ and $S = 2$ states. After irradiation by a 505 nm light for 7 h at 10 K, an LIESST effect was observed. The high-spin state showed no significant decay after the light was turned off. An alternating current (AC) magnetic susceptibility measurement was performed using a SQUID magnetometer, which revealed slow magnetic relaxation behavior in the high-spin Fe(II) state under an external magnetic field (H_{dc} = 2000 Oe) in the temperature range of 1.9–5.0 K, with an effective barrier of 22 K. An analysis of EPR spectra demonstrated that the anisotropy parameters of the light-induced phase of $[Fe(1\text{-propyltetrazole})_6](BF_4)_2$ agree with the SMM behavior, with a total spin reversal barrier of 72 K. The effective barrier of 22 K from ac magnetic susceptibility is lower than that of the total spin reversal barrier due to the presence of vibronic coupling, typically observed in mononuclear SMMs. In the same year, Smith's group reported another iron(II) molecular nanomagnet with photomagnetic properties, $PhB(MesIm)_3Fe^{II}\text{-}N = PPh_3$ (Figure 4b). After 18 h of white light irradiation at 10 K, an LIESST effect was observed. Magnetic susceptibility measurements under light irradiation indicated a transition from the low-spin state to the high-spin state. The high-spin complex exhibited slow magnetic relaxation behavior under an external magnetic field (H_{dc} = 1000 Oe) in the temperature range of 1.8–4.6 K, with an effective barrier of 22 K [32].

Cyanide-bridged SCO complexes have shown intriguing photoinduced magnetic properties. These complexes generally consist of divalent or trivalent iron/cobalt ions linked by cyanide ligands, resulting in either chain or three-dimensional framework structures. From 2013 onwards, Liu's group reported a series of molecular nanomagnets with photomagnetic properties, including $\{[Fe^{III}(Tp^*)(CN)_3]_2Fe^{II}(bpmh)\}\cdot 2H_2O$, (bpmh, *N,N′*-bis-pyridin-4ylmethylene hydrazine), $\{[Fe^{III}(bpy)(CN)_4]_2[Co^{II}(phpy)_2]\}\cdot 2H_2O$, (bpy, 2,2′-bipyridine, phpy, 4-phenylpyridine), $\{[W(CN)_8][(Fe^{II})(bib)_2](bibH)\}\cdot 2CH_3OH$, $\{[(pzTp)Fe^{III}(CN)_3]_4Fe^{II}_2(Pmat)_4\}_n\cdot 12H_2O$, $\{[pzTpFe^{II}(CN)_3]_4Co^{II}_2(Bib)_4\}\cdot 3H_2O$, and $\{[pzTpFe^{II}(CN)_3]_2Co^{II}(Bpi)_2\}\cdot CH_3CN\cdot 4H_2O$ [15,33,34,102,103] (Figure 4c–g). $\{[Fe^{III}(Tp^*)(CN)_3]_2Fe^{II}(bpmh)\}\cdot 2H_2O$ is a double zigzag chain that includes Fe(II) spin-crossover units and well-isolated paramagnetic Fe(III) ions. This chain exhibited thermally reversible SCO at the Fe(II) sites. Following 12 h of irradiation with 473 nm blue light at 5 K, an SCM behavior was triggered, accompanied by the LIESST effect at the Fe(II) sites and synergistic ferromagnetic coupling between the photoinduced high-spin Fe(II) and low-spin Fe(III) ions [33] (Figure 5).

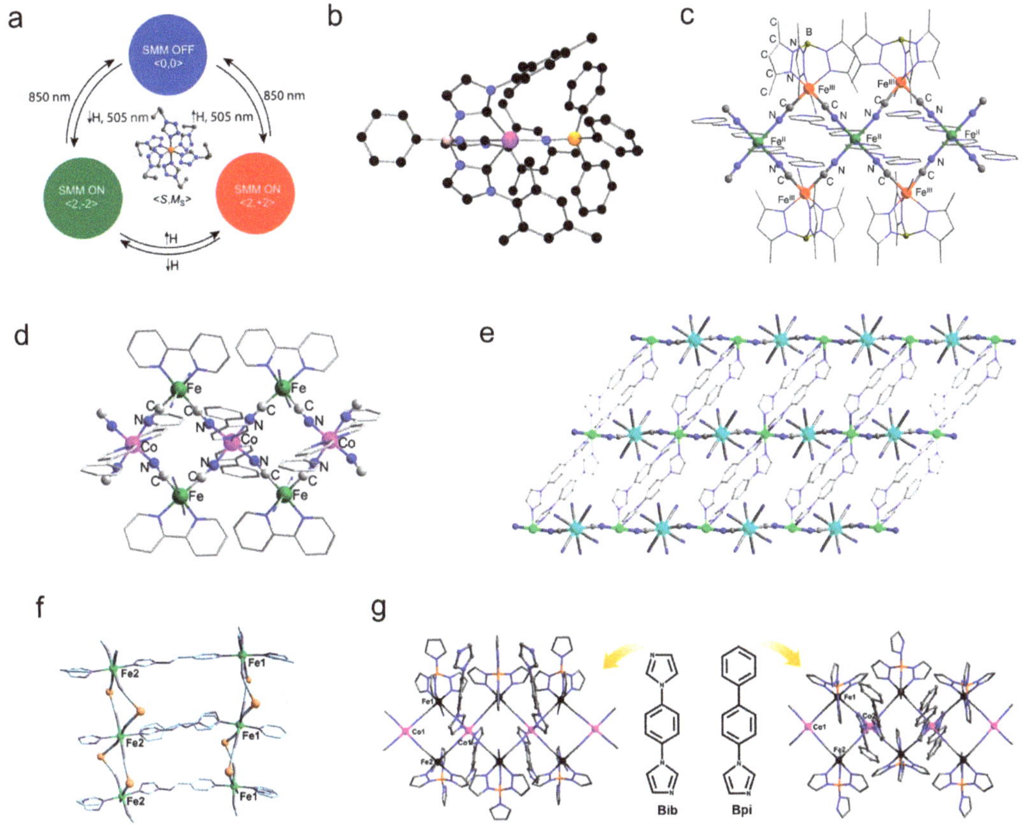

Figure 4. Transition metal molecular nanomagnets with photomagnetic properties. (**a**) $[Fe^{II}(1\text{-propyltetrazole})_6](BF_4)_2$; (**b**) $PhB(MesIm)_3Fe^{II}\text{-}N = PPh_3$; (**c**) $\{[Fe^{III}(Tp^*)(CN)_3]_2Fe^{II}(bpmh)\}\cdot 2H_2O$; (**d**) $\{[Fe^{III}(bpy)(CN)_4]_2[Co^{II}(phpy)_2]\}\cdot 2H_2O$; (**e**) $\{[W(CN)_8][(Fe^{II})(bib)_2](bibH)\}\cdot 2CH_3OH$; (**f**) $\{[(pzTp)Fe^{III}(CN)_3]_4Fe^{II}_2(Pmat)_4\}_n\cdot 12H_2O$; (**g**) $\{[pzTpFe^{II}(CN)_3]_4Co^{II}_2(Bib)_4\}\cdot 3H_2O$ and $\{[pzTpFe^{II}(CN)_3]_2Co^{II}(Bpi)_2\}\cdot CH_3CN\cdot 4H_2O$. Reprinted with permission from Refs. [15,31,32,102] Copyright 2021 Nature Chemistry, 2013 American Chemical Society, 2013 American Chemical Society, and 2021 Angewandte Chemie International Edition. Reprinted from Refs. [33,34,103].

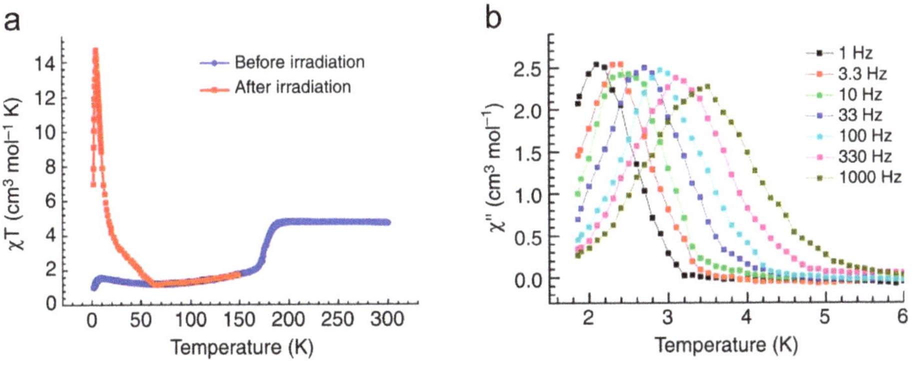

Figure 5. The χT–T (**a**) and χ''–T (**b**) of $\{[Fe^{III}(Tp^*)(CN)_3]_2Fe^{II}(bpmh)\}\cdot 2H_2O$. Reprinted from Ref. [33].

In 2021, Liu's group reported a molecular nanomagnet with photomagnetic properties in the form of {[W(CN)$_8$][(Fe^{II})(bib)$_2$](bibH)}·2CH$_3$OH, which exhibited a reversible LIESST effect when subjected to alternating infrared (808 nm) and blue (473 nm) light, enabling reversible switching between paramagnetic high-spin and diamagnetic low-spin states of the Fe(II) center [15]. At 1.8 K, irradiation with infrared light generated SCM behavior due to the magnetic interactions between the photogenerated high-spin Fe(II) centers and tungsten(V) centers, accompanied by wide magnetic hysteresis loops and a large coercive field of 19 kOe (Figure 6a). Irradiation by blue light resulted in isolated Fe(II)–W(V) species exhibiting SMM behavior. Magnetic hysteresis loops measured at 3.3 K showed that the high-spin Fe(II) state under infrared light maintained a relatively narrow hysteresis loop, while the low-spin Fe(II) state under blue light led to the closure of the hysteresis loop (Figure 6b). This represents an interesting example of reversible switching between SCM and SMM accompanied by light-controlled magnetic hysteresis loop switching.

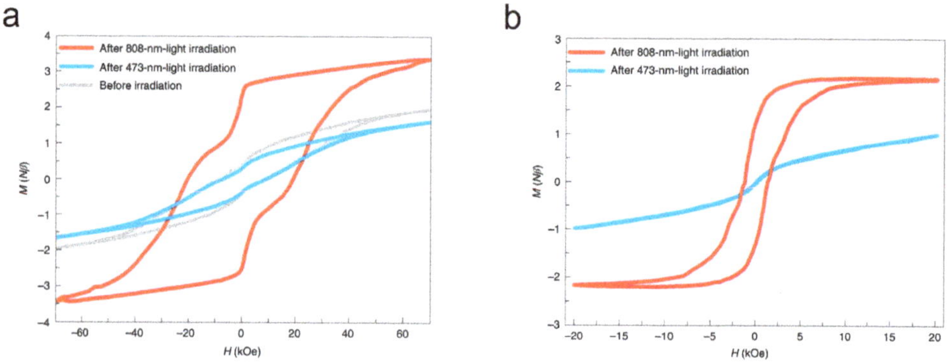

Figure 6. Magnetic hysteresis loops of {[W(CN)$_8$][(Fe^{II})(bib)$_2$](bibH)}·2CH$_3$OH at 1.8 K (**a**) and 3.3 K (**b**) after irradiation. Reprinted with permission from Ref. [15]. Copyright 2021 Nature Chemistry.

3.2. Photomagnetic Properties of Other Transition Metal Molecular Nanomagnets

Dithienylethene is a photochromic molecule that undergoes reversible ring-opening and ring-closing reactions when exposed to light of different wavelengths, resulting in significant alterations to its electronic and optical properties. In 2009, Irie's group reported a chain-like molecular nanomagnet based on closed-ring dithienylethene, [Mn$_4$(hmp)$_6$(dae-c)$_2$(H$_2$O)$_2$](ClO$_4$)$_2$·CH$_3$CN·4H$_2$O (Mn$_4$-c; Hhmp = 2-hydroxymethylpyridine, H$_2$dae = 1,2-bis(5-carboxyl-2-methyl-3-thienyl)perfluorocyclopentene) [104]. Upon irradiation with visible light ($\lambda > 480$ nm), the closed-ring Mn$_4$-c underwent an open-ring reaction, which enhanced the magnetic interactions between the [Mn$_4$] units and altered its slow magnetic relaxation behavior. In 2019, Yamashita's group reported zero-dimensional molecular nanomagnets with open-ring and closed-ring dithienylethene, [Mn$_2$(saltmen)$_2$(dae-o)] (Mn$_2$-o) and [Mn(saltmen)(dae-c)]·H$_2$O·Et$_3$N (Mn-c) (H$_2$saltmen = 2,2′-((1E,1′E)-((2,3-dimethylbutane-2,3-diyl)bis(azaneylylidene))bis(methaneylylidene))diphenol, H$_2$dae = 1,2-bis(5-carboxyl-2-methyl-3-thienyl)perfluorocyclopentene) [105]. Under an external magnetic field, the photoproducts Mn$_2$-o-UV and Mn-c-vis exhibited a change in AC magnetic susceptibilities compared to Mn$_2$-o and Mn-c. The effective energy barriers were 15.89, 18.10, 13.96, and 19.91 K for Mn$_2$-o, Mn$_2$-o-UV, Mn-c, and Mn-c-vis, respectively.

Cyano-bridged heterometallic {FeCo} chain systems provide the opportunity to achieve photoswitchable SCMs via light-induced MMET [110,111]. A cyanide-bridged chain {[(Tp)Fe(CN)$_3$]$_2$Co(BIT)}·2CH$_3$OH (Tp = hydrotris(pyrazolyl)borate, BIT = 3,4-bis-(1H-imidazol-1-yl)thiophene) exhibits reversible, multi-phase transitions driven by

MMET [110]. Light-irradiated single-crystal X-ray diffraction reveals that MMET occurs exclusively between the Co center and one Fe site, switching the chain between diamagnetic $[(Fe1)^{II}{}_{LS}–Co^{III}{}_{LS}–(Fe2)^{III}{}_{LS}]$ and metastable ferromagnetic $[(Fe1)^{III}{}_{LS}–Co^{II}{}_{HS}–(Fe2)^{III}{}_{LS}]$ properties at low temperatures. At 10 K, alternating 946 and 532 nm illumination enables the reversible on–off switching of SCM behavior. Moreover, rapid thermal quenching traps a distinct supercooled phase (HT^{sc}, $P2_12_12_1$), which differs from both the low-temperature phase (LT, *Pnma*) and the light-induced high-temperature phase (HT*, $P2_1/n$). In addition, removal of lattice methanol upon desolvation suppresses MMET and SCM behavior, underscoring the essential role of hydrogen bonding in tuning redox potentials. These findings establish $\{[(Tp)Fe(CN)_3]_2Co(BIT)\}\cdot 2CH_3OH$ as an uncommon platform that integrates light-, thermal-, and solvent-responsive SCM.

4. Photomagnetic Properties of Lanthanide Molecular Nanomagnets

The combination of photoresponsive ligands with lanthanide ions could result in photoresponsive molecular nanomagnets, which have important applications for high-density data storage, quantum information processing, and optical switches. This strategy combines the strong magnetic anisotropy of lanthanide ions with the dynamic characteristics of photoresponsive ligands. In recent years, researchers have assembled lanthanide ions with photocyclization organic compounds and radical precursors as photoresponsive units to construct various lanthanide molecular nanomagnets with photomagnetic properties [27–30,112–122] (Figure 7 and Table 1).

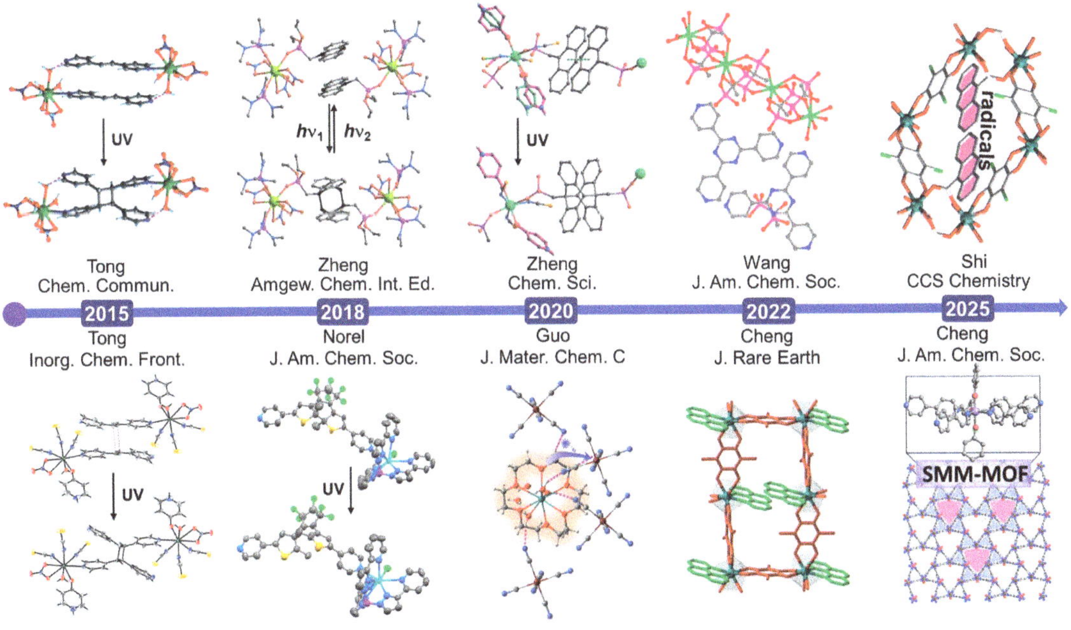

Figure 7. Selected examples of lanthanide molecular nanomagnets with photomagnetic properties. Reprinted with permission from Refs. [27–30,112,113,116,120,121]. Copyright 2015 Chemical Communications, 2018 Angewandte Chemie International Edition, 2019 American Chemical Society, 2020 American Chemical Society, 2017 Inorganic Chemistry Frontiers, 2021 Chemical Science, 2022 Journal of Materials Chemistry C, 2025 Journal of Rare Earths, 2025 American Chemical Society. Reprinted from Ref. [122].

Table 1. The static and dynamic magnetic properties of selected photoresponsive lanthanide molecular nanomagnets discussed in this review.

Complex/ Photoproduct	Photochemical Reaction	$\chi_M T$ at RT cm^3 mol^{-1} K	Hysteresis Loop at 2 K H_c (Oe)	U_{eff} (K) H_{dc} (Oe)	Ref.
$[Dy(bpe)(H_2O)_4(NO_3)_2](NO_3)\cdot 2bpe$	[2,2] cycloaddition	13.9	-	- (slow magnetization relaxation)	[27]
$[Dy_2(tpcb)(H_2O)_8(NO_3)_4](NO_3)_2\cdot 2bpe\cdot tpcb$		14.2	-	-	
$(Hbpe)_2[Dy(bpe)(H_2O)(4\text{-}pyO)(NO_3)(SCN)_3]SCN$ with bpe	[2,2] cycloaddition	13.9	- (butterfly-shaped)	153.8 (H_{dc} = 0) 201.9 (H_{dc} = 1500)	[112]
$(Hbpe)_2(H_2tpcb)[Dy_2(tpcb)_2(H_2O)_2(4\text{-}pyO)_2(NO_3)_2(SCN)_6](SCN)_2$		13.9	- (smaller butterfly-shaped)	205.5 (H_{dc} = 0) 234.5 (H_{dc} = 1500)	
$Dy^{III}(depma)(NO_3)_3(hmpa)_2$	[4 + 4] cycloaddition	13.8	-	20.4 (H_{dc} = 500)	[28]
$Dy^{III}_2(depma_2)(NO_3)_6(hmpa)_4$		14.1	-	43.2 (H_{dc} = 500)	
$Dy^{III}(SCN)_3(depma)_2(4\text{-}hpy)_2$ (**1**)	[4 + 4] cycloaddition	14.1	- (butterfly-shaped)	141 (H_{dc} = 0)	[113]
1UV		14.0	-	101 (H_{dc} = 0)	
1R		14.1	- (butterfly-shaped)	153 (H_{dc} = 0)	
$[Dy(Tppy)F(L_c)]PF_6$ (**1c**)	ring opening	13.9	- (butterfly-shaped)	225.9 (H_{dc} = 0)	[29]
1o		13.8	- (smaller butterfly-shaped)	225.9 (H_{dc} = 0)	
$[Dy_3(H\text{-}HEDP)_3(H_2\text{-}HEDP)_3]\cdot 2H_3\text{-}TPT\cdot H_4\text{-}HEDP\cdot 10H_2O$ (**QDU-1**)	photogenerated radical	42.7	-	-(SMM off)	[30]
QDU-1a		43.7	-	108.1 (H_{dc} = 0)	
QDU-1b		42.7	-	-(SMM off)	
$[Ln_3(H\text{-}HEDP)_2(H_2\text{-}HEDP)_2(H_3\text{-}HEDP)_2]\cdot H_3\text{-}TPP\cdot 11H_2O$ (Ln = Dy (**1-Dy**))	photogenerated radical	42.6	-	16.9 (H_{dc} = 2000)	[115]
1a-Dy		41.8	-	-(SMM off)	
$[Ln_3(H\text{-}HEDP)_2(H_2\text{-}HEDP)_2(H_3\text{-}HEDP)_2]\cdot H_3\text{-}TPP\cdot 11H_2O$ (Ln = Tb (**2-Tb**))		35.2	-	12.1 (H_{dc} = 2000)	
2a-Tb		33.3	-	-(SMM off)	
$[Dy^{III}(18C6)(H_2O)_3]Fe^{III}(CN)_6\cdot 2H_2O$ (**DyFe18C6**)	photogenerated radical	14.5	-	18.4 (H_{dc} = 800)	[116]
DyFe18C6-UV		17.1	-	17.0 (H_{dc} = 800)	
$[Er(CA)_{1.5}(phen)(DMF)]_n$ (**Er(phen)**)	photogenerated radical	11.5	-	66.4 (H_{dc} = 600)	[120]
Er(phen)-UV		14.4	-	72.7 (H_{dc} = 600)	
$\{[Dy_{1.5}(OPh)_2Cl(BPy)_3(THF)_{1.5}][(BPh_4)_{1.5}]\cdot 0.5THF\}_n$ ((**Dy(BPy)**)	photogenerated radical	20.57	H_c = 4500 Oe	822 (H_{dc} = 0, T = 12–40 K) 1048 (H_{dc} = 0, T = 53–71 K)	[121]
Dy(BPy)-UV		20.63	H_c = 1300 Oe	641 (H_{dc} = 0, T = 12–40 K) 1000 (H_{dc} = 0, T = 53–71 K)	
$[Yb(CA)(ACA)(DMF)(H_2O)]_n$ (**Yb(ACA)**)	photogenerated radical	2.65	-	94 (H_{dc} = 800)	[122]
Yb@Y(ACA)-UV		4.83	-	111 (H_{dc} = 800)	

4.1. *Photomagnetic Properties of Photocyclization-Type Lanthanide Molecular Nanomagnets*

Photocyclization-type organic ligands include diarylethene, 9-anthrylmethylphosphonic acids, dithienylethenes, spiropyrans, fulgides, and so on [123–127], which have been employed as photoresponsive units in lanthanide molecular nanomagnets. Diarylethenes and

9-anthrylmethylphosphonic ligands can undergo [2 + 2] and [4 + 4] cycloaddition reactions, respectively, under ultraviolet light [27,28,112,113]. Dithienylethene can undergo cycloaddition reaction when exposed to the light of a specific wavelength [29]. It is worth noting that 9-diethylphosphonomethylanthracene (depma) can undergo photoinduced or thermally induced ring-opening reactions, enabling reversible structural conversions [28,113].

To date, the photomagnetic properties of photocyclization-type lanthanide molecular nanomagnets are primarily focused on [2 + 2] and [4 + 4] cycloaddition reactions induced by light based on the alterations of the coordination environment of the metal centers [27,28,112,113].

The photoinduced cyclization process of 1,2-bis(4-pyridyl)ethylene (bpe) primarily occurs through a [2 + 2] cycloaddition reaction. Under illumination, the double bonds of bpe interact with another molecule, leading to the formation of a cyclobutane structure. In 2015, Tong's group reported a mononuclear Dy(III) complex with bpe, $[Dy(bpe)(H_2O)_4(NO_3)_2](NO_3)\cdot 2bpe$ [27], where the alkene bonds were arranged parallel at a distance of 3.67 Å. Upon UV irradiation, a [2 + 2] cycloaddition reaction occurred, yielding a binuclear Dy(III) complex, $[Dy_2(tpcb)(H_2O)_8(NO_3)_4](NO_3)_2\cdot 2bpe\cdot tpcb$ (tpcb = tetrakis(4-pyridyl)cyclobutane) (Figure 8a,b). Under a zero-dc field, the mononuclear complex exhibited slow magnetic relaxation, while the photoproduced binuclear complex displayed only a very weak temperature-dependent magnetic relaxation (Figure 8c,d).

In 2017, another mononuclear Dy(III) complex $(Hbpe)_2[Dy(bpe)(H_2O)(4\text{-}pyO)(NO_3)(SCN)_3]SCN$ with bpe [112] was reported, where the alkene bonds were parallel with a distance of 3.44 Å. UV irradiation induced a [2 + 2] cycloaddition reaction, yielding a binuclear Dy(III) complex $(Hbpe)_2(H_2tpcb)[Dy_2(tpcb)_2(H_2O)_2(4\text{-}pyO)_2(NO_3)_2(SCN)_6](SCN)_2$ (tpcb = tetrakis(4-pyridyl)cyclobutane, 4-pyO = 4-(1H)-pyridone) (Figure 8e,f). Subtle changes in the coordination environment around the Dy(III) ions increased the effective energy barrier from 153.8 K (H_{dc} = 0 Oe) and 201.9 K (H_{dc} = 1500 Oe) in the mononuclear complex to 205.5 K (H_{dc} = 0 Oe) and 234.5 K (H_{dc} = 1500 Oe) in the binuclear complex (Figure 8g,h) [128]. However, in the temperature range dominated by the Raman process, the magnetic relaxation time of the binuclear complex under the 1500 Oe field was significantly shorter compared to the mononuclear complex.

Similarly to the [2,2] cycloaddition of bpe, 9-diethylphosphonomethylanthracene (depma) can undergo a [4,4] cycloaddition reaction upon illumination. Under ultraviolet light, the π bonds of two anthracene rings interacted to form a dimer with an eight-membered ring structure. In 2018, Zheng's group reported a mononuclear Dy(III) complex, $Dy^{III}(depma)(NO_3)_3(hmpa)_2$ (hmpa = hexamethylphosphoramide) [28], where the anthracene rings were arranged parallel to each other with an interplanar distance of 3.44 Å. UV irradiation at 365 nm induced a [4 + 4] cycloaddition reaction, forming a binuclear $Dy^{III}_2(depma_2)(NO_3)_6(hmpa)_4$ (Figure 9a,b). Under a 500 Oe field, the effective energy barriers of 20.4 K for the mononuclear complex and 43.2 K for the binuclear complex were obtained (Figure 9c,d). Exposure to 254 nm of UV irradiation or thermal treatment reverted the binuclear complex to the mononuclear complex, leading to a reduced effective energy barrier of 18.0 K. This light-driven reversibility highlights the wavelength-dependent switching capabilities.

In 2020, another depma-based mononuclear Dy(III) complex, $Dy^{III}(SCN)_3(depma)_2(4\text{-}hpy)_2$ (**1**; 4-hpy = 4-hydroxypyridine) [113], was reported, where the anthracene rings were arranged parallel to each other with an interplanar distance of 3.49 Å at 300 K (**1RT**) and 3.52/3.56 Å at 193 K (**1LT**). UV irradiation at 365 nm induced a [4 + 4] cycloaddition reaction, leading to a one-dimensional Dy(III) complex **1UV**. **1UV** reverted to **1** upon heating at 120 °C (Figure 9e). The one-dimensional Dy(III) complex had a different dc magnetic susceptibility compared to the mononuclear complex, and heating at **1UV** yielded **1R**, which exhibited the same DC magnetic susceptibility as **1** (Figure 9f). Under zero field, the effective energy barriers of **1**, **1UV**, and **1R** were 141, 101, and 153 K, respectively (Figure 9g–i). This work demonstrates the feasibility of switching magnetic properties via light and heat.

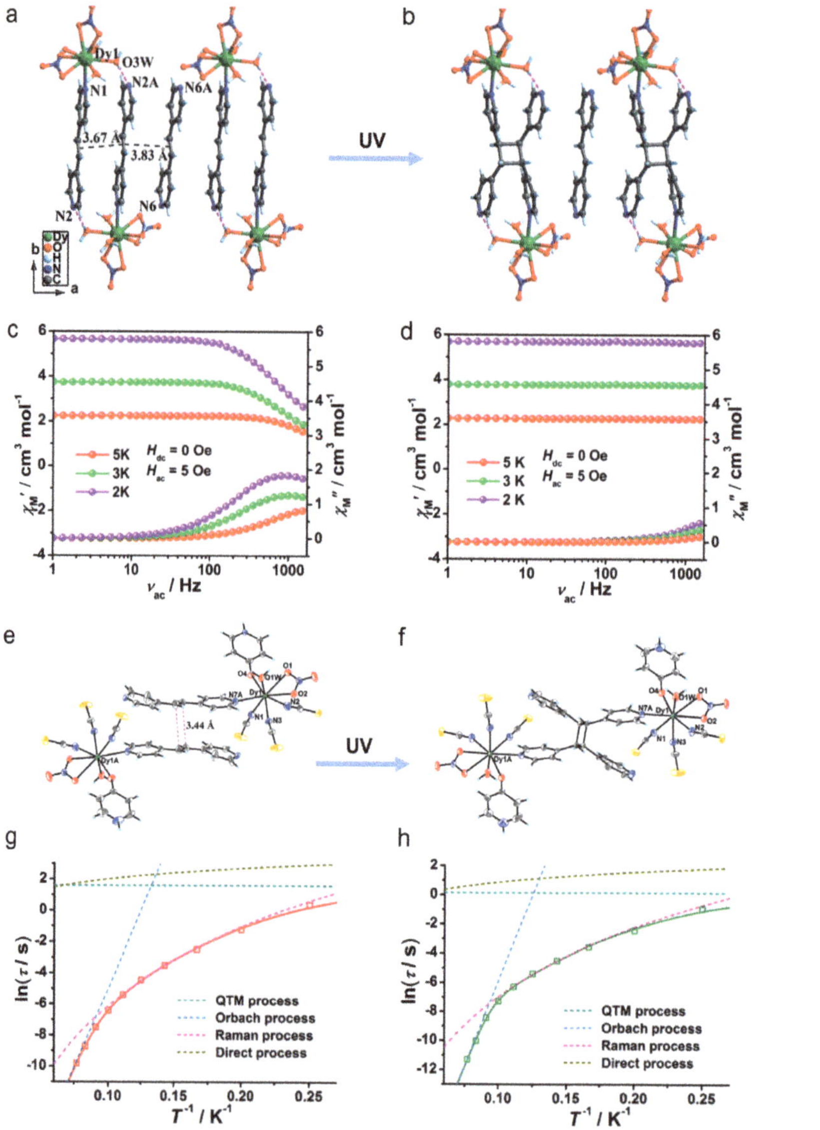

Figure 8. The photochemical reaction from $[Dy(bpe)(H_2O)_4(NO_3)_2](NO_3)\cdot 2bpe$ (**a**) to $[Dy_2(tpcb)(H_2O)_8(NO_3)_4](NO_3)_2\cdot 2bpe\cdot tpcb$ (**b**) via [2,2] photocyclization and their frequency-dependent ac susceptibilities (**c**,**d**). Reprinted with permission from Ref. [27]. Copyright 2015 Chemical Communications. The photochemical reaction from $(Hbpe)_2[Dy(bpe)(H_2O)(4\text{-}pyO)(NO_3)(SCN)_3]SCN$ (**e**) to $(Hbpe)_2(H_2tpcb)[Dy_2(tpcb)_2(H_2O)_2(4\text{-}pyO)_2(NO_3)_2(SCN)_6](SCN)_2$ (**f**) via a [2,2] photocyclization and their ln(τ) vs. T^{-1} plots (**g**,**h**). Reprinted with permission from Ref. [112]. Copyright 2017 Inorganic Chemistry Frontiers.

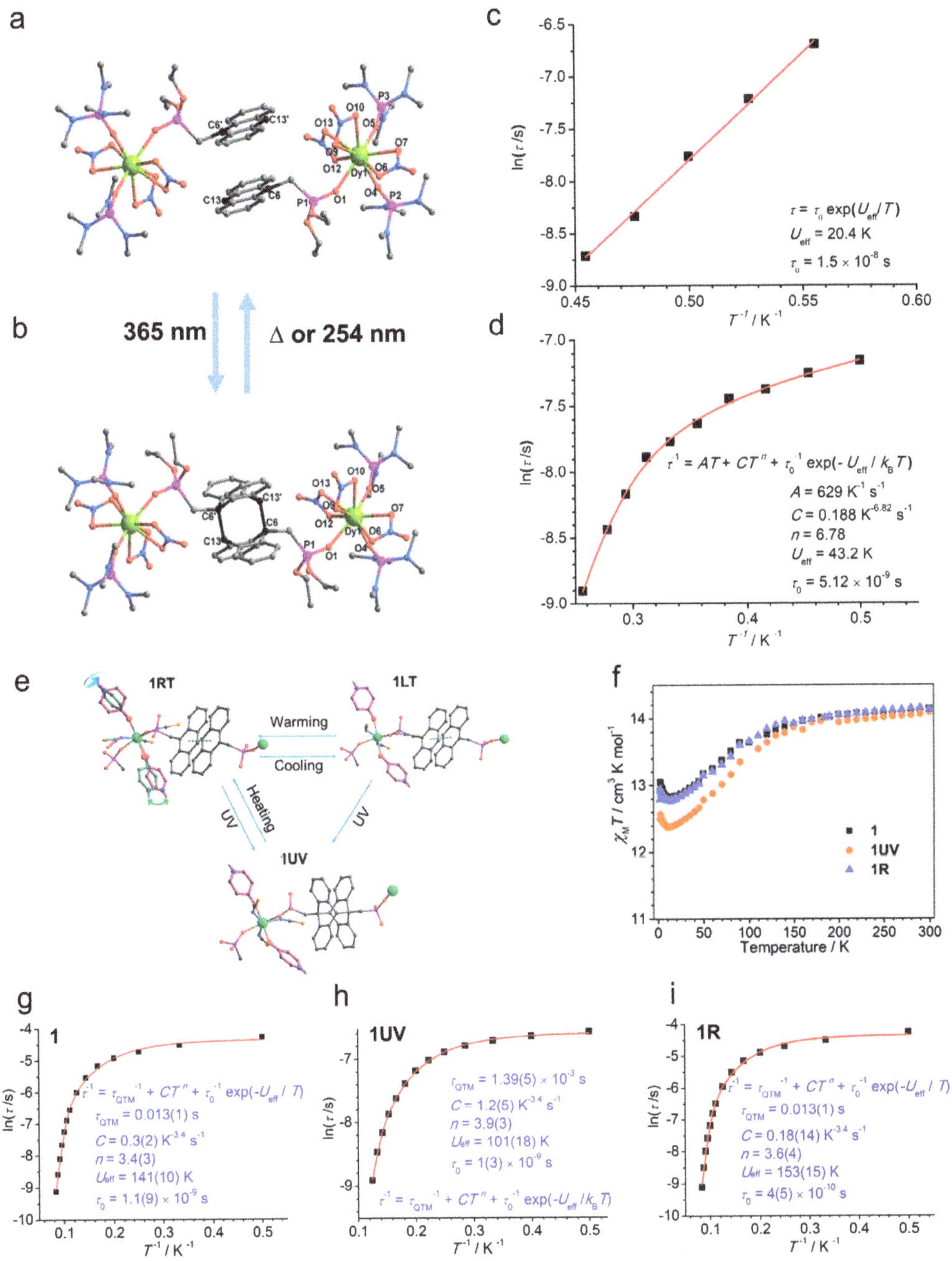

Figure 9. The photochemical reaction from $Dy^{III}(depma)(NO_3)_3(hmpa)_2$ (**a**) to $Dy^{III}_2(depma_2)(NO_3)_6(hmpa)_4$ (**b**) via [4,4] photocyclization and their ln(τ) vs. T^{-1} plots (**c**,**d**). Reprinted with permission from Ref. [28]. Copyright 2018 Angewandte Chemie International Edition. Structural transformation among **1RT**, **1LT**, and **1UV** (**e**). The $\chi_M T$ vs. T (**f**) and ln(τ) vs. T^{-1} plots (**g**–**i**) for **1**, **1UV**, and **1R**. Reprinted with permission from Ref. [113]. Copyright 2021 Chemical Science.

Dithienylethene is another photoresponsive ligand. The double bond of ethylene could react with two thiophene groups to form a closed-loop structure under ultraviolet irradiation. Upon exposure to visible light or heating, the closed-loop structure can revert to the open-loop structure, which is a reversible process. In 2020, Bernot's group reported a one-dimensional Dy(III) complex [Dy(Tppy)F(L_c)]PF_6 (**1c**; Tppy = tris(3-(2-pyridyl)pyrazolyl)hydroborate, L = bispyridyl dithienylethene) with a bispyridyl dithienylethene ligand [29]. Upon irradiation using 532 nm green light, the ligand underwent a transformation from a closed-ring to an open-ring structure, resulting in the formation of the open-ring one-dimensional complex **1o** (Figure 10a,b). An in-depth study on the multiple relaxation processes of molecular nanomagnets is crucial for clarifying their dominant mechanisms across different temperature ranges. **1o** and **1c** exhibited Raman, Orbach, and quantum tunneling relaxation processes with the same effective energy barrier of 225.9 K. Here, $\tau^{-1} = \tau_{QTM}{}^{-1} + \tau_0{}^{-1}\exp(-U_{eff}/k_BT) + BT^n$ was used to fit the relaxation data [19]. **1o** and **1c** showed the Orbach process at a high temperature range and QTM at a low temperature range. However, **1o** (τ_{QTM} = 0.0015 s) demonstrated a faster quantum tunneling rate compared to **1c** (τ_{QTM} = 0.0337 s), and exhibited narrower hysteresis loops (Figure 10c,d). Ab initio calculations indicated that the photomodulation of the magnetic properties originated from the change in crystal packing rather than ligand isomerization-induced crystal field splitting.

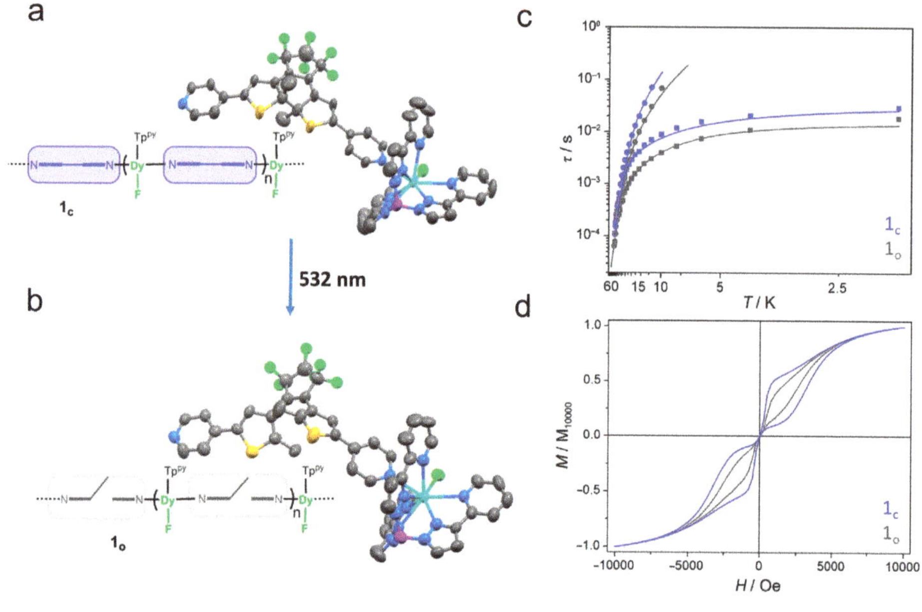

Figure 10. The photochemical reaction from **1c** (**a**) to **1o** (**b**) via photocyclization and their τ–T (**c**) and magnetic hysteresis loops (**d**). Reprinted with permission from Ref. [29]. Copyright 2019 American Chemical Society.

4.2. Photomagnetic Properties of Photogenerated Radical-Type Lanthanide Molecular Nanomagnets

Compared to photocyclization-type systems, photogenerated radical families usually exhibit faster response and minimal structural changes. For example, pyridyls and 9-anthracene carboxylic acids are used for lanthanide molecular nanomagnets with photogenerated radicals. If a system employs pyridyl derivatives as electron acceptors and hydroxyethylidene diphosphonate (HEDP) as electron donors, light irradiation induces electron transfer between them, resulting in the generation of stable radicals [30,115]. The aromatic structure could stabilize photogenerated radicals as well. These radicals have the capacity to modulate the coordination field strength of metal centers and/or the magnetic interaction between the metal centers and the radicals.

Wang's group reported a Dy(III) molecular nanomagnet, $[Dy_3(H\text{-}HEDP)_3(H_2\text{-}HEDP)_3]\cdot 2H_3\text{-}TPT\cdot H_4\text{-}HEDP\cdot 10H_2O$ (**QDU-1**; TPT = 2,4,6-tri(4-pyridyl)-1,3,5-triazine) [30] (Figure 11a). At room temperature and under UV light, colorless **QDU-1** transformed into blue **QDU-1a**, simultaneously generating stable radicals, with a sharp EPR signal at $g = 2.0004$. Heating at 100 °C for 30 min restored **QDU-1** to its colorless state (**QDU-1b**). Before irradiation, **QDU-1** was in a paramagnetic state; after irradiation, the stable radicals induced ferromagnetic interactions with Dy(III) ions at low temperatures, and **QDU-1a** exhibited SMM behavior (Figure 11b–d), with an effective energy barrier of 108.1 K. Through heating, **QDU-1a** can be reverted to its initial state. This study represents the observation of room-temperature-reversible photo-thermochromic SMM on/off properties in lanthanide complexes.

In 2021, two photoresponsive lanthanide complexes, $[Ln_3(H\text{-}HEDP)_2(H_2\text{-}HEDP)_2(H_3\text{-}HEDP)_2]\cdot H_3\text{-}TPP\cdot 11H_2O$ (Ln = Dy (**1-Dy**), Tb (**2-Tb**) and TPP = 2,4,6-tris(4-pyridyl)pyridine) [115], were reported (Figure 11e). At room temperature and under irradiation with a xenon lamp for 40 min, yellow **1-Dy** and **2-Tb** were transformed to purple **1a-Dy** and **2a-Tb** with EPR signals at $g = 2.014$ and 1.996, respectively, from the generation of stable radicals. At low temperatures, antiferromagnetic coupling existed between the lanthanide ions and the photogenerated radicals (Figure 11f). **1-Dy** and **2-Tb** exhibited SMM behavior, while **1a-Dy** and **2a-Tb** were paramagnetic (Figure 11g–j). Reversible photo-thermochromic properties and controlled quenching and recovery of SMM behaviors were observed.

The photogenerated crown radicals were proven to be a means to modulate the magnetization of lanthanide complexes. Guo's group reported a crown-based complex, $[Dy^{III}(18C6)(H_2O)_3]Fe^{III}(CN)_6\cdot 2H_2O$ (**DyFe18C6**) [116] (Figure 12a). At room temperature and under UV irradiation, yellow **DyFe18C6** transformed into orange **DyFe18C6-UV**, accompanied by the generation of radicals. Heating at 75 °C for 2 h or storing in darkness for 7 days restored the complex to its initial state. The obtained compound exhibited photochromic and photomagnetic behaviors at room temperature, demonstrating a transition from paramagnetic to ferromagnetic states and a remarkable enhancement of $\chi_M T$ by 20.9% (Figure 12b). **DyFe18C6** and its photoproducts had effective energy barriers of 18.4 and 17.0 K, respectively (Figure 12c,d).

Due to the advantages of combining the intrinsic properties of photochromic molecules with those of coordination polymers (CPs)/metal–organic frameworks (MOFs), such as modularity, porosity, and structural adjustability, the application of photoresponsive CPs/MOFs in gas storage and separation has been studied [129–135]. Since 2013, we have reported a series of SMM-based CPs/MOFs [11,120–122,136–149]. However, studies on photoresponsive SMMs in CPs/MOFs remain rare.

In 2023, two Er(III) CPs $[Er(CA)_{1.5}(bpy)(DMF)]_n$ (**Er(bpy)**) and $[Er(CA)_{1.5}(phen)(DMF)]_n$ (**Er(phen)**) (H_2CA = 2,5-dichloro-3,6-dihydroxy-p-quinone, bpy = 2,2′-bipyridine, phen = 1, 10-phenanthroline) and their light-converted products **Er(bpy)-UV** and **Er(phen)-UV**, exhibiting field-induced magnetization dynamics, were reported [120]. The structure of **Er(phen)** is shown in Figure 13a. The room temperature $\chi_M T$ of **Er(phen)-UV** increased 25.7% compared with that of **Er(phen)** (Figure 13b). The EPR signal at $g = 2.002$ further supported the generation of radicals, influencing the magnetization dynamics. The magnetization dynamics of **Er(phen)** and **Er(phen)-UV** were dominated by both Raman and Orbach processes, and **Er(phen)-UV** showed an enhanced effective energy barrier value compared with that of **Er(phen)** (Figure 13c,d). It represents an example where both static and dynamic magnetizations of an SMM-based CP were enhanced by UV illumination.

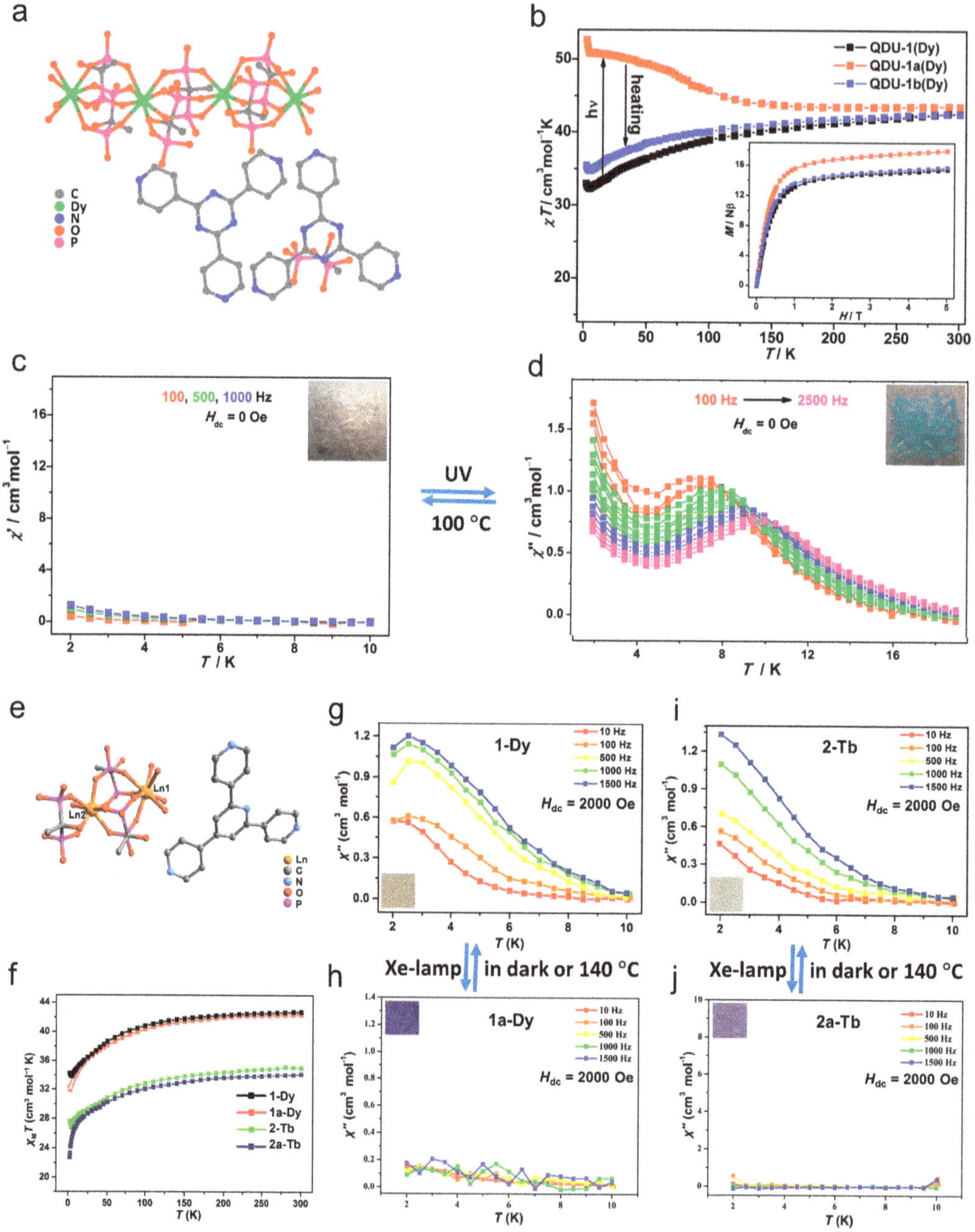

Figure 11. The structure of **QDU-1** (**a**). The $\chi_M T$–T (**b**) and ac susceptibilities (**c**,**d**) of **QDU-1**, **QDU-1a**, and **QDU-1b**. Reprinted with permission from Ref. [30]. Copyright 2020 American Chemical Society. The structure of **1-Ln** (**e**). The $\chi_M T$–T (**f**) and ac susceptibilities (**g**–**j**) of **1-Dy**, **1a-Dy**, **2-Tb**, and **2a-Tb**. Reprinted with permission from Ref. [115]. Copyright 2022 Science China Materials.

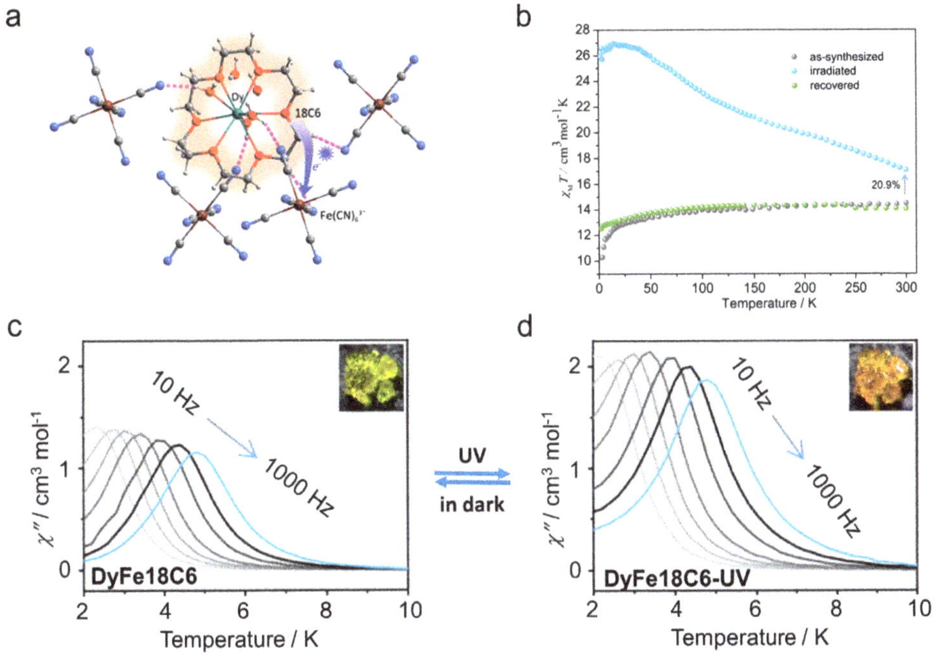

Figure 12. Crystal structure of **DyFe18C6** (**a**). The $\chi_M T$–T (**b**) and ac susceptibilities (**c**,**d**) of **DyFe18C6** and **DyFe18C6-UV**. Reprinted with permission from Ref. [116]. Copyright 2022 Journal of Materials Chemistry C.

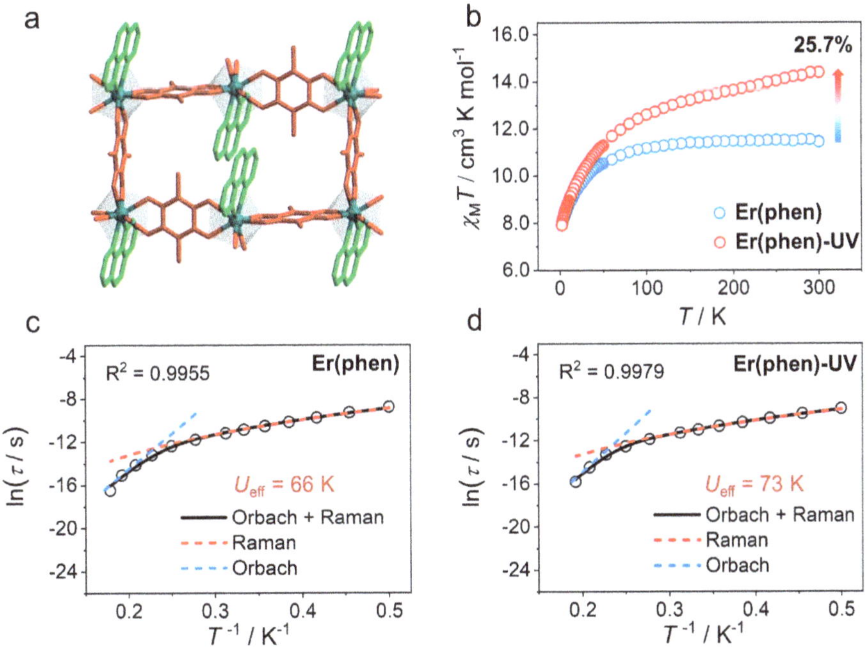

Figure 13. The structure of **Er(phen)** (**a**). The $\chi_M T$–T (**b**) and $\ln(\tau)$–T^{-1} plots (**c**,**d**) for **Er(phen)** and **Er(phen)-UV**. Reprinted with permission from Ref. [120]. Copyright 2025 Journal of Rare Earths.

Recently, a SMM-MOF $\{[Dy_{1.5}(OPh)_2Cl(BPy)_3(THF)_{1.5}][(BPh_4)_{1.5}]\cdot 0.5THF\}_n$ **(Dy(BPy)**, BPy = 4,4′-bipyridine, OPh^- = phenoxide, BPh_4^- = tetraphenylborate) was reported, which features a pentagonal bipyramid SMM as the node and the photosensitive ligand BPy as the linker [121] (Figure 14a–c). The precise synthesis ensured the equatorial arrangement of BPy and the axial coordination of PhO^-, thereby enhancing the vertical orientation of the main magnetic axis of Dy^{III} ions across all Kagomé layers. **Dy(BPy)** displayed photochromic properties when irradiated with UV light at room temperature, generating radicals, with an EPR signal at g = 2.0000. In comparison to **Dy(BPy)**, **Dy(BPy)-UV** demonstrates faster relaxation kinetics in the temperature range of 12–70 K and a lower divergence temperature of 6 K in the field cooling and zero-field cooling curves (9 K for **Dy(BPy)**). The energy barrier of **Dy(BPy)-UV** was 641 K, which is lower than that of 822 K of **Dy(BPy)** in the low-temperature domain. The coercivity field sharply decreases from 4500 Oe for **Dy(BPy)** to 1300 Oe for **Dy(BPy)-UV** at 2 K, while the hysteresis loop opening temperature diminishes from 20 K for **Dy(BPy)** to 14 K for **Dy(BPy)-UV**(Figure 14d,e). A study of the mechanism revealed that the modulated magnetic properties are attributed to the formation of photogenerated radicals and alterations in crystal packing.

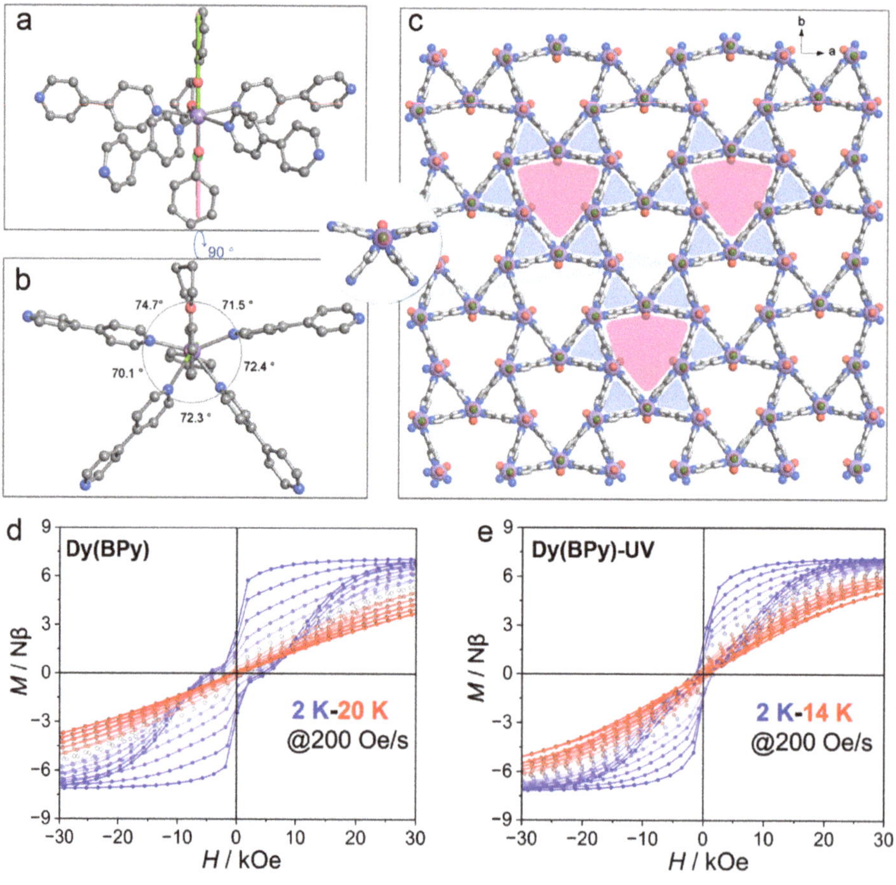

Figure 14. Structure of $[Dy(OPh)X(BPy)_4(THF)]^+$ (X = Cl/PhO^-) cation in **Dy(BPy)** viewed from the side (**a**) and from above (**b**). Kagomé layer of **Dy(BPy)** (**c**). The magnetic hysteresis loops of **Dy(BPy)** (**d**) and **Dy(BPy)-UV** (**e**). Reprinted with permission from Ref. [121]. Copyright 2025 American Chemical Society.

9-anthracene carboxylic acid (HACA) serves as a valuable unit for the construction of molecular switches, attributed to its ability to transform into a photogenerated radical. Stable two-dimensional Yb(III) CPs (**Yb(ACA)**) and their Y(III)-diluted analog **Yb@Y(ACA)**, along with their photogenerated radical forms, **Yb(ACA)-UV** and **Yb@Y(ACA)-UV**, were reported [122]. For **Yb(ACA)** (Figure 15a,b), the formation of the radical leads to an 82.3% enhancement of the static magnetic susceptibilities of **Yb(ACA)-UV** at room temperature compared to those of **Yb(ACA)**, supported by high-level calculation results, which are among the highest values reported so far (Figure 15c). Unexpected increases in the effective energy barrier for magnetization reversal of **Yb(ACA)-UV** (111 K) compared with **Yb(ACA)** (94 K) were observed. This study not only establishes a strategy for designing photomagnetic materials via SMM-based CP platform but also reveals the influence of photogenerated radicals on the static and dynamic magnetizations of lanthanide CPs.

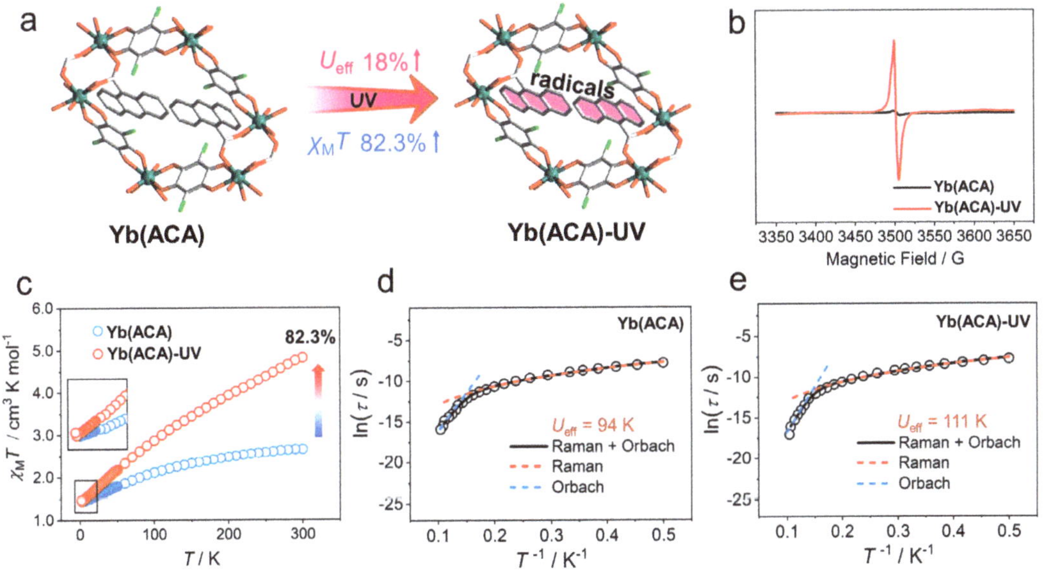

Figure 15. The photochemical reaction from **Yb(ACA)** to **Yb(ACA)-UV** (**a**). EPR spectra of **Yb(ACA)** and **Yb(ACA)-UV** (**b**). The $\chi_M T$–T (**c**) and $\ln(\tau)$–T^{-1} plots (**d**,**e**) of **Yb(ACA)** and **Yb(ACA)-UV**. Reprinted from Ref. [122].

5. Conclusions and Outlook

The recent advances of molecular nanomagnets with photomagnetic properties are summarized. Researchers have applied various strategies to introduce photoresponsive groups into molecular nanomagnets featuring magnetic bistability. Although molecular nanomagnets with photomagnetic properties have received increasing attention in recent years, this field is still in its early stages. Challenges primarily include two aspects: (i) how to rationally synthesize photoresponsive molecular nanomagnets with modulated static and dynamic magnetic properties and (ii) how to achieve reversible switching of the magnetic properties using light. To date, the photomagnetic properties of transition metal molecular nanomagnets are limited mainly to SCO complexes, and the lanthanide molecular nanomagnets are usually connected with photocyclization ligands and photogenerated radicals. Generally, the cycloaddition reaction is restricted by the crystal lattice, leading to slow or even irreversible reactions; photogenerated radicals may not be highly stable, making it challenging to observe the influence of these radicals on magnetic properties quantitatively.

Currently, the static and dynamic magnetic properties of most molecular nanomagnets induced by light decrease in terms of the energy barrier and hysteresis window, and only a limited number of examples demonstrate reversible capability. In addition to the discrete system, CPs and MOFs are a promising platform for the study of molecular nanomagnets with photomagnetic properties, which have the advantages of high stability and the structural flexibility of metal nodes, organic linkers, and guest molecules. Based on the fast development of coordination chemistry and MOF chemistry, research into photoresponsive magnetic CPs and MOFs may develop at a fast pace in the future.

Author Contributions: Conceptualization, W.S., P.C., and X.G.; writing—original draft preparation, X.G.; writing—review and editing, P.C., W.S., X.G., and X.S.; visualization, X.G. and W.S.; funding acquisition, P.C., W.S., and X.G. All authors have read and agreed to the published version of the manuscript.

Funding: This work was supported by the National Natural Science Foundation of China (Grant Nos. 22435002 and 62401563).

Institutional Review Board Statement: Not applicable.

Informed Consent Statement: Not applicable.

Data Availability Statement: No new data were created or analyzed in this study. Data sharing is not applicable to this article.

Conflicts of Interest: The authors declare no conflicts of interest.

References

1. Ferreira, F.; Moreira, S.; Zhao, M.; Barriga, E.H. Stretch-induced endogenous electric fields drive directed collective cell migration in vivo. *Nat. Mater.* **2025**, *24*, 462–470. [CrossRef]
2. Ghitani, N.; von Buchholtz, L.J.; MacDonald, D.I.; Falgairolle, M.; Nguyen, M.Q.; Licholai, J.A.; Ryba, N.J.P.; Chesler, A.T. A distributed coding logic for thermosensation and inflammatory pain. *Nature* **2025**, *642*, 1016–1023. [CrossRef] [PubMed]
3. Li, Z.; Wu, Q.-Q.; Chen, M.-H.; Xu, J.-J.; Chen, H.-Y.; Zhao, W.-W. Organic photoelectrochemical memtransistor. *eScience* **2025**, *5*, 100374. [CrossRef]
4. Wang, Y.-X.; Su, D.; Ma, Y.; Sun, Y.; Cheng, P. Electrical detection and modulation of magnetism in a Dy-based ferroelectric single-molecule magnet. *Nat. Commun.* **2023**, *14*, 7901. [CrossRef]
5. Wang, Y.-H.; Yan, S.-K.; Liang, S.; Zhao, Y.-R.; Zhang, J.; Wang, G.-M.; Hu, J.-X. Synergistic photomagnetic and photomechanical dynamics in a dysprosium-based smart molecule. *Adv. Sci.* **2025**, e09088. [CrossRef] [PubMed]
6. Raju, M.S.; Paillot, K.; Breslavetz, I.; Novitchi, G.; Rikken, G.L.J.A.; Train, C.; Atzori, M. Optical readout of single-molecule magnets magnetic memories with unpolarized light. *J. Am. Chem. Soc.* **2024**, *146*, 23616–23624. [CrossRef]
7. Wang, J.; Zakrzewski, J.J.; Zychowicz, M.; Xin, Y.; Tokoro, H.; Chorazy, S.; Ohkoshi, S. Desolvation-induced highly symmetrical terbium(III) single-molecule magnet exhibiting luminescent self-monitoring of temperature. *Angew. Chem. Int. Ed.* **2023**, *62*, e202306372. [CrossRef]
8. Magott, M.; Brzozowska, M.; Baran, S.; Vieru, V.; Pinkowicz, D. An intermetallic molecular nanomagnet with the lanthanide coordinated only by transition metals. *Nat. Commun.* **2022**, *13*, 2014. [CrossRef]
9. Gatteschi, D.; Sessoli, R.; Villain, J. *Molecular Nanomagnets*; Oxford University Press: Oxford, UK, 2006.
10. Rodolphe, C.; Hitoshi, M.; Masahiro, Y.; Claude, C. Evidence for single-chain magnet behavior in a Mn(III)-Ni(II) chain designed with high spin magnetic units: A route to high temperature metastable magnets. *J. Am. Chem. Soc.* **2002**, *124*, 12837–12844. [CrossRef]
11. Liu, X.; Feng, X.; Meihaus, K.R.; Meng, X.; Zhang, Y.; Li, L.; Liu, J.L.; Pedersen, K.S.; Keller, L.; Shi, W.; et al. Coercive fields above 6T in two cobalt(II)-radical chain compounds. *Angew. Chem. Int. Ed.* **2020**, *59*, 10610–10618. [CrossRef]
12. Wu, Y.; Ruan, Z.-Y.; Zhang, C.; Chen, J.-N.; Wang, X.-Y.; Chen, Z.; Gou, X.; Lan, W.; Kong, X.-J.; Liu, J.-L.; et al. Strong and switchable magneto-optical effect in air-stable chiral Dy^{III} complexes with magnetic anisotropy. *J. Am. Chem. Soc.* **2025**, *147*, 20799–20806. [CrossRef] [PubMed]

13. Marin, R.; Brunet, G.; Murugesu, M. Shining new light on multifunctional lanthanide single-molecule magnets. *Angew. Chem. Int. Ed.* **2021**, *60*, 1728–1746. [CrossRef] [PubMed]
14. Gou, X.; Wu, Y.; Wang, M.; Liu, N.; Lan, W.; Zhang, Y.-Q.; Shi, W.; Cheng, P. The influence of light on the field-induced magnetization dynamics of two Er(III) coordination polymers with different halogen substituents. *Dalton Trans.* **2024**, *53*, 148–152. [CrossRef]
15. Zhao, L.; Meng, Y.S.; Liu, Q.; Sato, O.; Shi, Q.; Oshio, H.; Liu, T. Switching the magnetic hysteresis of an [Fe(II)-NC-W(V)]-based coordination polymer by photoinduced reversible spin crossover. *Nat. Chem.* **2021**, *13*, 698–704. [CrossRef]
16. Errulat, D.; Harriman, K.L.M.; Gálico, D.A.; Kitos, A.A.; Mansikkamäki, A.; Murugesu, M. A trivalent 4f complex with two bis-silylamide ligands displaying slow magnetic relaxation. *Nat. Chem.* **2023**, *15*, 1100–1107. [CrossRef]
17. Meihaus, K.R.; Minasian, S.G.; Lukens, W.W., Jr.; Kozimor, S.A.; Shuh, D.K.; Tyliszczak, T.; Long, J.R. Influence of pyrazolate vs N-heterocyclic carbene ligands on the slow magnetic relaxation of homoleptic trischelate lanthanide(III) and uranium(III) complexes. *J. Am. Chem. Soc.* **2014**, *136*, 6056–6068. [CrossRef]
18. Gao, C.; Genoni, A.; Gao, S.; Jiang, S.; Soncini, A.; Overgaard, J. Observation of the asphericity of 4f-electron density and its relation to the magnetic anisotropy axis in single-molecule magnets. *Nat. Chem.* **2019**, *12*, 213–219. [CrossRef]
19. Guo, F.S.; Day, B.M.; Chen, Y.C.; Tong, M.L.; Mansikkamäki, A.; Layfield, R.A. Magnetic hysteresis up to 80 kelvin in a dysprosium metallocene single-molecule magnet. *Science* **2018**, *362*, 1400–1403. [CrossRef]
20. Goodwin, C.A.P.; Ortu, F.; Reta, D.; Chilton, N.F.; Mills, D.P. Molecular magnetic hysteresis at 60 kelvin in dysprosocenium. *Nature* **2017**, *548*, 439–442. [CrossRef] [PubMed]
21. Li, L.; Fang, Y.; Liu, S.; Hu, M.; Wang, W. Slow magnetic relaxation in a 3D dysprosium(III)-fluoro-oxalate framework containing zig-zag [Dy-F] chains. *J. Rare Earths* **2023**, *41*, 100–107. [CrossRef]
22. Liao, X.; Chen, Y.; Xie, T.; Sun, R.J.; Yang, L.Z.; Liu, C.F.; Wang, R.; Klyatskaya, S.; Ruben, M.; Zhang, W.; et al. Spatially dependent f-pi exchange interaction within a single-molecule magnet $TbPc_2$. *Nat. Commun.* **2025**, *16*, 6263. [CrossRef]
23. Liu, Y.; Luo, Q.; Jin, P.; Zhai, Y.; Zheng, Y. Fine tuning dynamic magnetism of dysprosiacarboranyl sandwiches. *J. Rare Earths* **2025**, *43*, 1205–1212. [CrossRef]
24. Khariushin, I.V.; Ovsyannikov, A.S.; Islamov, D.R.; Samigullina, A.I.; Solovieva, S.E.; Zakrzewski, J.J.; Chorazy, S.; Ferlay, S. Tuning crystal packing and magnetic properties in a series of [Dy_{12}] metallocubanes based on azobenzene derivatives of salicylic acid. *Inorg. Chem.* **2023**, *62*, 10548–10558. [CrossRef] [PubMed]
25. Li, G.-M.; Xu, F.; Han, S.-D.; Pan, J.; Wang, G.-M. Hybrid photochromic lanthanide phosphonate with multiple photoresponsive functionalities. *Inorg. Chem.* **2022**, *61*, 8379–8385. [CrossRef]
26. Bartolomé, E.; Arauzo, A.; Luzón, J.; Gasque, L. Light-induced magnetic switching in a coumarin-based Tb single molecule magnet. *J. Mater. Chem. C* **2025**, *13*, 831–841. [CrossRef]
27. Wang, L.-F.; Qiu, J.-Z.; Liu, J.-L.; Chen, Y.-C.; Jia, J.-H.; Jover, J.; Ruiz, E.; Tong, M.-L. Modulation of single-molecule magnet behaviour via photochemical [2 + 2] cycloaddition. *Chem. Commun.* **2015**, *51*, 15358–15361. [CrossRef]
28. Huang, X.-D.; Xu, Y.; Fan, K.; Bao, S.-S.; Kurmoo, M.; Zheng, L.-M. Reversible SC-SC Transformation involving [4 + 4] cycloaddition of anthracene: A single-ion to single-molecule magnet and yellow-green to blue-white emission. *Angew. Chem. Int. Ed.* **2018**, *57*, 8577–8581. [CrossRef]
29. Hojorat, M.; Al Sabea, H.; Norel, L.; Bernot, K.; Roisnel, T.; Gendron, F.; Guennic, B.L.; Trzop, E.; Collet, E.; Long, J.R.; et al. Hysteresis photomodulation via single-crystal-to-single-crystal isomerization of a photochromic chain of dysprosium single-molecule magnets. *J. Am. Chem. Soc.* **2020**, *142*, 931–936. [CrossRef] [PubMed]
30. Ma, Y.-J.; Hu, J.-X.; Han, S.-D.; Pan, J.; Li, J.-H.; Wang, G.-M. Manipulating on/off single-molecule magnet behavior in a Dy(III)-based photochromic complex. *J. Am. Chem. Soc.* **2020**, *142*, 2682–2689. [CrossRef]
31. Feng, X.; Mathoniere, C.; Jeon Ie, R.; Rouzieres, M.; Ozarowski, A.; Aubrey, M.L.; Gonzalez, M.I.; Clerac, R.; Long, J.R. Tristability in a light-actuated single-molecule magnet. *J. Am. Chem. Soc.* **2013**, *135*, 15880–15884. [CrossRef] [PubMed]
32. Mathonière, C.; Lin, H.J.; Siretanu, D.; Clerac, R.; Smith, J.M. Photoinduced single-molecule magnet properties in a four-coordinate iron(II) spin crossover complex. *J. Am. Chem. Soc.* **2013**, *135*, 19083–19086. [CrossRef]
33. Liu, T.; Zheng, H.; Kang, S.; Shiota, Y.; Hayami, S.; Mito, M.; Sato, O.; Yoshizawa, K.; Kanegawa, S.; Duan, C. A light-induced spin crossover actuated single-chain magnet. *Nat. Commun.* **2013**, *4*, 2826. [CrossRef]
34. Jiang, W.; Jiao, C.; Meng, Y.; Zhao, L.; Liu, Q.; Liu, T. Switching single chain magnet behavior via photoinduced bidirectional metal-to-metal charge transfer. *Chem. Sci.* **2018**, *9*, 617–622. [CrossRef]
35. Zhou, X.; Yan, F.; Lyubartsev, A.; Shen, B.; Zhai, J.; Conesa, J.C.; Hedin, N. Efficient production of solar hydrogen peroxide using piezoelectric polarization and photoinduced charge transfer of nanopiezoelectrics sensitized by carbon quantum dots. *Adv. Sci.* **2022**, *9*, 2105792. [CrossRef] [PubMed]
36. Homer, M.K.; Kuo, D.-Y.; Dou, F.Y.; Cossairt, B.M. Photoinduced charge transfer from quantum dots measured by cyclic voltammetry. *J. Am. Chem. Soc.* **2022**, *144*, 14226–14234. [CrossRef] [PubMed]

37. Lu, Z.; Wu, X.; Chen, N.; Cao, M.; Sartin, M.M.; Ren, B. Photoinduced charge transfer from a semiconductor to a metal probed at the single-nanoparticle level. *ACS Energy Lett.* **2021**, *6*, 3473–3480. [CrossRef]
38. Li, X.; Yu, J.; Lu, Z.; Duan, J.; Fry, H.C.; Gosztola, D.J.; Maindan, K.; Rajasree, S.S.; Deria, P. Photoinduced charge transfer with a small driving force facilitated by exciplex-like complex formation in metal-organic frameworks. *J. Am. Chem. Soc.* **2021**, *143*, 15286–15297. [CrossRef]
39. Cammarata, M.; Zerdane, S.; Balducci, L.; Azzolina, G.; Mazerat, S.; Exertier, C.; Trabuco, M.; Levantino, M.; Alonso-Mori, R.; Glownia, J.M.; et al. Charge transfer driven by ultrafast spin transition in a CoFe Prussian blue analogue. *Nat. Chem.* **2020**, *13*, 10–14. [CrossRef]
40. Ahmad, I.; Akhtar, M.S.; Ahmed, E.; Ahmad, M. Aluminium and cerium co-doped ZnO nanoparticles: Facile and inexpensive synthesis and visible light photocatalytic performances. *J. Rare Earths* **2021**, *39*, 151–159. [CrossRef]
41. Ishikawa, T.; Hayes, S.A.; Keskin, S.; Corthey, G.; Hada, M.; Pichugin, K.; Marx, A.; Hirscht, J.; Shionuma, K.; Onda, K.; et al. Direct observation of collective modes coupled to molecular orbital-driven charge transfer. *Science* **2015**, *350*, 1501–1505. [CrossRef]
42. Kaneva, N.; Bachvarova-Nedelcheva, A. Photocatalytic efficiency of pure and palladium Co-catalytic modified binary system. *Inorganics* **2025**, *13*, 161. [CrossRef]
43. Zoppellaro, G.; Medved, M.; Hrubý, V.; Zbořil, R.; Otyepka, M.; Lazar, P. Solvent controlled generation of spin active polarons in two-dimensional material under UV light irradiation. *J. Am. Chem. Soc.* **2024**, *146*, 15010–15018. [CrossRef] [PubMed]
44. Widel, Z.X.W.; Alatis, J.A.; Stephenson, R.H.; Mastrocinque, F.; Wilcox, A.C.; Bullard, G.; Olivier, J.-H.; Bai, Y.; Zhang, P.; Beratan, D.N.; et al. Fluence-dependent photoinduced charge transfer dynamics in polymer-wrapped semiconducting single-walled carbon nanotubes. *J. Am. Chem. Soc.* **2024**, *146*, 31169–31176. [CrossRef] [PubMed]
45. Shu, A.; Qin, C.; Li, M.; Zhao, L.; Shangguan, Z.; Shu, Z.; Yuan, X.; Zhu, M.; Wu, Y.; Wang, H. Electric effects reinforce charge carrier behaviour for photocatalysis. *Energy Environ. Sci.* **2024**, *17*, 4907–4928. [CrossRef]
46. Royakkers, J.; Yang, H.; Gillett, A.J.; Eisner, F.; Ghosh, P.; Congrave, D.G.; Azzouzi, M.; Andaji-Garmaroudi, Z.; Leventis, A.; Rao, A.; et al. Synthesis of model heterojunction interfaces reveals molecular-configuration-dependent photoinduced charge transfer. *Nat. Chem.* **2024**, *16*, 1453–1461. [CrossRef] [PubMed]
47. Pearce, N.; Reynolds, K.E.A.; Kayal, S.; Sun, X.Z.; Davies, E.S.; Malagreca, F.; Schürmann, C.J.; Ito, S.; Yamano, A.; Argent, S.P.; et al. Selective photoinduced charge separation in perylenediimide-pillar[5]arene rotaxanes. *Nat. Commun.* **2022**, *13*, 415. [CrossRef]
48. Zheng, F.; Dong, F.; Zhou, L.; Yu, J.; Luo, X.; Zhang, X.; Lv, Z.; Jiang, L.; Chen, Y.; Liu, M. Cerium and carbon-sulfur codoped mesoporous TiO_2 nanocomposites for boosting visible light photocatalytic activity. *J. Rare Earths* **2023**, *41*, 539–549. [CrossRef]
49. Pajuelo-Corral, O.; García, J.A.; Castillo, O.; Luque, A.; Rodríguez-Diéguez, A.; Cepeda, J. Single-ion magnet and photoluminescence properties of lanthanide(III) coordination polymers based on pyrimidine-4,6-dicarboxylate. *Magnetochemistry* **2021**, *7*, 8. [CrossRef]
50. Fu, C.; Li, F.; Zhang, J.; Li, D.; Qian, K.; Liu, Y.; Tang, J.; Fan, F.; Zhang, Q.; Gong, X.Q.; et al. Site sensitivity of interfacial charge transfer and photocatalytic efficiency in photocatalysis: Methanol oxidation on anatase TiO_2 nanocrystals. *Angew. Chem. Int. Ed.* **2021**, *60*, 6160–6169. [CrossRef] [PubMed]
51. Graf, M.; Vonbun-Feldbauer, G.B.; Koper, M.T.M. Direct and broadband plasmonic charge transfer to enhance water oxidation on a gold electrode. *ACS Nano* **2021**, *15*, 3188–3200. [CrossRef]
52. Wang, M.; Zhang, G.; Guan, Z.; Yang, J.; Li, Q. Spatially separating redox centers and photothermal effect synergistically boosting the photocatalytic hydrogen evolution of $ZnIn_2S_4$ Nanosheets. *Small* **2021**, *17*, 2006952. [CrossRef]
53. Delgado, T.; Pelé, A.-L. Measurement of reverse-light-induced excited spin state trapping in spin crossover systems: A study case with $Zn_{1-x}Fe_x(6\text{-mepy})_3tren(PF_6)_2{\cdot}CH_3CN$; x = 0.5%. *Crystals* **2024**, *14*, 210. [CrossRef]
54. Zhu, H.-L.; Meng, Y.-S.; Hu, J.-X.; Oshio, H.; Liu, T. Photoinduced magnetic hysteresis in a cyanide-bridged two-dimensional $[Mn_2W]$ coordination polymer. *Inorg. Chem. Front.* **2022**, *9*, 4974–4981. [CrossRef]
55. Su, S.-Q.; Wu, S.-Q.; Huang, Y.-B.; Xu, W.-H.; Gao, K.-G.; Okazawa, A.; Okajima, H.; Sakamoto, A.; Kanegawa, S.; Sato, O. Photoinduced persistent polarization change in a spin transition crystal. *Angew. Chem. Int. Ed.* **2022**, *61*, e202208771. [CrossRef] [PubMed]
56. Nadeem, M.; Cruddas, J.; Ruzzi, G.; Powell, B.J. Toward high-temperature light-induced spin-state trapping in spin-crossover materials: The interplay of collective and molecular effects. *J. Am. Chem. Soc.* **2022**, *144*, 9138–9148. [CrossRef] [PubMed]
57. Benchohra, A.; Li, Y.; Chamoreau, L.M.; Baptiste, B.; Elkaïm, E.; Guillou, N.; Kreher, D.; Lescouëzec, R. The atypical hysteresis of $[Fe(C_6F_5Tp)_2]$: Overlay of spin-crossovers and symmetry-breaking phase transition. *Angew. Chem. Int. Ed.* **2021**, *60*, 8803–8807. [CrossRef]
58. Wen, W.; Meng, Y.-S.; Jiao, C.-Q.; Liu, Q.; Zhu, H.-L.; Li, Y.-M.; Oshio, H.; Liu, T. A mixed-valence $\{Fe_{13}\}$ cluster exhibiting metal-to-metal charge-transfer-switched spin crossover. *Angew. Chem. Int. Ed.* **2020**, *59*, 16393–16397. [CrossRef]

59. Ossinger, S.; Näther, C.; Buchholz, A.; Schmidtmann, M.; Mangelsen, S.; Beckhaus, R.; Plass, W.; Tuczek, F. Spin transition of an iron(II) organoborate complex in different polymorphs and in vacuum-deposited thin films: Influence of cooperativity. *Inorg. Chem.* **2020**, *59*, 7966–7979. [CrossRef] [PubMed]
60. Zhang, K.; Ash, R.; Girolami, G.S.; Vura-Weis, J. Tracking the metal-centered triplet in photoinduced spin crossover of fe(phen)$_3^{2+}$ with tabletop femtosecond M-Edge X-ray absorption Near-Edge structure spectroscopy. *J. Am. Chem. Soc.* **2019**, *141*, 17180–17188. [CrossRef]
61. Fatur, S.M.; Shepard, S.G.; Higgins, R.F.; Shores, M.P.; Damrauer, N.H. A synthetically tunable system to control MLCT excited-state lifetimes and spin states in iron(II) polypyridines. *J. Am. Chem. Soc.* **2017**, *139*, 4493–4505. [CrossRef]
62. Wu, D.-Y.; Sato, O.; Einaga, Y.; Duan, C.-Y. A spin-crossover cluster of iron(II) exhibiting a mixed-spin structure and synergy between spin transition and magnetic interaction. *Angew. Chem. Int. Ed.* **2009**, *48*, 1475–1478. [CrossRef]
63. Halder, G.J.; Chapman, K.W.; Neville, S.M.; Moubaraki, B.; Murray, K.S.; Létard, J.F.; Kepert, C.J. Elucidating the mechanism of a two-step spin transition in a nanoporous metal-organic framework. *J. Am. Chem. Soc.* **2009**, *130*, 17552–17562. [CrossRef] [PubMed]
64. Agustí, G.; Ohtani, R.; Yoneda, K.; Gaspar, A.B.; Ohba, M.; Sánchez-Royo, J.F.; Muñoz, M.C.; Kitagawa, S.; Real, J.A. Oxidative addition of halogens on open metal sites in a microporous spin-crossover coordination polymer. *Angew. Chem. Int. Ed.* **2009**, *48*, 8944–8947. [CrossRef] [PubMed]
65. Chen, T.; Yang, Q.; Fang, C.; Deng, S.; Xu, B. Advanced design for stimuli-reversible chromic wearables with customizable functionalities. *Adv. Mater.* **2024**, *37*, 2413665. [CrossRef]
66. He, Y.; Dang, T.; Leach, A.G.; Zhang, Z.-Y.; Li, T. Photoswitchable azobispyrazole crystals achieving near-quantitative crystalline-state bidirectional $E \leftrightarrows Z$ conversions. *J. Am. Chem. Soc.* **2024**, *146*, 29237–29244. [CrossRef]
67. Zhao, H.; Liao, J.; Fu, S.; Zi, Y.; Bai, X.; Ci, Y.; Zhang, Y.; Cai, X.; Li, Y.; Cun, Y.; et al. Direct 3D lithography of reversible photochromic patterns with tunable luminescence in amorphous transparent media. *ACS Energy Lett.* **2025**, *10*, 1235–1244. [CrossRef]
68. Dong, G.; Lin, H.; Yang, H.; Hu, T.; Huang, Q.; Han, X.; Li, X.; Wang, P.; Xu, J.; Cheng, Y.; et al. Twin-perovskite structured $Ba_8Ti_3Nb_4O_{24}$: Pr^{3+}: A fast photochromics with millisecond-level response speed for optical storage. *ACS Energy Lett.* **2024**, *10*, 484–495. [CrossRef]
69. Hou, J.; Long, G.; Zhao, W.; Zhou, G.; Liu, D.; Broer, D.J.; Feringa, B.L.; Chen, J. Phototriggered complex motion by programmable construction of light-driven molecular motors in liquid crystal networks. *J. Am. Chem. Soc.* **2022**, *144*, 6851–6860. [CrossRef]
70. Song, Y.-P.; Zhang, J.-N.; Wang, J.-R.; Li, K.; Yuan, Y.-X.; Li, B.; Zang, S.-Q. A fast responsive photochromic SCC-MOF for photoswitching and information encryption. *Sci. China Mater.* **2024**, *67*, 698–704. [CrossRef]
71. Lu, H.; Xie, J.; Wang, X.-Y.; Wang, Y.; Li, Z.-J.; Diefenbach, K.; Pan, Q.-J.; Qian, Y.; Wang, J.-Q.; Wang, S.; et al. Visible colorimetric dosimetry of UV and ionizing radiations by a dual module photochromic nanocluster. *Nat. Commun.* **2021**, *12*, 2798. [CrossRef]
72. Roubinet, B.; Weber, M.; Shojaei, H.; Bates, M.; Bossi, M.L.; Belov, V.N.; Irie, M.; Hell, S.W. Fluorescent photoswitchable diarylethenes for biolabeling and single-molecule localization microscopies with optical superresolution. *J. Am. Chem. Soc.* **2017**, *139*, 6611–6620. [CrossRef]
73. Zhang, X.; Hou, L.; Samorì, P. Coupling carbon nanomaterials with photochromic molecules for the generation of optically responsive materials. *Nat. Commun.* **2016**, *7*, 11118. [CrossRef] [PubMed]
74. Babii, O.; Afonin, S.; Garmanchuk, L.V.; Nikulina, V.V.; Nikolaienko, T.V.; Storozhuk, O.V.; Shelest, D.V.; Dasyukevich, O.I.; Ostapchenko, L.I.; Iurchenko, V.; et al. Direct photocontrol of peptidomimetics: An alternative to oxygen-dependent photodynamic cancer therapy. *Angew. Chem. Int. Ed.* **2016**, *55*, 5493–5496. [CrossRef] [PubMed]
75. Rice, A.M.; Martin, C.R.; Galitskiy, V.A.; Berseneva, A.A.; Leith, G.A.; Shustova, N.B. Photophysics modulation in photoswitchable metal-organic frameworks. *Chem. Rev.* **2019**, *120*, 8790–8813. [CrossRef]
76. Bisoyi, H.K.; Li, Q. Light-directing chiral liquid crystal nanostructures: From 1D to 3D. *Acc. Chem. Res.* **2014**, *47*, 3184–3195. [CrossRef]
77. Kole, G.K.; Vittal, J.J. Solid-state reactivity and structural transformations involving coordination polymers. *Chem. Soc. Rev.* **2013**, *42*, 1755–1775. [CrossRef]
78. Cao, L.-H.; Xu, X.-Q.; Tang, X.-H.; Yang, Y.; Liu, J.; Yin, Z.; Zang, S.-Q.; Ma, Y.-M. Controllable strategy for metal-organic framework light-driven [2 + 2] cycloaddition reactions via solvent-assisted linker exchange. *Inorg. Chem.* **2021**, *60*, 2117–2121. [CrossRef]
79. Ou, Y.-C.; Liu, W.-T.; Li, J.-Y.; Zhang, G.-G.; Wang, J.; Tong, M.-L. Solvochromic and photodimerization behaviour of 1D coordination polymer via single-crystal-to-single-crystal transformation. *Chem. Commun.* **2011**, *47*, 9384–9386. [CrossRef]

80. Wang, X.-Y.; Wang, Z.-M.; Gao, S. A pillared layer MOF with anion-tunable magnetic properties and photochemical [2 + 2] cycloaddition. *Chem. Commun.* **2007**, 1127–1129. [CrossRef] [PubMed]
81. Hutchins, K.M.; Rupasinghe, T.P.; Ditzler, L.R.; Swenson, D.C.; Sander, J.R.G.; Baltrusaitis, J.; Tivanski, A.V.; MacGillivray, L.R. Nanocrystals of a metal-organic complex exhibit remarkably high conductivity that increases in a single-crystal-to-single-crystal transformation. *J. Am. Chem. Soc.* **2014**, *136*, 6778–6781. [CrossRef]
82. Park, I.H.; Medishetty, R.; Kim, J.Y.; Lee, S.S.; Vittal, J.J. Distortional supramolecular isomers of polyrotaxane coordination polymers: Photoreactivity and sensing of nitro compounds. *Angew. Chem. Int. Ed.* **2014**, *53*, 5591–5595. [CrossRef]
83. Zouev, I.; Cao, D.K.; Sreevidya, T.V.; Botoshansky, M.; Kaftory, M. Photodimerization of anthracene derivatives in their neat solid state and in solid molecular compounds. *CrystEngComm* **2011**, *13*, 4376–4381. [CrossRef]
84. Becker, H.D. Unimolecular photochemistry of anthracenes. *Chem. Rev.* **1993**, *93*, 145–172. [CrossRef]
85. Furlong, B.J.; Katz, M.J. Bistable dithienylethene-based metal-organic framework illustrating optically induced changes in chemical separations. *J. Am. Chem. Soc.* **2017**, *139*, 13280–13283. [CrossRef]
86. Kobatake, S.; Takami, S.; Muto, H.; Ishikawa, T.; Irie, M. Rapid and reversible shape changes of molecular crystals on photoirradiation. *Nature* **2007**, *446*, 778–781. [CrossRef] [PubMed]
87. Yang, X.; Du, M.; Chu, Z.; Li, C. Synchronizing multicolor changes and shape deformation into structurally homogeneous hydrogels via a single photochromophore. *Adv. Mater.* **2025**, *37*, 2500857. [CrossRef]
88. Williams, D.E.; Martin, C.R.; Dolgopolova, E.A.; Swifton, A.; Godfrey, D.C.; Ejegbavwo, O.A.; Pellechia, P.J.; Smith, M.D.; Shustova, N.B. Flipping the switch: Fast photoisomerization in a confined environment. *J. Am. Chem. Soc.* **2018**, *140*, 7611–7622. [CrossRef] [PubMed]
89. Du, Y.; Liao, W.-Q.; Li, Y.; Huang, C.-R.; Gan, T.; Chen, X.-G.; Lv, H.-P.; Song, X.-J.; Xiong, R.-G.; Wang, Z.-X. A homochiral fulgide organic ferroelectric crystal with photoinduced molecular orbital breaking. *Angew. Chem. Int. Ed.* **2023**, *62*, e202315189. [CrossRef]
90. Du, Y.; Huang, C.-R.; Xu, Z.-K.; Hu, W.; Li, P.-F.; Xiong, R.-G.; Wang, Z.-X. Photochromic single-component organic fulgide ferroelectric with photo-triggered polarization response. *JACS Au* **2023**, *3*, 1464–1471. [CrossRef] [PubMed]
91. Wu, Y.; Schneider, S.; Yuan, Y.; Young, R.M.; Francese, T.; Mansoor, I.F.; Dudenas, P.J.; Lei, Y.; Gomez, E.D.; DeLongchamp, D.M.; et al. Twisted A-D-A type acceptors with thermally-activated delayed crystallization behavior for efficient nonfullerene organic solar cells. *Adv. Energy Mater.* **2022**, *12*, 2103957. [CrossRef]
92. Pugliese, E.; Vo, N.T.; Boussac, A.; Banse, F.; Mekmouche, Y.; Simaan, J.; Tron, T.; Gotico, P.; Sircoglou, M.; Halime, Z.; et al. Photocatalytic generation of a non-heme Fe(III)-hydroperoxo species with O_2 in water for the oxygen atom transfer reaction. *Chem. Sci.* **2022**, *13*, 12332–12339. [CrossRef]
93. Hein, R.; Docker, A.; Davis, J.J.; Beer, P.D. Redox-switchable chalcogen bonding for anion recognition and sensing. *J. Am. Chem. Soc.* **2022**, *144*, 8827–8836. [CrossRef]
94. Min, H.; Han, Z.; Sun, T.; Wang, K.; Xu, J.; Yao, P.; Yang, S.; Cheng, P.; Shi, W. Dynamic-static coupled sensing of trace biomarkers by molecularly imprinted metal-organic frameworks. *Sci. China Chem.* **2023**, *66*, 3511–3517. [CrossRef]
95. Han, Z.; Wang, K.; Chen, Y.; Li, J.; Teat, S.J.; Yang, S.; Shi, W.; Cheng, P. A multicenter metal-organic framework for quantitative detection of multicomponent organic mixtures. *CCS Chem.* **2022**, *4*, 3238–3245. [CrossRef]
96. Burton, S.T.; Lee, G.; Moore, C.E.; Sevov, C.S.; Turro, C. Cyclometallated Co(III) complexes with lowest-energy charge transfer excited states accessible with visible light. *J. Am. Chem. Soc.* **2025**, *147*, 13315–13327. [CrossRef]
97. Martin, C.R.; Kittikhunnatham, P.; Leith, G.A.; Berseneva, A.A.; Park, K.C.; Greytak, A.B.; Shustova, N.B. Let the light be a guide: Chromophore communication in metal-organic frameworks. *Nano Res.* **2020**, *14*, 338–354. [CrossRef]
98. Wang, S.; Sun, H.; Erdmann, T.; Wang, G.; Fazzi, D.; Lappan, U.; Puttisong, Y.; Chen, Z.; Berggren, M.; Crispin, X.; et al. A chemically doped naphthalenediimide-bithiazole polymer for n-type organic thermoelectrics. *Adv. Mater.* **2018**, *30*, 1801898. [CrossRef]
99. Xiong, Z.; Li, Y.; Yuan, Z.; Liang, J.; Wang, S.; Yang, X.; Xiang, S.; Lv, Y.; Chen, B.; Zhang, Z. Switchable anisotropic/isotropic photon transport in a double-dipole metal-organic framework via radical-controlled energy transfer. *Adv. Mater.* **2024**, *36*, 2314005. [CrossRef]
100. Liao, P.Y.; Li, J.X.; Liu, J.C.; Xiong, Q.; Ruan, Z.Y.; Li, T.; Deng, W.; Jiang, S.D.; Jia, J.H.; Tong, M.L. Radical-induced photochromic silver(I) metal-organic frameworks: Alternative topology, dynamic photoluminescence and efficient photothermal conversion modulated by anionic guests. *Angew. Chem. Int. Ed.* **2024**, *63*, e202401448. [CrossRef]
101. Gong, T.; Sui, Q.; Li, P.; Meng, X.F.; Zhou, L.J.; Chen, J.; Xu, J.; Wang, L.; Gao, E.-Q. Versatile and switchable responsive properties of a lanthanide-viologen metal-organic framework. *Small* **2019**, *15*, 1803468. [CrossRef] [PubMed]
102. Liu, Q.; Hu, J.-X.; Meng, Y.-S.; Jiang, W.-J.; Wang, J.-L.; Wen, W.; Wu, Q.; Zhu, H.-L.; Zhao, L.; Liu, T. Asymmetric coordination toward a photoinduced single-chain magnet showing high coercivity values. *Angew. Chem. Int. Ed.* **2021**, *60*, 10537–10541. [CrossRef]

103. Hu, J.-X.; Zhu, H.-L.; Meng, Y.-S.; Pang, J.; Li, N.; Liu, T.; Bu, X.-H. Ligand modified and light switched on/off single-chain magnets of {Fe_2Co} coordination polymers via metal-to-metal charge transfer. *CCS Chem.* **2023**, *5*, 865–875. [CrossRef]
104. Morimoto, M.; Miyasaka, H.; Yamashita, M.; Irie, M. Coordination assemblies of [Mn_4] single-molecule magnets linked by photochromic ligands: Photochemical control of the magnetic properties. *J. Am. Chem. Soc.* **2009**, *131*, 9823–9835. [CrossRef] [PubMed]
105. Fetoh, A.; Cosquer, G.; Morimoto, M.; Irie, M.; El-Gammal, O.; Abu El-Reash, G.M.; Breedlove, B.K.; Yamashita, M. Synthesis, structures, and magnetic properties of two coordination assemblies of Mn(III) single molecule magnets bridged via photochromic diarylethene ligands. *Inorg. Chem.* **2019**, *58*, 2307–2314. [CrossRef]
106. Jiao, T.; Wu, C.-H.; Zhang, Y.-S.; Miao, X.; Wu, S.; Jiang, S.-D.; Wu, J. Solution-phase synthesis of Clar's goblet and elucidation of its spin properties. *Nat. Chem.* **2025**, *17*, 924–932. [CrossRef]
107. Eaton, S.S.; Yamabayashi, T.; Horii, Y.; Yamashita, M.; Eaton, G.R. Anisotropy of spin-lattice relaxation time (T_1) for oxo-vanadium(IV) and nitrido chromium(V) porphyrins. *J. Am. Chem. Soc.* **2025**, *147*, 13815–13823. [CrossRef] [PubMed]
108. Graham, M.J.; Krzyaniak, M.D.; Wasielewski, M.R.; Freedman, D.E. Probing nuclear spin effects on electronic spin coherence via EPR measurements of vanadium(IV) complexes. *Inorg. Chem.* **2017**, *56*, 8106–8113. [CrossRef]
109. Tesi, L.; Lucaccini, E.; Cimatti, I.; Perfetti, M.; Mannini, M.; Atzori, M.; Morra, E.; Chiesa, M.; Caneschi, A.; Sorace, L.; et al. Quantum coherence in a processable vanadyl complex: New tools for the search of molecular spin qubits. *Chem. Sci.* **2016**, *7*, 2074–2083. [CrossRef] [PubMed]
110. Yao, N.-T.; Zhao, L.; Yi, C.; Liu, Q.; Li, Y.-M.; Meng, Y.-S.; Oshio, H.; Liu, T. Manipulating selective metal-to-metal electron transfer to achieve multi-phase transitions in an asymmetric [Fe_2Co]-assembled mixed-valence chain. *Angew. Chem. Int. Ed.* **2022**, *61*, e202115367. [CrossRef]
111. Shang, M.-J.; Lu, H.-H.; Liu, Q.; Zhou, R.-H.; Sun, H.-Y.; Shao, Z.; Zhao, L.; Meng, Y.-S.; Liu, T. Reversible on–off switching of a single-chain magnet via single-crystal-to-single-crystal transition and light-induced metal-to-metal electron transfer. *Inorg. Chem. Front.* **2024**, *11*, 8704–8714. [CrossRef]
112. Wang, L.-F.; Qiu, J.-Z.; Chen, Y.-C.; Liu, J.-L.; Li, Q.-W.; Jia, J.-H.; Tong, M.-L. [2 + 2] Photochemical modulation of the Dy(III) single-molecule magnet: Opposite influence on the energy barrier and relaxation time. *Inorg. Chem. Front.* **2017**, *4*, 1311–1318. [CrossRef]
113. Huang, X.-D.; Wen, G.-H.; Bao, S.-S.; Jia, J.-G.; Zheng, L.-M. Thermo- and light-triggered reversible interconversion of dysprosium-anthracene complexes and their responsive optical, magnetic and dielectric properties. *Chem. Sci.* **2021**, *12*, 929–937. [CrossRef]
114. Li, J.; Kong, M.; Yin, L.; Zhang, J.; Yu, F.; Ouyang, Z.-W.; Wang, Z.; Zhang, Y.-Q.; Song, Y. Photochemically tuned magnetic properties in an erbium(III)-based easy-plane single-molecule magnet. *Inorg. Chem.* **2019**, *58*, 14440–14448. [CrossRef]
115. Zhang, Q.; Han, S.-D.; Li, Q.; Hu, J.-X.; Wang, G.-M. Electron transfer photochromism of Ln-based (Ln = Dy, Tb) coordinated polymers for reversibly switching off/on single-molecule magnetic behavior. *Sci. China Mater.* **2022**, *65*, 788–794. [CrossRef]
116. Cai, L.-Z.; Guo, P.-Y.; Wang, M.-S.; Guo, G.-C. Photoinduced magnetic phase transition and remarkable enhancement of magnetization for a photochromic single-molecule magnet. *J. Mater. Chem. C* **2021**, *9*, 2231–2235. [CrossRef]
117. Tian, H.; Su, J.-B.; Bao, S.-S.; Kurmoo, M.; Huang, X.-D.; Zhang, Y.-Q.; Zheng, L.-M. Reversible ON-OFF switching of single-molecule-magnetism associated with single-crystal-to-single-crystal structural transformation of a decanuclear dysprosium phosphonate. *Chem. Sci.* **2018**, *9*, 6424–6433. [CrossRef] [PubMed]
118. Selvanathan, P.; Dorcet, V.; Roisnel, T.; Bernot, K.; Huang, G.; Le Guennic, B.; Norel, L.; Rigaut, S. Trans to cis photo-isomerization in merocyanine dysprosium and yttrium complexes. *Dalton. Trans.* **2018**, *47*, 4139–4148. [CrossRef] [PubMed]
119. Cosquer, G.; Kamila, M.; Li, Z.-Y.; Breedlove, B.; Yamashita, M. Photo-modulation of single-molecule magnetic dynamics of a dysprosium dinuclear complex via a diarylethene bridge. *Inorganics* **2018**, *6*, 9. [CrossRef]
120. Gou, X.; Chen, Z.; Jiang, J.; Liu, N.; Lan, W.; Wu, Y.; Cheng, P.; Shi, W. Influence of chelating ligands on photomagnetic properties of two erbium(III) coordination polymers. *J. Rare Earths* **2025**, *43*, 552–555. [CrossRef]
121. Wang, X.-Q.; Geng, Y.-W.; Wang, Z.; Xie, C.; Han, T.; Cheng, P. Two-dimensional metal-organic framework with high-performance single-molecule magnets as nodes showing magnetic coercivity photomodulation. *J. Am. Chem. Soc.* **2025**, *147*, 18044–18053. [CrossRef]
122. Gou, X.; Wu, Y.; Wen, H.; Li, L.; Lan, W.; Liu, N.; Zhang, S.; Ungur, L.; Cheng, P.; Shi, W. Modulating static and dynamic magnetizations of ytterbium(III) coordination polymers by light-induced radicals. *CCS Chem.* **2025**, 1–9. [CrossRef]
123. Zhang, P.; Zhang, Y.; Wu, F.; Xiao, W.; Hua, W.; Tang, Z.; Liu, W.; Chen, S.; Wang, Y.; Wu, W.; et al. Photoisomerization-mediated tunable pore size in metal organic frameworks for U(VI)/V(V) selective separation. *Nat. Commun.* **2025**, *16*, 2361. [CrossRef]
124. Wu, L.; Liu, Y.; Yang, W.; Liu, Z.; Liu, C.; Yuan, X.; Zhang, L.; Ju, J.; Yao, X. In situ stimuli transfer in multi-environment shape-morphing hydrogels based on the copolymer between spiropyran and acrylic acid. *Adv. Sci.* **2025**, *12*, 2416173. [CrossRef]

125. Tan, X.; Wang, Y.; Li, H.; Duan, Y.; Wen, B.; Zhao, J.; Kim, H.; Lee, J.Y.; Zhou, L.; Cheng, H.-B.; et al. Supramolecular synthesis of dithienylethene-albumin complexes for enhanced photoswitching in photoacoustic imaging-guided near-infrared photothermal therapy. *Small* **2025**, *21*, 2409027. [CrossRef]
126. Jin, T.; Zhang, Z.; He, S.; Kaledin, A.L.; Xu, Z.; Liu, Y.; Zhang, P.; Beratan, D.N.; Lian, T. Shell thickness and heterogeneity dependence of triplet energy transfer between core-shell quantum dots and adsorbed molecules. *J. Am. Chem. Soc.* **2025**, *147*, 16282–16292. [CrossRef]
127. Malkmus, S.; Koller, F.O.; Draxler, S.; Schrader, T.E.; Schreier, W.J.; Brust, T.; DiGirolamo, J.A.; Lees, W.J.; Zinth, W.; Braun, M. All-optical operation cycle on molecular bits with 250-GHz clock-rate based on photochromic fulgides. *Adv. Funct. Mater.* **2007**, *17*, 3657–3662. [CrossRef]
128. Guo, F.-S.; Benjamin, M.D.; Chen, Y.-C.; Tong, M.-L.; Mansikkamäki, A.; Richard, A.L. A Dysprosium metallocene single-molecule magnet functioning at the axial limit. *Angew. Chem. Int. Ed.* **2017**, *56*, 11445–11449. [CrossRef]
129. Ou, R.; Zhang, H.; Zhao, C.; Hegab, H.M.; Jiang, L.; Truong, V.X.; Wang, H. Photoresponsive styrylpyrene-modified MOFs for gated loading and release of cargo molecules. *Chem. Mater.* **2020**, *32*, 10621–10627. [CrossRef]
130. Arima, H.; Hiraide, S.; Nagano, H.; Abylgazina, L.; Senkovska, I.; Auernhammer, G.K.; Fery, A.; Kaskel, S.; Watanabe, S. Atomic force microscopy strategies for capturing guest-induced structural transitions in single flexible metal-organic framework particles. *J. Am. Chem. Soc.* **2025**, *147*, 14491–14503. [CrossRef] [PubMed]
131. Han, Z.; Zhang, R.; Jiang, J.; Chen, Z.; Ni, Y.; Xie, W.; Xu, J.; Zhou, Z.; Chen, J.; Cheng, P.; et al. High-efficiency lithium-ion transport in a porous coordination chain-based hydrogen-bonded framework. *J. Am. Chem. Soc.* **2023**, *145*, 10149–10158. [CrossRef] [PubMed]
132. Han, Z.; Li, J.; Lu, W.; Wang, K.; Chen, Y.; Zhang, X.; Lin, L.; Han, X.; Teat, S.J.; Frogley, M.D.; et al. A {Ni_{12}}-wheel-based metal-organic framework for coordinative binding of sulphur dioxide and nitrogen dioxide. *Angew. Chem. Int. Ed.* **2022**, *61*, e202115585. [CrossRef]
133. Sun, T.; Min, H.; Han, Z.; Wang, L.; Cheng, P.; Shi, W. Rapid detection of nanoplastic particles by a luminescent Tb-based coordination polymer. *Chin. Chem. Lett.* **2024**, *35*, 108718. [CrossRef]
134. Han, Z.; Yan, Z.; Wang, K.; Kang, X.; Lv, K.; Zhang, X.; Zhou, Z.; Yang, S.; Shi, W.; Cheng, P. Observation of oxygen evolution over a {Ni_{12}}-cluster-based metal-organic framework. *Sci. China Chem.* **2022**, *65*, 1088–1093. [CrossRef]
135. Jing, P.; Wu, B.; Han, Z.; Shi, W.; Cheng, P. An efficient Ag/MIL-100(Fe) catalyst for photothermal conversion of CO_2 at ambient temperature. *Chin. Chem. Lett.* **2021**, *32*, 3505–3508. [CrossRef]
136. Gou, X.; Liu, N.; Wu, Y.; Lan, W.; Wang, M.; Shi, W.; Cheng, P. Modulation of magnetization dynamics of an Er(III) coordination polymer by the conversion of a ligand to a radical using UV light. *Dalton Trans.* **2023**, *52*, 10372–10377. [CrossRef] [PubMed]
137. Liu, X.; Wang, Y.-X.; Han, Z.-S.; Han, T.; Shi, W.; Cheng, P. Tuning the magnetization dynamics of Tb^{III}-based single-chain magnets through substitution on the nitronyl nitroxide radical. *Dalton Trans.* **2019**, *48*, 8989–8994. [CrossRef] [PubMed]
138. Zhang, X.; Xu, N.; Shi, W.; Wang, B.-W.; Cheng, P. The influence of an external magnetic field and magnetic-site dilution on the magnetization dynamics of a coordination network based on ferromagnetic coupled dinuclear dysprosium(III) units. *Inorg. Chem. Front.* **2018**, *5*, 432–437. [CrossRef]
139. Wang, M.; Meng, X.; Song, F.; He, Y.; Shi, W.; Gao, H.; Tang, J.; Cheng, P. Reversible structural transformation induced switchable single-molecule magnet behavior in lanthanide metal-organic frameworks. *Chem. Commun.* **2018**, *54*, 10183–10186. [CrossRef]
140. Liu, X.; Zhang, Y.; Shi, W.; Cheng, P. Rational design and synthesis of a chiral lanthanide-radical single-chain magnet. *Inorg. Chem.* **2018**, *57*, 13409–13414. [CrossRef]
141. Song, F.; Zhang, Y.; Chen, J.; Han, T.; Cheng, P. Syntheses, crystal structures and magnetic properties of three lanthanide-nitronyl nitroxide complexes. *J. Rare Earth* **2017**, *35*, 24–27. [CrossRef]
142. Zhang, S.; Li, H.; Duan, E.; Han, Z.; Li, L.; Tang, J.; Shi, W.; Cheng, P. A 3D heterometallic coordination polymer constructed by trimeric {$NiDy_2$} single-molecule magnet units. *Inorg. Chem.* **2016**, *55*, 1202–1207. [CrossRef]
143. Zhang, X.; Vieru, V.; Feng, X.; Liu, J.L.; Zhang, Z.; Na, B.; Shi, W.; Wang, B.-W.; Powell, A.K.; Chibotaru, L.F.; et al. Influence of guest exchange on the magnetization dynamics of dilanthanide single-molecule-magnet nodes within a metal-organic framework. *Angew. Chem. Int. Ed.* **2015**, *54*, 9861–9865. [CrossRef]
144. Liu, K.; Li, H.; Zhang, X.; Shi, W.; Cheng, P. Constraining and tuning the coordination geometry of a lanthanide ion in metal-organic frameworks: Approach toward a single-molecule magnet. *Inorg. Chem.* **2015**, *54*, 10224–10231. [CrossRef]
145. Li, L.; Liu, S.; Li, H.; Shi, W.; Cheng, P. Influence of external magnetic field and magnetic-site dilution on the magnetic dynamics of a one-dimensional Tb(III)-radical complex. *Chem. Commun.* **2015**, *51*, 10933–10936. [CrossRef]
146. Chen, D.-M.; Ma, X.-Z.; Zhang, X.-J.; Xu, N.; Cheng, P. Switching a 2D Co(II) layer to a 3D Co_7-cluster-based metal-organic framework: Syntheses, crystal structures, and magnetic properties. *Inorg. Chem.* **2015**, *54*, 2976–2982. [CrossRef]

147. Zou, J.-Y.; Shi, W.; Gao, H.-L.; Cui, J.-Z.; Cheng, P. Spin canting and metamagnetism in 3D pillared-layer homospin cobalt(II) molecular magnetic materials constructed via a mixed ligands approach. *Inorg. Chem. Front.* **2014**, *1*, 242–248. [CrossRef]
148. Han, T.; Shi, W.; Niu, Z.; Na, B.; Cheng, P. Magnetic blocking from exchange interactions: Slow relaxation of the magnetization and hysteresis loop observed in a dysprosium-nitronyl nitroxide chain compound with an antiferromagnetic ground state. *Chem. Eur. J.* **2013**, *19*, 994–1001. [CrossRef] [PubMed]
149. Zhang, S.; Wang, B.; Xu, N.; Shi, W.; Gao, S.; Cheng, P. Ferromagnetic interactions in EO-azido-bridged binuclear transition metal(II) systems: Syntheses, crystal structures and magnetostructural correlations. *Sci. China Chem.* **2012**, *55*, 942–950. [CrossRef]

magnetochemistry

Article

Dysprosium Complexes Incorporating Halogen-Substituted Anthracene: Piezochromism and Single-Molecule Magnet Properties

Ye-Hui Qin, Qian-Qian Su, Song-Song Bao and Li-Min Zheng *

State Key Laboratory of Coordination Chemistry, School of Chemistry and Chemical Engineering, Collaborative Innovation Center of Advanced Microstructures, Nanjing University, Nanjing 210023, China; yehuiqin@smail.nju.edu.cn (Y.-H.Q.); 602022240101@smail.nju.edu.cn (Q.-Q.S.); baososo@nju.edu.cn (S.-S.B.)
* Correspondence: lmzheng@nju.edu.cn

Abstract: Lanthanide-based single-molecule magnets (Ln-SMMs) showing stimuli-responsive changes in photoluminescence (PL) and magnetic properties are attractive for their potential applications in information storage and molecular devices. In this work, we report two mononuclear complexes, namely, $Dy(SCN)_2(NO_3)(Cl\text{-}depma)_2(4\text{-}hpy)_2$ (**Dy-Cl**) and $Dy(SCN)_2(NO_3)(Br\text{-}depma)_2(4\text{-}hpy)_2$ (**Dy-Br**), where X-depma represents 10-X-9-diethylphosphinomethylanthracene (X = Cl, Br) and 4-hpy is 4-hydroxypyridine. Both contain face-to-face π-π-interacted anthracene rings and exhibit yellow-green excimer emission. Unlike the other related Dy–anthracene complexes without a halogen substituent, **Dy-Cl** and **Dy-Br** cannot undergo photocycloaddition reaction under UV-light irradiation. However, they exhibited remarkable grinding-induced changes in luminescence. Magnetic studies revealed that **Dy-Cl** and **Dy-Br** show SMM behavior under zero dc field with the effective energy barriers (U_{eff}/k_B) of 259 K and 264 K, respectively. We also investigated the effect of pressure on the magnetic properties of **Dy-Br** and observed a reduction in the magnetization value, narrowing of the butterfly-shaped hysteresis loop, and acceleration of the magnetic relaxation under 1.09 GPa. The results demonstrate that introducing a halogen substituent into an anthracene group may pose significant influences on the photophysical and photochemical properties of the complexes. In addition, pressure may be a promising external stimulus to modulate the PL and SMM behaviors of Dy–anthracene complexes.

Keywords: single-molecule magnet; dysprosium(III); anthracene; halogen substitution; photoluminescence; piezochromism; pressure

1. Introduction

Luminescent lanthanide-based single-molecule magnets (Ln-SMMs) have received great attention in recent years owing to their potential applications in high-density data storage, molecular devices, and sensors [1–4]. Of particular interest are those that undergo synergistic changes in magnetic and photoluminescence (PL) properties in response to external stimuli [5–12]. By utilizing reversible [4 + 4] photocycloaddition of anthracene groups, we and others have successfully isolated a series of lanthanide–anthracene complexes, including mononuclear [13–15] and polynuclear complexes [16–20] and coordination polymers [21,22], which exhibit photoresponsive PL and/or SMM properties. Introducing substituents into anthracene may alter its electronic structure and thus affect the photophysical properties of the complex [23–25]. The photochemical properties may also vary due to the different stacking patterns and electronic structures of the anthracene groups

 https://doi.org/10.3390/magnetochemistry11120102

brought about by the substituents [26,27]. For example, Bardeen and co-workers found that 10-halogen (F, Cl, Br)-substituted 9-anthracenecarboxylate acid (9AC) showed a similar stacking mode, but only F-substituted 9AC exhibited photoreactivity [26]. Noting that brominated anthracene encapsulated in a metal–organic cage experienced a photocycloaddition reaction [28], we were curious whether the same would happen with lanthanide complexes containing halogen-substituted anthracene moieties.

Pressure is another promising stimulus for inducing alterations in luminescence and magnetic properties [29–32]. However, although numerous efforts have been devoted to the study of pressure-induced magnetic changes in molecular systems, especially the spin crossover compounds [33–35], investigations into pressure-induced magnetic changes in lanthanide complexes are still rare [36–40]. In particular, there are only a handful of examples of lanthanide complexes reported in the literature that exhibit both piezochromism and SMM behavior [40,41]. For lanthanide complexes containing π-π-interacted anthracene groups, pressure may change not only the stacking pattern of the anthracene groups but also the coordination environment of the lanthanide ion, thus modulating both the luminescent and magnetic properties, but similar studies have not been reported.

In this work, we report two new complexes $Dy(SCN)_2(NO_3)(Cl\text{-}depma)_2(4\text{-}hpy)_2$ (**Dy-Cl**) and $Dy(SCN)_2(NO_3)(Br\text{-}depma)_2(4\text{-}hpy)_2$ (**Dy-Br**), where X-depma represents 10-X-9-diethylphosphinomethylanthracene (X = Cl, Br) (Scheme 1). Their structures are very similar to the known compound $Dy(SCN)_2(NO_3)(depma)_2(4\text{-}hpy)_2$ (**Dy-H**, depma = 9-diethylphosphinomethylanthracene), in which the anthracene groups are face-to-face π-π-interacted [14]. All exhibit excimer emission arising from the anthracene pair and SMM behavior under zero *dc* field. But the PL and SMM properties of **Dy-Cl** and **Dy-Br** are slightly different from those of **Dy-H**. The main difference in the **Dy-Cl** and **Dy-Br** compounds compared to **Dy-H**, which undergoes photoinduced single-crystal-to-single-crystal (SC-SC) structural transformations, is that they do not experience photochemical reaction under 365 nm light irradiation. However, both **Dy-Cl** and **Dy-Br** show remarkable changes in photoluminescence caused by grinding. Magnetic studies on compound **Dy-Br** revealed a decrease in magnetization value under high pressure (1.09 GPa). These results indicate that the introduction of halogen substituents to anthracene significantly affects the photochemical properties of the resulting dysprosium complexes, as well as their photophysical and magnetic properties. More importantly, pressure can be a potential stimulus to modulate the PL and SMM properties of Dy–anthracene systems. This work provides useful clues for the design and synthesis of stimuli-responsive lanthanide complexes based on anthracene derivatives.

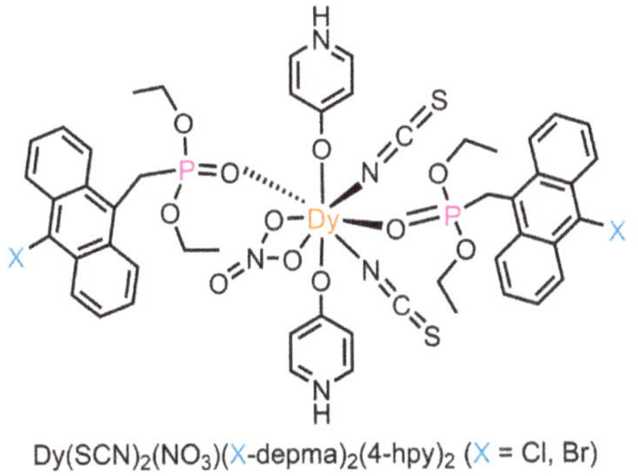

Scheme 1. Molecular structures of **Dy-X** (X = Cl, Br).

2. Experimental Section

2.1. General Information

10-Chloro-9-diethylphosphinomethylanthracene (Cl-depma) was synthesized from 10-chloro-9-anthracenecarboxaldehyde and characterized by ^{1}H NMR (Figures S1 and S2) (see details in the Supporting Information) [42]. 10-Bromo-9-diethylphosphinomethylanthracene (Br-depma) was synthesized according to the literature [43]. CH_2Cl_2 and EtOAc were obtained from Yasheng Chemical Co., Ltd., Wuxi, China, and other starting reagents and solvents were obtained from Aladdin, Shanghai, China. All were used directly. All other starting reagents and solvents were obtained from commercial sources and used directly. The elemental analysis for C, H, and N was performed using an Elementar Vario Macro cube, Elementar Analysensysteme GmbH, Langenselbold, German. The infrared (IR) spectra were measured in the range 4000–400 cm^{-1} using KBr pellets on a Bruker Tensor 27 spectrometer, Bruker Corporation, Billerica, MA, USA. Powder X-ray diffraction (PXRD) data were recorded on a Bruker D8 advance diffractometer (Bruker Corporation, Billerica, MA, USA), with Cu-Kα radiation in a range of 5–50° at room temperature. The differential scanning calorimetry (DSC) measurements was were conducted on a Mettler DSC823e instrument (Mettler-Toledo, Zurich, Switzerland). The UV/Vis spectra were measured on a Perkin Elmer Lambda 950 UV/VIS/NIR spectrometer (PerkinElmer, Inc., Waltham, MA, USA), using powder samples. The steady fluorescence spectra were recorded using PerkinElmer LS55 spectrometer (PerkinElmer, Inc., Waltham, MA, USA). Time-resolved fluorescence and quantum yield measurements were performed on an Edinburgh FLS 980 (Edinburgh Instruments, Livingston, UK), at room temperature. The 1H NMR spectra were recorded on a BRUKER AVANCE III 400 MHz spectrometer (Bruker Corporation, Billerica, MA, USA). The dc and ac magnetic susceptibility data were collected on polycrystalline samples by a Quantum Design vibrating sample magnetometer (VSM) (Quantum Design Inc., San Diego, CA, USA). A drop of Paraffin smaller than 2 mg was added to avoid the movement and reorientation of samples during measurement. The dc susceptibilities were corrected for diamagnetic contributions of holder, paraffin, and sample.

*2.2. Synthesis of $Dy(SCN)_2(NO_3)(Cl\text{-}depma)_2(4\text{-}hpy)_2$ (**Dy-Cl**)*

$Dy(NO_3)_3{\cdot}6H_2O$ (228 mg, 0.5 mmol) and KSCN (107 mg, 1.1 mmol) were dissolved in 10 mL CH_3CN and stirred for 30 min. After filtration of the white precipitate, 1 mL of the filtrate was taken and 4-hydroxypyridine (4-hpy, 10.4 mg, 0.11 mmol), Cl-depma (37.0 mg, 0.1 mmol), and 5 mL CH_3CN were added and stirred for about 20 min. After filtration, the clear solution was slowly evaporated for one week to give yellow crystals, the purity of which was verified by PXRD measurements (Figure S3). Yield: 35.0 mg (55%). Elemental anal. calcd (%): C, 47.80; H, 4.01; N, 5.64. Found (%): C, 47.45; H, 4.03; N, 5.50. IR (KBr): ν(C≡N): 2086, 2064 cm^{-1}; ν(P-O): 1024, 1056, 1185 cm^{-1} (Figure S4).

*2.3. Synthesis of $Dy(SCN)_2(NO_3)(Br\text{-}depma)_2(4\text{-}hpy)_2$ (**Dy-Br**)*

Compound **Dy-Br** was synthesized using a method similar to **Dy-Cl**, except that the ligand Cl-depma was replaced with Br-depma (40.7 mg, 0.1 mmol). The purity was verified by PXRD measurements (Figure S3). The infrared spectrum of **Dy-Br** is also similar to that of **Dy-Cl** (Figure S4). Yield: 38.5 mg (56%). Elemental anal. calcd (%): C, 44.64; H, 3.75; N, 5.21. Found (%): C, 44.92; H, 4.05; N, 5.12. IR (KBr): ν(C≡N): 2086, 2064 cm^{-1}; ν(P-O): 1025, 1054, 1185 cm^{-1} (Figure S4).

2.4. *Single-Crystal X-Ray Crystallography*

Single crystals were selected for data collection on a Bruker D8 diffractometer with a Metal Jet (Ga-Kα radiation, λ = 1.34139 Å). The data were integrated using the SAINT program [44]. Multiscan absorption corrections were applied. The structures were solved by intrinsic phasing methods and refined on F^2 by full-matrix least squares using SHELXTL [45]. All the non-hydrogen atoms were refined anisotropically. All H atoms were placed in theoretical positions and refined isotropically. The residual electron densities were of no chemical significance.

2.5. *High-Pressure Magnetic Measurement*

High-pressure magnetic measurements were performed using a Quantum Design HMD high-pressure cell made of BeCu. The data were collected on polycrystalline samples by a Quantum Design vibrating sample magnetometer (VSM). A small amount of lead (<0.2 mg) was used as a pressure calibrant, and mineral oil served as the pressure-transmitting medium. Before each measurement, the magnet reset option was executed to eliminate any remnant field from the superconducting magnet. The background signal was collected at ambient pressure with samples loaded in the cell, and pressurized magnetic data was collected under pressure. The applied pressure was determined by measuring the superconducting transition temperature of the lead calibrant.

3. Results and Discussion

3.1. *Crystal Structure*

Single-crystal structural analysis shows that **Dy-Cl** crystallizes in the monoclinic $P2_1/m$ space group at 193 K (Table 1 and Table S1). The asymmetric unit consists of half of a Dy^{III} ion, one SCN^-, 0.5 NO_3^-, one Cl-depma, and two halves of 4-hpy, respectively. A symmetry plane is present in the structure, where one 4-hpy ligand and NO_3^- ligand are located on this symmetry plane. Each Dy^{III} ion is eight-coordinated with two N atoms (N3, N3A) from two SCN^-, two O atoms (O6, O6A) from two Cl-depma ligands, two O atoms (O3, O4) from chelated NO_3^-, and two O atoms (O1, O2) from two 4-hpy, respectively, forming an eight-coordinated $[DyO_6N_2]$ configuration (Figure 1a, Table S2). According to the continuous shape measure (CShM) analysis [46], the geometry of the $[DyO_6N_2]$ polyhedron in **Dy-Cl** is a distorted triangular dodecahedron (CShM = 1.856, D_{2d}) (Table S3). However, if we consider the chelating NO_3^- anion as occupying one position, the coordination geometry of the Dy^{III} ion has an approximately D_{5h} symmetry. The two oxygen atoms from 4-hpy (O1, O2) occupy the axial positions, while the remaining atoms occupy the equatorial plane. The Dy1−O/N bond lengths in the equatorial plane range from 2.337(4) to 2.572(5) Å, and the axial Dy1−O1 and Dy1−O2 bond lengths are 2.251(6) and 2.246(6) Å, respectively (Table 1 and Table S2). One of the two axial 4-hpy (N1) ligands is disordered over two sites. The protonated pyridine nitrogen atoms of 4-hpy are involved in the intermolecular hydrogen bonding with the uncoordinated nitrate oxygen atom (O5) and the sulfur atom (S1) from the NCS^- anion (Table S4), forming a supramolecular layer in the *ac* plane (Figure S5a). Neighboring anthracene groups are face-to-face stacked through $\pi-\pi$ interactions, thus connecting the supramolecular layers into a three-dimensional (3D) network (Figure 1b and Figure S5b). The center-to-center distance (d_{cc}) and plane-to-plane distance (d_{pp}) of the anthracene pair are 3.758 and 3.372 Å, respectively, which satisfy Schmidt's rule for photodimerization [47]. The slip angle is 23.10° and the central C2−C9A distance ($d_{C2\text{-}C9A}$) is 3.743 Å. In addition, the Cl···S distance between the S atom in SCN^- and Cl atom in the Cl-depma ligand of another molecule is 3.378 Å, which is less than the sum of the van der Waals radius (3.55 Å) of one Cl atom and one S atom [48], indicating the presence of halogen bonds.

Table 1. The unit cell and structural parameters and PL and magnetic properties of **Dy-Cl**, **Dy-Br**, and **Dy-H**.

	Compound	Dy-Cl	Dy-Br	Dy-H [a]
Cell parameters	Temperature	193 K	193 K	193 K
	Crystal system	Monoclinic	Monoclinic	Monoclinic
	Space group	$P2_1/m$	$P2_1/n$	$P2_1/m$
	a/Å	9.736(1)	9.754(1)	9.605(3)
	b/Å	25.723(1)	25.833(1)	25.654(10)
	c/Å	11.251(1)	21.114(1)	11.161(4)
	β/°	111.4(1)	94.739(1)	112.1(1)
	V/Å^3	2623.2(2)	5302.0(4)	2548.5(16)
	Z	2	4	2
	CCDC number	2475934	2475970	2233761
Structural parameters	Dy1-O(axial)/Å	2.251(5)/2.246(5)	2.260(4)/2.235(5)	2.248(8)/2.213(9)
	Dy1-O(equatorial)/Å	2.337(4)-2.572(5)	2.328(3)-2.574(4)	2.319(7)-2.571(8)
	Dy1-N/Å	2.450(5)	2.455(4)/2.440(4)	2.426(8)
	O1-Dy1-O2(axial)/°	163.9(2)	163.7(2)	163.9(4)
	$d_{Dy\text{-}Dy}$/Å	9.736	9.754	9.605
	d_{cc}/Å	3.758	3.810	3.753
	d_{pp}/Å	3.372	3.408	3.409
	slip angle/°	23.10	23.61/23.06	22.36
	$d_{C2\text{-}C9A}$/Å	3.743	-	3.754
	$d_{C2\text{-}C28}$, $d_{C9\text{-}C21}$/Å	-	3.819/3.811	-
PL properties	Excimer peak/nm	550	535	540
	Lifetime/ns	54.1	7.11	41.7
	PLQY/%	10.31	1.76	12.69
Magnetic properties	$\chi_M T$/cm^3 K mol^{-1}	14.77	14.92	14.18
	$M/N\beta$ (70 kOe)	5.65	5.54	5.22
	U_{eff}/k_B/K	259(12)	264(9)	277
	τ_0/s	2.77(222) × 10^{-11}	4.46(243) × 10^{-11}	7.16 × 10^{-11}
	C/K^{-n} s^{-1}	2.1(5) × 10^{-3}	9.1(7) × 10^{-4}	1.17 × 10^{-3}
	n	4.37(10)	4.80(8)	4.42
	τ_{QTM}/s	0.241(7)	0.316(8)	0.188

[a] The parameters for **Dy-H** are cited from [14].

Compound **Dy-Br** crystallizes in the space group $P2_1/n$ at 193 K (Tables 1 and S1). The asymmetric unit consists of one Dy^{III} ion, two SCN^- ions, one NO_3^-, two Br-depma, and two 4-hpy. It has a similar mononuclear structure to **Dy-Cl**. However, the replacement of Cl by Br brought slight changes in structural parameters (Tables 1 and S5). The $[DyO_6N_2]$ polyhedron is close to a distorted triangular dodecahedron, with a CShM value (1.831) slightly smaller than that in **Dy-Cl** (CShM = 1.856) (Table S3). Again, the coordination geometry of Dy^{III} can be viewed as having a pseudo-D_{5h} symmetry if the chelating NO_3^- anion is considered to occupy one position. The axial Dy1–O1 and Dy1–O2 bond lengths are 2.260(4) and 2.235(5) Å, respectively, while the equatorial Dy1-O/N bond lengths are 2.328(3)–2.574(4) Å. The O1–Dy1–O2 angle is 163.7(2)°, close to that in **Dy-Cl** [163.9(2)°]. The Br···S halogen bonds are observed between the molecules with the distances of 3.337 and 3.346 Å, less than the sum of the van der Waals radius of Br and S (3.65 Å) [48]. These halogen bonds, together with the N–H···O(S) hydrogen bonds, connect the adjacent molecules, forming a 3D supramolecular network (Table S4, Figure S6).

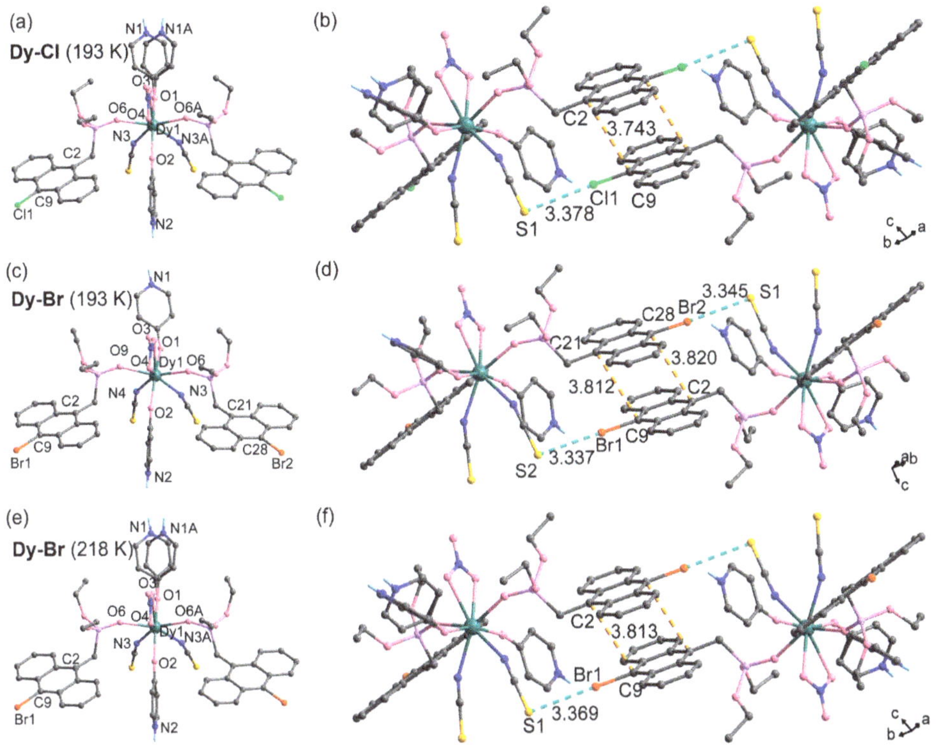

Figure 1. (**a**,**c**,**e**) The molecular structures of **Dy-Cl** (**a**), **Dy-Br (193K)** (**c**), and **Dy-Br (218K)** (**e**); (**b**,**d**,**f**) fragments of the supramolecular chains formed by halogen bonds and face-to-face π-π interactions of anthracene rings in compounds **Dy-Cl** (**b**), **Dy-Br (193K)** (**d**), and **Dy-Br (218K)** (**f**).

Compared to the analog compound $[Dy(SCN)_2(NO_3)(depma)_2(4\text{-}hpy)_2]$ (**Dy-H**), **Dy-Cl** and **Dy-Br** have several distinct structural features. First, there exist halogen bonds in the structures of **Dy-Cl** and **Dy-Br**, while not in **Dy-H**. The halogen bond in **Dy-Br** is stronger, with shorter Br···S distances (3.337 and 3.346 Å) than the Cl···S distance (3.378 Å) in **Dy-Cl**. We expect that the intermolecular halogen bonds will slightly alter the packing of the anthracene groups. Indeed, the slip angles, defined as the angle between the centroid–centroid line and the vertical line of the anthracene pair, follow the order **Dy-Br** (23.61/23.06°) > **Dy-Cl** (23.10°) > **Dy-H** (22.36°) (Figures S5c and S6c), which is consistent with the decreasing strength of the intermolecular halogen bonds. A similar trend is found for the center-to-center distance (d_{cc}) of the anthracene pair. Notably, the plane-to-plane distance (d_{pp}) of **Dy-Cl** is the shortest among the three complexes. Secondly, the coordination environments of the dysprosium ions are slightly different. The axial Dy1–O1 bond length in **Dy-Cl** [2.251(5) Å] is shorter than in **Dy-Br** [2.260(4) Å], but comparable to **Dy-H** [2.248(8) Å], whereas the axial Dy1–O2 bond length in **Dy-Cl** [2.246(5) Å] is longer than in **Dy-Br** [2.235(5) Å] and **Dy-H** [2.213(9) Å]. The equatorial Dy1–N bond lengths are elongated in **Dy-Cl** [2.450(5) Å] and **Dy-Br** [2.455(4) and 2.440(4) Å] compared to **Dy-H** [2.426(8) Å], potentially arising from the halogen bond between halogen atom of the anthracene ligand and sulfur atom of the NCS^- anion. Thirdly, the Dy···Dy separation is expanded following the sequence **Dy-Br** (9.754 Å) > **Dy-Cl** (9.736 Å) > **Dy-H** (9.605 Å), consistent with the decreasing atomic radii of Br, Cl, and H. All these structural differences will be reflected by their optical and magnetic properties.

It is noteworthy that both 4-hpy ligands in **Dy-Br** are ordered at 193 K, unlike for **Dy-Cl** at the same temperature, in which one of the two 4-hpy ligands is disordered over two sites. To examine whether an order–disorder phase transition occurred during the temperature change, we performed DSC measurements on **Dy-Br** in the temperature range of 153–303 K. Unfortunately, after several attempts, we were unable to observe an endothermic or exothermic peak in the DSC curves, possibly due to the very small entropy change during the phase transition. Nevertheless, single-crystal structural analysis measured at 218 K revealed that one of the two 4-hpy in the structure of **Dy-Br** was disordered (Table S6, Figures 1e,f and S7), akin to **Dy-Cl**. The fact that the order–disorder phase transition occurred above 193 K for **Dy-Br** but not for **Dy-Cl** and **Dy-H** should be related to their structures, but the underlying mechanism is not clear to us.

3.2. Photophysical Properties and External Stimulus Effects

To investigate the photophysical properties of compounds **Dy-Cl** and **Dy-Br**, we first measured their UV-vis diffuse reflectance spectra in the solid state. As shown in Figure S8, the two compounds display intense and broad absorption bands in the range of 200–600 nm, corresponding to the $\pi^*\leftarrow\pi$ transitions of the X-depma and 4-hpy ligands. Furthermore, small and narrow absorption peaks appear at 754 and 800 nm for both compounds, originating from the characteristic 4f–4f transitions of the Dy^{III} ion from the ground state ($^6H_{15}/_2$) to the excited states ($^6F_3/_2$ and $^6F_5/_2$).

When polycrystalline samples of **Dy-Cl** and **Dy-Br** were irradiated with 365 nm UV light, they showed bright yellow luminescence. We then measured the photoluminescence (PL) spectra of **Dy-Cl** and **Dy-Br**. Upon excitation at 365 nm, **Dy-Cl** and **Dy-Br** exhibited broad emission bands peaking at 550 nm and 535 nm, respectively (Figure 2), attributed to the excimer emission due to the face-to-face $\pi-\pi$-interacted anthracene pair. Compared to **Dy-H** (540 nm), the emission peaks of **Dy-Cl** and **Dy-Br** are either red-shifted or blue-shifted. According to the structural description, **Dy-Cl** has the shortest anthracene plane spacing (d_{pp}) (3.372 Å) compared to **Dy-Br** (3.408 Å) and **Dy-H** (3.409 Å). This result suggests that the $\pi-\pi$-interactions between the anthracene groups in **Dy-Cl** are enhanced, which may account for the longest wavelength of the excimer emission peak in **Dy-Cl**. In contrast, although **Dy-Br** shows nearly the same d_{pp} distance as that of **Dy-H**, the slip angle of the anthracene pair in **Dy-Br** (23.61/23.06°) is larger than that in **Dy-H** (22.36°), which may account for the blue-shift in the excimer emission in **Dy-Br**. We notice that the slip angle of the anthracene pair in **Dy-Cl** (23.10°) is also larger than that in **Dy-H** (22.36°), but the emission peak appears at a longer wavelength. Obviously, compared to the slip angle, the shortening of anthracene plane spacing plays a more important role in the red-shift in the excimer emission of **Dy-Cl**. Moreover, we measured the photoluminescent lifetimes of the two compounds. As expected, **Dy-Cl** displays the longest lifetime (54.1 ns) among the three compounds (**Dy-Br**: 7.11 ns; **Dy-H**: 41.7 ns). The quantum yield (PLQY) was also obtained as 10.31% for **Dy-Cl**, 1.76% for **Dy-Br**, and 12.69% for **Dy-H**. The extremely short fluorescence lifetime and low quantum yield of **Dy-Br** can be ascribed to the heavy atom effect of the Br atom.

Considering that the anthracene groups in **Dy-Cl** and **Dy-Br** are face-to-face $\pi-\pi$-interacted with separation satisfying Schmidt's rule for the photocycloaddition reaction, we further studied the photochemical properties of the two complexes. We irradiated the two samples with 365 nm UV light (power density: ca. 100 mW cm^{-2}) for 1 h, but observed almost no change in their PL spectra (Figure S9). The IR spectra of **Dy-Cl** and **Dy-Br** before and after light irradiation are also identical (Figure S10). The results indicate that **Dy-Cl** and **Dy-Br** cannot undergo photocycloaddition reactions, which is similar to the case of 10-X-9-anthracenecarboxylate acid (X = Cl, Br) in its solid state [26]. The reason for this could be two-fold: first, the larger size of the halogen atom increases the spatial repulsion around the anthracene ligand and prevents photocycloaddition reactions [26]; second, the presence of heavy atoms such as Br and Cl facilitates the formation of triplet states by intersystem crossing and is not conducive to the formation of a singlet excited state, which is the reactive species in the anthracycline photodimerization reaction [27].

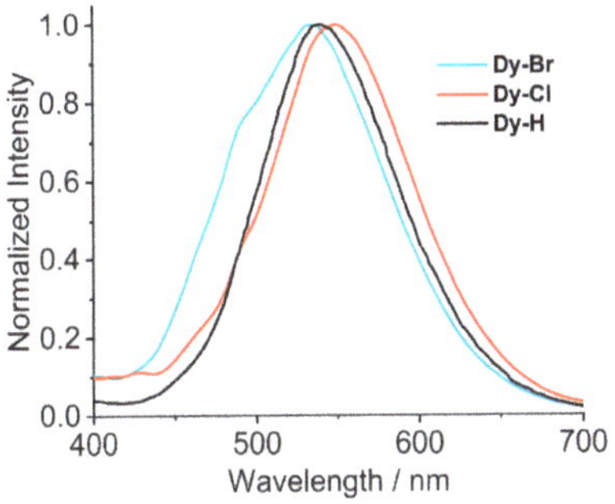

Figure 2. Normalized PL spectra of **Dy-Cl**, **Dy-Br**, and **Dy-H** (λ_{ex} = 365 nm).

Applying mechanical force may cause a change in the stacking mode of anthracene rings in compounds **Dy-Cl** and **Dy-Br**, thereby affecting their PL properties [49–53]. We next investigated the PL spectra of **Dy-Cl** and **Dy-Br** before and after grinding. Figure S11 shows clearly that after 5 min of grinding, the emission color of **Dy-Cl** (named as **Dy-Cl-ground**) changed from vivid yellow to yellow-green, and the emission color of **Dy-Br** (named as **Dy-Br-ground**) changed from yellow-green to blue-green. By heating the grinding products at 100 °C for 10 min (named as **Dy-Cl-heat** and **Dy-Br-heat**), the emission colors returned to their initial color. The PL spectra can more accurately display the changes in luminescence before and after grinding. As shown in Figure 3a,c, both **Dy-Cl-ground** and **Dy-Br-ground** exhibited broad bands with maxima blue-shifted to 535 nm and 490 nm, respectively, which are 15 nm and 45 nm less than the pristine samples. The blue-shift of the emission peak after grinding can be attributed to the formation of a more pronounced displaced stacking structure of the anthracene pair induced by external forces. After heating at 100 °C for 10 min, the PL spectra of **Dy-X-heat** matched those of **Dy-X** (X = Cl, Br), suggesting that the luminescence changes induced by grinding are reversible. Apparently, thermal treatment promotes the recovery of anthracene stacking to its initial state. The PXRD and IR measurements confirmed that compounds **Dy-X** (X = Cl, Br) experienced grinding-induced amorphization and thermo-induced crystalline-phase recovery (Figures 3b,d and S12). We also performed five cycles of grinding/heating measurements on compounds **Dy-Cl** and **Dy-Br**. The PL spectra showed that the luminescence changes and recovery remained reversible throughout the entire process (Figures S13 and S14).

It is noteworthy that compound **Dy-Br** shows a significantly greater blue-shift after grinding than **Dy-Cl**. This phenomenon may be attributed to the fact that the d_{cc} and d_{pp} distances between the anthracene rings in **Dy-Br** are larger and the interactions are weaker, which makes it easier for them to be moved by mechanical forces, ultimately leading to a more pronounced blue-shift.

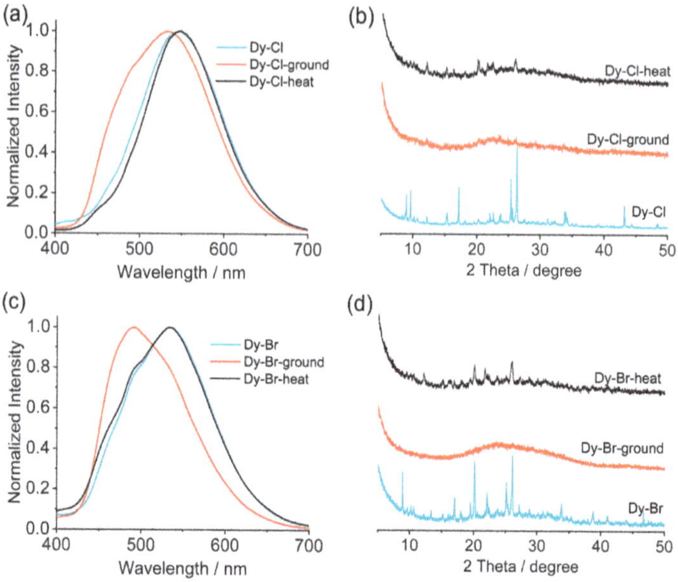

Figure 3. (**a**) Normalized PL spectra (λ_{ex} = 365 nm) and (**b**) PXRD patterns for **Dy-Cl**, **Dy-Cl-ground**, and **Dy-Cl-heat**; (**c**) normalized PL spectra (λ_{ex} = 365 nm) and (**d**) PXRD patterns for **Dy-Br**, **Dy-Br-ground**, and **Dy-Br-heat**.

3.3. Magnetic Properties

Structural description has demonstrated that the introduction of a halogen atom causes slight changes in the coordination environment of the Dy^{III} ion in **Dy-Cl** and **Dy-Br**. Since the magnetic properties of lanthanide complexes are sensitive to environmental changes, we studied their magnetic properties. Figure S15 shows the direct current (*dc*) magnetic susceptibilities of **Dy-Cl** and **Dy-Br** measured under a 1 kOe *dc* field in the temperature range of 2–300 K. At 300 K, the $\chi_M T$ values were 14.77 cm^3 K mol^{-1} and 14.92 cm^3 K mol^{-1} for **Dy-Cl** and **Dy-Br**, respectively, slightly larger than that of an isolated Dy^{III} ion (14.17 cm^3 K mol^{-1}, $^6H_{15}/_2$, $S = 5/2$, $L = 5$, $g_J = 4/3$). Upon cooling, the $\chi_M T$ values slowly decreased until 10 K, at which point the values were 12.32 cm^3 K mol^{-1} for **Dy-Cl** and 12.31 cm^3 K mol^{-1} for **Dy-Br**. Below 10 K, the $\chi_M T$ value showed an upturn and reached a maximum at 5 K, which may be ascribed to the weak intermolecular ferromagnetic dipole–dipole interactions. Finally, below 5 K, the $\chi_M T$ values dropped sharply, reaching 8.89 cm^3 K mol^{-1} for **Dy-Cl** and 8.75 cm^3 K mol^{-1} for **Dy-Br** at 2 K. This dramatic decrease is attributed to the blocking of magnetic moment [54]. Notably, while 4f–4f interactions are typically negligible or exhibit only weak antiferromagnetic behavior in mononuclear systems, weak ferromagnetic dipole–dipole interactions have been observed in certain complexes. These include the **Dy-H** complex with the shortest Dy···Dy distance of 9.605 Å [14], a mononuclear complex [$Dy(NTA)_3L$] with the shortest Dy···Dy distance of 9.729 Å [55], and a Dy^{III}-Cd^{II}-phthalocyaninato sextuple-decker complex (Dy_2Cd_3) with a long intramolecular Dy···Dy distance of 13.047 Å and intermolecular Dy···Dy distance of 11.246 Å [56]. We also measured the field-dependent magnetization curves of the two compounds. The *M* values at 2 K and 70 kOe were 5.65 $N\beta$ and 5.54 $N\beta$ for **Dy-Cl** and **Dy-Br**, respectively, lower than the theoretical value of 10 $N\beta$ for a single Dy^{III} ion, ascribed to strong magnetic anisotropy (Figure S16a,c). The non-superposition of the *M* vs. H/T curves provides further evidence for the presence of strong magnetic anisotropy (Figure S16b,d). Notably, the *M* vs. H/T curves of both compounds at 2 K show a clear sigmoid shape, a phenomenon also observed in the single spin-based SMMs associated with the magnetic blocking [57]. The hysteresis measurements confirmed the magnetic blocking, with butterfly-shaped hysteresis loops for both compounds (Figures 4a and S17). However, the coercive field and remnant magnetization were both zero at H_{dc} = 0 Oe, indicating the presence of strong quantum tunneling of magnetization (QTM).

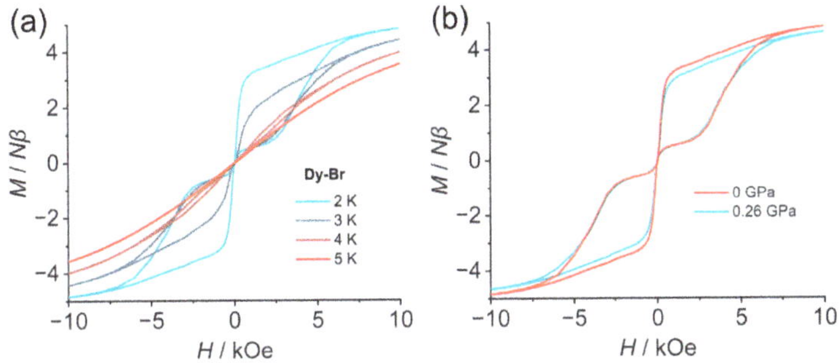

Figure 4. (**a**) The magnetic hysteresis loops for **Dy-Br** measured at the indicated temperatures; (**b**) the magnetic hysteresis loops for **Dy-Br** measured under 0 GPa and 1.09 GPa at 2 K.

We subsequently measured the alternating current (*ac*) magnetic susceptibility of **Dy-Cl** and **Dy-Br** to investigate their magnetic dynamics (Figure 5). Both the in-phase (χ') and out-of-phase (χ'') *ac* susceptibilities exhibited strong frequency dependence under zero *dc* field below 16 K for **Dy-Cl** and 18 K for **Dy-Br** (Figure 5a,b,d,e), respectively, which is characteristic of SMM behavior. The Cole–Cole plots displayed a distinct semicircle, indicating a single-step magnetic relaxation process (Figure S18). These Cole–Cole plots can be fitted by a generalized Debye model to extract the relaxation time (τ) (Tables S7 and S8) [58]. The plot of ln(τ) versus T^{-1} could then be fitted using Equation (1), which accounts for the QTM, Raman, and Orbach relaxation processes, where U_{eff} represents the effective energy barrier for the Orbach process (Figure 5c,f). The best fit gave the parameters U_{eff}/k_B = 259(12) K, τ_0 = 2.77(222) × 10^{-11} s, C = 2.1(5) × 10^{-3} $K^{-4.37}$ s^{-1}, n = 4.37(10), and τ_{QTM} = 0.241(7) s for **Dy-Cl,** and U_{eff}/k_B = 264(9) K, τ_0 = 4.46(243) × 10^{-11} s, C = 9.1 × 10^{-4} $K^{-4.80}$ s^{-1}, n = 4.80(8), and τ_{QTM} = 0.361(8) s for **Dy-Br**.

$$\tau^{-1} = \tau_{QTM}^{-1} + CT^n + \tau_0^{-1}\left(\frac{-U_{eff}}{k_B T}\right) \tag{1}$$

Clearly, the U_{eff}/k_B value of **Dy-Cl** (259 K) is slightly smaller than those of **Dy-Br** (264 K) and **Dy-H** (277 K). By comparing the structures of **Dy-Cl** and **Dy-H**, we found that the average equatorial Dy1-O/N bond lengths in **Dy-Cl** [2.447 Å/2.450(5) Å] are longer than those in **Dy-H** [2.437 Å/2.426(8) Å], suggesting that **Dy-Cl** has a weaker equatorial ligand field. In addition, the axial Dy1–O1 distance and O1–Dy1–O2 angle in **Dy-Cl** [2.251(5) Å, 163.9(2)°] are close to those in **Dy-H** [2.248(8) Å, 163.9(4)°]. However, the axial Dy1–O2 distance in **Dy-Cl** is significantly longer [2.246(5) Å vs. 2.213(9) Å in **Dy-H**]. Apparently, the axial Dy1–O2 distance plays a key role in determining the magnetic anisotropy of the material. Compound **Dy-H**, with a much shorter Dy1–O2 distance, shows a higher energy barrier than **Dy-Cl** with a longer Dy1–O2 distance. For **Dy-Br**, both the axial and equatorial Dy1–O/N bond lengths are elongated compared to **Dy-H**; this may explain why **Dy-Br** exhibits a lower energy barrier than **Dy-H**. Notably, applying an external field (0–500 Oe) cannot enhance the slow magnetic relaxation (Figures S19 and S20).

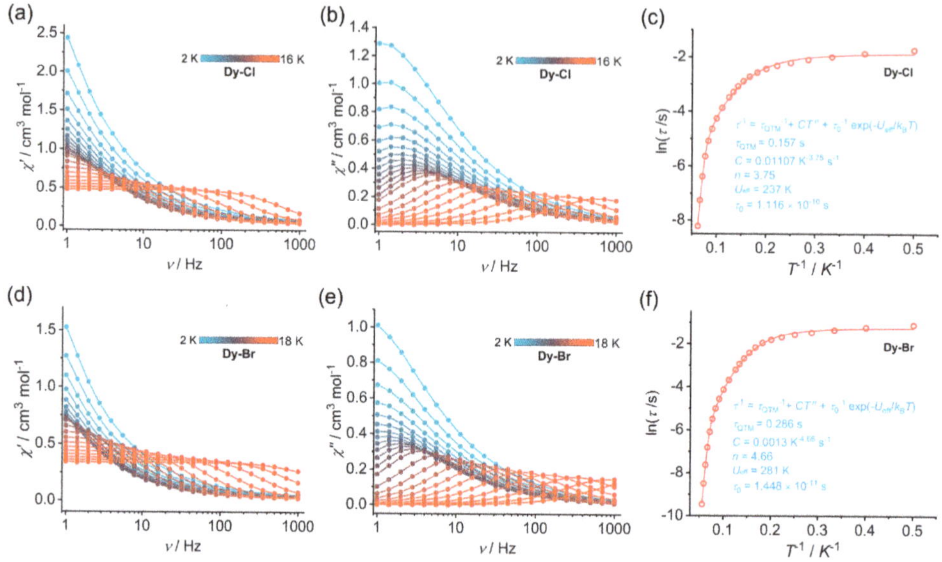

Figure 5. (**a**,**d**) The in-phase *ac* susceptibilities for **Dy-Cl** (**a**) and **Dy-Br** (**d**); (**b**,**e**) the out-of-phase *ac* susceptibilities for **Dy-Cl** (**b**) and **Dy-Br** (**e**); (**c**,**f**) the plot of $\ln\tau$ vs. T^{-1} for **Dy-Cl** (**c**) and **Dy-Br** (**f**).

In addition to the different effective energy barriers of the Orbach process, we also found that the Raman process exponents of **Dy-Cl** (n = 4.37) and **Dy-Br** (n = 4.80) are different. Both theoretical and experimental studies have demonstrated that Raman relaxation is associated with low-energy phonons, which dominate at low temperatures [59]. Molecular vibrations beyond the ion's first coordination shell play a significant role in spin–phonon coupling through the electrostatic polarization effect [60]. The fact that **Dy-Br** has a higher n value than **Dy-Cl** may be due to the presence of stronger halogen bonds between the molecules.

3.4. *Effect of High Pressure on the Magnetic Properties of* ***Dy-Br***

As already described, both **Dy-Cl** and **Dy-Br** show grinding-induced emission color change due to the displaced stacking structure of their anthracene pairs. To examine whether an external pressure may affect their magnetic properties, we measured the magnetic hysteresis loops of the ground samples (**Dy-Cl-ground** and **Dy-Br-ground**) at 2 K. As shown in Figure S21, a pronounced narrowing of the loops was observed for the ground samples, in contrast to the pristine samples, indicating enhancement of QTM effect after grinding.

We then selected compound **Dy-Br** for high-pressure magnetic measurements using a pressurized BeCu device. By measuring the superconducting transition temperature of the lead calibrant, the pressure was determined to be 1.09 GPa (Figure S22). At 300 K, the $\chi_M T$ value under pressure was 11.97 cm^3 K mol^{-1}, much lower than the value of 14.92 cm^3 K mol^{-1} at ambient pressure (Figure S23). The temperature dependence of $\chi_M T$ under pressure followed a trend similar to that at ambient pressure, reaching a minimum value of 7.67 cm^3 K mol^{-1} at 2 K, which was also slightly lower than the value at ambient pressure (8.75 cm^3 K mol^{-1}). Magnetization curves at 2 K under both high pressure and ambient pressure exhibited a sigmoid shape (Figures S24 and S25). The magnetization value at 70 kOe was 4.40 $N\beta$ under pressure, lower than the 5.54 $N\beta$ observed at ambient pressure. Hysteresis loops at 2 K remained butterfly-shaped under pressure but were significantly narrower compared to those at ambient pressure (Figures 4b and S26). Overgaard and co-workers found that for a pseudo-pentagonal bipyramidal Dy^{III} complex, the axial O-Dy-O angle decreased under pressure, leading to a drop in the magnetization at zero field [38]. We speculate that the loss of magnetization in **Dy-Br** may be related to the distortion of the pseudo-pentagonal bipyramidal geometry of Dy^{III} ions, leading to a decrease in magnetic anisotropy, as well as the shortening of the Dy···Dy distance, which enhances QTM and/or Raman relaxation processes [40].

The severe distortion of *ac* magnetic susceptibility at high frequencies (>100 Hz) arises from eddy currents induced in the BeCu device under alternating magnetic fields (Figure S27a,b). These eddy currents become particularly intense at elevated frequencies, generating substantial background signals that are challenging to fully compensate. Consequently, *ac* susceptibility data become unavailable in the higher-frequency regime. Additionally, this effect contributes to discontinuous jumps in data points observed at lower frequencies. Fortunately, for compound **Dy-Br**, the susceptibility data in the range of 2–12 K are still available in the lower-frequency range (<50 Hz) under high pressure. As shown in Figure S27b, the peak signals in out-of-phase *ac* susceptibility curves clearly shift towards lower frequencies upon cooling. The Cole–Cole plot exhibits a semicircular shape, indicating a single-step relaxation process (Figure S27c). The Cole–Cole plots can be fitted by a generalized Debye model to extract the relaxation time (τ) (Table S9). However, the plot of ln(τ) versus T^{-1} could not be adequately fitted by considering any combination of the QTM, Raman, and Orbach processes (Figure S27d). Nevertheless, a linear fit with the Orbach process using Equation (2) in the high-temperature region provides a reasonable approximation for estimating the effective energy barrier. The fit gives an energy barrier of 30(3) K, which is significantly lower than that obtained under ambient pressure [264(5) K]. This result suggests that the application of high pressure accelerates the magnetic relaxation and reduces the energy barrier in **Dy-Br**.

$$ln(\tau) = ln(\tau_0) + \left(\frac{U_{\text{eff}}}{k_B}\right) \times \frac{1}{T} \quad (2)$$

4. Conclusions

We report two new compounds, namely $[Dy(SCN)_2(NO_3)(Cl\text{-}depma)_2(4\text{-}hpy)_2]$ (**Dy-Cl**) and $[Dy(SCN)_2(NO_3)(Br\text{-}depma)_2(4\text{-}hpy)_2]$ (**Dy-Br**). Both contain anthracene rings with face-to-face π-π interactions and exhibit yellow-green excimer emission. The halogen substituent has an effect on the stacking modes of the anthracene pairs, such as the plane-to-plane distance and the slip angle of the anthracene rings. As a result, the emission peak in **Dy-Cl** is red-shifted by 15 nm compared to **Dy-Br**. Moreover, **Dy-Cl** shows a significantly longer fluorescence lifetime and higher quantum yield than **Dy-Br**, attributed to the heavy atom effect of Br. Interestingly, neither **Dy-Cl** nor **Dy-Br** undergoes the [4 + 4] photocycloaddition reaction, though the stacking of anthracene pairs in the two compounds meet the criteria for Schmidt's rule. However, they show piezochromism upon grinding, with the emission peak blue-shifted by 15 nm for **Dy-Cl** and 45 nm for **Dy-Br**, respectively. This color change is reversible after thermal annealing. In addition, both compounds exhibit single-molecule magnet behavior, with energy barriers of 259 K for **Dy-Cl** and 264 K for **Dy-Br**. High-pressure magnetic studies on **Dy-Br** revealed a decrease in magnetization value and the acceleration of the magnetic relaxation at 1.09 GPa. This work offers new insights into the design and synthesis of stimulus-responsive anthracene-based single-molecule magnets.

Supplementary Materials: The following supporting information can be downloaded at: https://www.mdpi.com/article/10.3390/magnetochemistry11120102/s1. Scheme S1: Synthetic procedures for Cl-depma. Figure S1: The ^1H NMR spectra of 10-Chloro-9-anthracenemethanol;

Figure S2: The ^{1}H NMR spectra of Cl-depma; Figure S3: The PXRD patterns for **Dy-Cl** and **Dy-Br**; Figure S4: The FT-IR spectra in the range of 4000 cm^{-1} to 400 cm^{-1} (a) and 1800 cm^{-1} to 600 cm^{-1} (b) of **Dy-Cl** and **Dy-Br**; Figure S5: (a) The 2D supramolecular layer in the *ac* plane in **Dy-Cl**; (b) the 3D supramolecular structure viewed from the *bc* plane in **Dy-Cl**; Figure S6: (a) The 2D supramolecular layer in the ac plane in **Dy-Br (193K)**; (b) the 3D supramolecular structure viewed from the *bc* plane in **Dy-Br (193K)**; Figure S7: (a) The 2D supramolecular layer in the *ac* plane in **Dy-Br (218K)**; (b) the 3D supramolecular structure viewed from the *bc* plane in **Dy-Br (218K)**; Figure S8: The solid-state UV-Vis absorption spectra of **Dy-Cl** (blue) and **Dy-Br** (red) at room temperature; Figure S9: The photoluminescence spectrum of **Dy-Cl** (a) and **Dy-Br** (b) before and after 1 h of 365 nm UV irradiation (λ_{ex} = 365 nm); Figure S10: The FT-IR spectra in the range of (a) 4000 cm^{-1} to 400 cm^{-1} and (b) 1800 cm^{-1} to 600 cm^{-1} for compounds **Dy-Cl** and **Dy-Br** before and after 1 h of 365 nm UV irradiation; Figure S11: Photoluminescence photograph of the samples under 365 nm UV light. (a) **Dy-Cl**; (b) **Dy-Cl-ground**; (c) **Dy-Cl-heat**; (d) **Dy-Br**; (e) **Dy-Br-ground**; (f) **Dy-Br-heat**; Figure S12: (a) The FT-IR spectra in the range of 1800 cm^{-1} to 600 cm^{-1} for **Dy-Cl**, **Dy-Cl-ground**, and **Dy-Cl-heat**; (b) The FT-IR spectra in the range of 1800 cm^{-1} to 600 cm^{-1} for **Dy-Br**, **Dy-Br-ground**, and **Dy-Br-heat**; Figure S13: Normalized photoluminescence spectra of **Dy-Cl-ground** and **Dy-Cl-heat** (a) and **Dy-Br-ground** and **Dy-Br-heat** (b) within five cycles of grinding/heating; Figure S14: Peak wavelength of photoluminescence spectra of **Dy-Cl-ground** and **Dy-Cl-heat** (a) and **Dy-Br-ground** and **Dy-Br-heat** (b) in cyclic grinding/heating test; Figure S15: The temperature-dependent $\chi_M T$ plots for **Dy-Cl** and **Dy-Br** within the range of 2–300 K; Figure S16: (a, c) The isothermal magnetization as a function of H at the depicted temperatures for **Dy-Cl** (a) and **Dy-Br** (c). (b, d) The isothermal magnetization as a function of HT^{-1} at the depicted temperatures for **Dy-Cl** (b) and **Dy-Br** (d); Figure S17: The magnetic hysteresis loops for **Dy-Cl** measured at the indicated temperatures; Figure S18: The Cole–Cole plots of **Dy-Cl** (a) and **Dy-Br** (b); Figure S19: The in-phase (a) and out-of-phase (b) *ac* magnetic susceptibilities of **Dy-Cl** under a 0–500 Oe *dc* field at 2 K; Figure S20: The in-phase (a) and out-of-phase (b) *ac* magnetic susceptibilities of **Dy-Br** under a 0–500 Oe *dc* field at 2 K; Figure S21: The magnetic hysteresis loops for **Dy-Cl** and **Dy-Cl-ground** (a) and **Dy-Br** and **Dy-Br-ground** (b) at 2 K; Figure S22: Plots of DC moment of lead to determine pressure; Figure S23: The temperature-dependent $\chi_M T$ plots for **Dy-Br** within the range of 2–300 K under 1.09 GPa and 0 GPa; Figure S24: (a) The isothermal magnetization as a function of H at the depicted temperatures for **Dy-Br** under 1.09 GPa; (b) the isothermal magnetization as a function of HT^{-1} at the depicted temperatures for **Dy-Br** under 1.09 GPa; Figure S25: The isothermal magnetization as a function of H at 2 K temperatures for **Dy-Br** under 1.09 GPa and 0 GPa; Figure S26: The magnetic hysteresis loops for **Dy-Br** under 1.09 GPa, measured at the indicated temperatures; Figure S27: (a, b) The in-phase (a) and out-of-phase (b) *ac* susceptibilities for **Dy-Br** under 1.09 GPa; (c) the Cole–Cole plot of **Dy-Br** under 1.09 GPa; (d) the plot of lnτ vs. T^{-1} for **Dy-Br** under 1.09 GPa; Table S1: The cell and structural parameters of **Dy-Cl**, **Dy-Br (193K)**, and **Dy-Br (218K)**; Table S2: Selected bond lengths (Å) and angles (°) for **Dy-Cl** at 193 K; Table S3: Continuous Shape Measure (CShM) analyses of dysprosium geometries for **Dy-Cl**, **Dy-Br (193K)**, and **Dy-Br (218K)** using the SHAPE2.1 software; Table S4: The parameters of H-bonding and halogen bonding for **Dy-Cl**, **Dy-Br (193K)**, and **Dy-Br (218K)**; Table S5: Selected bond lengths (Å) and angles (°) for **Dy-Br (193K)** measured at 193 K; Table S6: Selected bond lengths (Å) and angles (°) for **Dy-Br (218K)** measured at 218 K; Table S7: The fit parameters obtained from analyses of the *ac* susceptibilities of **Dy-Cl** under zero bias *dc* field; Table S8: The fit parameters obtained from analyses of the ac susceptibilities of **Dy-Br** under zero bias dc field; Table S9: The fit parameters obtained from analyses of the *ac* susceptibilities of **Dy-Br** under 1.09 GPa and zero bias *dc* field.

Author Contributions: Synthesizes, characterization, and writing, Y.-H.Q.; magnetic measurement and single-crystal structure analysis, Q.-Q.S.; single-crystal structure analysis, S.-S.B.; writing—review and editing, supervision, project administration, and funding acquisition, L.-M.Z. All authors have read and agreed to the published version of the manuscript.

Funding: This research was funded by the National Natural Science Foundation of China (No. 22273037, 21731003).

Informed Consent Statement: Not applicable.

Data Availability Statement: The original contributions presented in this study are included in the article/Supplementary Materials. Further inquiries can be directed to the corresponding author (L.-M.Z.).

Acknowledgments: We sincerely thank Dawid Pinkowicz of Jagiellonian University of Poland for the valuable discussions related to high-pressure magnetic measurements.

Conflicts of Interest: The authors declare no conflicts of interest.

References

1. Corredoira-Vázquez, J.; González-Barreira, C.; Fondo, M.; García-Deibe, A.M.; Sanmartín-Matalobos, J.; Gómez-Coca, S.; Ruiz, E.; Brites, C.D.S.; Carlos, L.D. An Air-Stable High-Performance Single-Molecule Magnet Operating as a Luminescent Thermometer below Its Blocking Temperature. *Inorg. Chem. Front.* **2025**, *12*, 5506–5516. [CrossRef]
2. Bispo, A.G., Jr.; Yeh, L.; Errulat, D.; Gálico, D.A.; Sigoli, F.A.; Murugesu, M. Improving the Performance of β-Diketonate-Based Dy^{III} Single-Molecule Magnets Displaying Luminescence Thermometry. *Chem. Commun.* **2023**, *59*, 8723–8726. [CrossRef]
3. Zanella, S.; Aragon-Alberti, M.; Brite, C.D.S.; Salles, F.; Carlos, L.D.; Long, J. Luminescent Single-Molecule Magnets as Dual Magneto-Optical Molecular Thermometers. *Angew. Chem. Int. Ed.* **2023**, *62*, e202306970. [CrossRef]
4. Dhbaibi, K.; Grasser, M.; Douib, H.; Dorcet, V.; Cador, O.; Vanthuyne, N.; Riobé, F.; Maury, O.; Guy, S.; Bensalah-Ledoux, A.; et al. Multifunctional Helicene-Based Ytterbium Coordination Polymer Displaying Circularly Polarized Luminescence, Slow Magnetic Relaxation and Room Temperature Magneto-Chiral Dichroism. *Angew. Chem. Int. Ed.* **2023**, *62*, e202215558. [CrossRef]
5. Zhang, Q.; Han, S.-D.; Li, Q.; Hu, J.-X.; Wang, G.-M. Electron Transfer Photochromism of Ln-Based (Ln = Dy, Tb) Coordinated Polymers for Reversibly Switching off/on Single-Molecule Magnetic Behavior. *Sci. China Mater.* **2022**, *65*, 788–794. [CrossRef]
6. Wang, J.; Zakrzewski, J.J.; Zychowicz, M.; Xin, Y.; Tokoro, H.; Chorazy, S.; Ohkoshi, S. Desolvation-Induced Highly Symmetrical Terbium(III) Single-Molecule Magnet Exhibiting Luminescent Self-Monitoring of Temperature. *Angew. Chem. Int. Ed.* **2023**, *62*, e202306372. [CrossRef] [PubMed]
7. Liao, P.-Y.; Liu, Y.; Ruan, Z.-Y.; Wang, H.-L.; Shi, C.-G.; Deng, W.; Wu, S.-G.; Jia, J.-H.; Tong, M.-L. Magnetic and Luminescent Dual Responses of Photochromic Hexaazamacrocyclic Lanthanide Complexes. *Inorg. Chem.* **2023**, *62*, 1075–1085. [CrossRef]
8. Münzfeld, L.; Dahlen, M.; Hauser, A.; Mahieu, N.; Kuppusamy, S.K.; Moutet, J.; Tricoire, M.; Köppe, R.; La Droitte, L.; Cador, O.; et al. Molecular Lanthanide Switches for Magnetism and Photoluminescence. *Angew. Chem. Int. Ed.* **2023**, *62*, e202218107. [CrossRef] [PubMed]
9. Cador, O.; Le Guennic, B.; Pointillart, F. Electro-Activity and Magnetic Switching in Lanthanide-Based Single-Molecule Magnets. *Inorg. Chem. Front.* **2019**, *6*, 3398–3417. [CrossRef]
10. Zhu, Z.; Li, X.-L.; Liu, S.; Tang, J. External Stimuli Modulate the Magnetic Relaxation of Lanthanide Single-Molecule Magnets. *Inorg. Chem. Front.* **2020**, *7*, 3315–3326. [CrossRef]
11. Zakrzewski, J.J.; Liberka, M.; Wang, J.; Chorazy, S.; Ohkoshi, S. Optical Phenomena in Molecule-Based Magnetic Materials. *Chem. Rev.* **2024**, *124*, 5930–6050. [CrossRef]
12. Bispo, A.G., Jr. Multifunctional Lanthanide(III) Single-Molecule Magnets Displaying Luminescence Thermometry: Progress, Challenges, and Future Directions. *Coord. Chem. Rev.* **2025**, *537*, 216685. [CrossRef]
13. Huang, X.-D.; Xu, Y.; Fan, K.; Bao, S.; Kurmoo, M.; Zheng, L. Reversible SC-SC Transformation Involving [4+4] Cycloaddition of Anthracene: A Single-Ion to Single-Molecule Magnet and Yellow-Green to Blue-White Emission. *Angew. Chem. Int. Ed.* **2018**, *57*, 8577–8581. [CrossRef] [PubMed]
14. Huang, X.-D.; Ma, X.; Zheng, L. Photo-responsive Single-Molecule Magnet Showing 0D to 1D Single-Crystal-to-Single-Crystal Structural Transition and Hysteresis Modulation. *Angew. Chem. Int. Ed.* **2023**, *62*, e202300088. [CrossRef]
15. Ma, X.-F.; Hou, X.-L.; Qin, Y.-H.; Teng, Q.; Bao, S.-S.; Tian, Y.-X.; Zheng, L.-M. A Thermally and Photoresponsive Luminescent Single-Molecule Magnet Based on Dysprosium–Anthracene: Effect of Temperature on Anthracene Photocycloaddition. *Chem. Sci.* **2025**, *16*, 10340–10348. [CrossRef]
16. Huang, X.-D.; Ma, X.-F.; Shang, T.; Zhang, Y.-Q.; Zheng, L.-M. Photocontrollable Magnetism and Photoluminescence in a Binuclear Dysprosium Anthracene Complex. *Inorg. Chem.* **2023**, *62*, 1864–1874. [CrossRef]
17. Ma, X.-F.; Huang, X.-D.; Zheng, L.-M. Tuning the Single-Molecule Magnet and Photoluminescence Properties of Binuclear Dysprosium Complexes by Light. *Cryst. Growth Des.* **2023**, *23*, 1095–1103. [CrossRef]
18. Wang, Y.-H.; Yan, S.-K.; Liang, S.; Zhao, Y.-R.; Zhang, J.; Wang, G.-M.; Hu, J.-X. Synergistic Photomagnetic and Photomechanical Dynamics in a Dysprosium-Based Smart Molecule. *Adv. Sci.* **2025**, *12*, e09088. [CrossRef]

19. Wang, Y.-H.; Gao, Z.-N.; Liang, S.; Wei, W.-J.; Han, S.-D.; Zhang, Y.-Q.; Hu, J.-X.; Wang, G.-M. Synergism of Light-Induced [4 + 4] Cycloaddition and Electron Transfer Toward Switchable Photoluminescence and Single-Molecule Magnet Behavior in a Dy4 Cubane. *Research* **2024**, *7*, 0411. [CrossRef] [PubMed]
20. Qin, Y.-H.; Ma, X.-F.; Hou, X.; Huang, X.-D.; Bao, S.-S.; Tian, Y.; Zhang, Y.-Q.; Zheng, L.-M. Photoresponsive luminescent single-molecule magnets based on dysprosium-anthracene complexes: Regulating the de-dimerization temperature of the photocycloaddition product by co-ligand. *Chem. Sci.* **2025**, *16*, 19806–19819. [CrossRef]
21. Huang, X.-D.; Hong, B.-K.; Wen, G.-H.; Li, S.-H.; Zheng, L.-M. Photo-Controllable Heterostructured Crystals of Metal–Organic Frameworks via Reversible Photocycloaddition. *Chem. Sci.* **2023**, *14*, 1852–1860. [CrossRef]
22. Gao, R.; Zou, Q.; Su, Q.-Q.; Ma, X.-F.; Qin, Y.-H.; Liao, R.; Bao, S.-S.; Zheng, L.-M. Photoresponsive Lanthanide-Dianthracene Framework: Introduction of Photoactive Anthracene Pairs by Controlling the Synthesis Temperature. *Chin. Chem. Lett.* **2024**, *36*, 110404. [CrossRef]
23. Abou-Hatab, S.; Spata, V.A.; Matsika, S. Substituent Effects on the Absorption and Fluorescence Properties of Anthracene. *J. Phys. Chem. A* **2017**, *121*, 1213–1222. [CrossRef]
24. Schillmöller, T.; Ruth, P.N.; Herbst-Irmer, R.; Stalke, D. Three Colour Solid-State Luminescence from Positional Isomers of Facilely Modified Thiophosphoranyl Anthracenes. *Chem. Commun.* **2020**, *56*, 7479–7482. [CrossRef]
25. Dai, Y.; Liu, H.; Geng, T.; Duan, R.; Li, X.; Liu, Y.; Liu, W.; He, B.-G.; Sui, L.; Wang, K.; et al. Piezochromic Fluorescence of Anthracene Derivative Crystals with Different Stacking Patterns Designed around Excimers. *J. Mater. Chem. C* **2023**, *11*, 4892–4898. [CrossRef]
26. Zhu, L.; Al-Kaysi, R.O.; Dillon, R.J.; Tham, F.S.; Bardeen, C.J. Crystal Structures and Photophysical Properties of 9-Anthracene Carboxylic Acid Derivatives for Photomechanical Applications. *Cryst. Growth Des.* **2011**, *11*, 4975–4983. [CrossRef]
27. Shu, Y.; Sun, J.; Yue, Y.; Ye, K.; Lu, R. Visible-Light-Triggered Actuators Based on the Molecular Crystals of Anthracenecarbonitrile Undergoing Reversible [4+4] Cycloaddition. *ChemPhotoChem* **2022**, *6*, e202200006. [CrossRef]
28. Gemen, J.; Białek, M.J.; Kazes, M.; Shimon, L.J.W.; Feller, M.; Semenov, S.N.; Diskin-Posner, Y.; Oron, D.; Klajn, R. Ternary Host-Guest Complexes with Rapid Exchange Kinetics and Photoswitchable Fluorescence. *Chem* **2022**, *8*, 2362–2379. [CrossRef]
29. Yang, X.; Dai, Y.; Liu, H.; Wang, K.; Yan, H.-L.; Yu, X.; Xia, Z.-A.; Wu, M.; Zhang, S.; Xiao, G.; et al. Antagonistic Effects of Distance and Overlap toward Anomalous Pressure-Induced Blueshift of π–π Excimer Fluorescence in 9-(2,2-Diphenylvinyl)Anthracene Crystals. *J. Am. Chem. Soc.* **2025**, *147*, 5300–5309. [CrossRef]
30. Chen, W.-T.; Chen, L.; Liang, Z.-Y.; Mo, Z.-W.; Ye, J.-W.; Chen, X.-M. Multiple Flexibilities Trigger Luminescent Piezochromism of Closely Packed Cu(I) Coordination Polymers. *Adv. Opt. Mater.* **2023**, *11*, 2202771. [CrossRef]
31. Yu, J.-X.; Christou, G.; Cheng, H.-P. Analysis of Exchange Interactions in Dimers of Mn3 Single-Molecule Magnets, and Their Sensitivity to External Pressure. *J. Phys. Chem. C* **2020**, *124*, 14768–14774. [CrossRef]
32. Craig, G.A.; Sarkar, A.; Woodall, C.H.; Hay, M.A.; Marriott, K.E.R.; Kamenev, K.V.; Moggach, S.A.; Brechin, E.K.; Parsons, S.; Rajaraman, G.; et al. Probing the Origin of the Giant Magnetic Anisotropy in Trigonal Bipyramidal Ni(II) under High Pressure. *Chem. Sci.* **2018**, *9*, 1551–1559. [CrossRef]
33. Pinkowicz, D.; Rams, M.; Mišek, M.; Kamenev, K.V.; Tomkowiak, H.; Katrusiak, A.; Sieklucka, B. Enforcing Multifunctionality: A Pressure-Induced Spin-Crossover Photomagnet. *J. Am. Chem. Soc.* **2015**, *137*, 8795–8802. [CrossRef]
34. Handzlik, G.; Dziubek, K.F.; Hanfland, M.; Pinkowicz, D. Simultaneous Manipulation of Iron(II) Spin Crossover and LIESST Behaviour Using Pressure, Temperature and Light. *Dalton Trans.* **2024**, *53*, 7677–7681. [CrossRef]
35. Liu, X.-L.; Zhao, Y.; Wang, T.; Wang, P.; Li, G.; Gao, B.-H.; Zhang, X.-L.; Wang, X.-Y. Fe^{III} Spin Crossover Complexes with a Dual-Substituted Hqsal Ligand: Anion, Solvent, and Pressure Effects. *Cryst. Growth Des.* **2025**, *25*, 3518–3528. [CrossRef]
36. Parmar, V.S.; Thiel, A.M.; Nabi, R.; Gransbury, G.K.; Norre, M.S.; Evans, P.; Corner, S.C.; Skelton, J.M.; Chilton, N.F.; Mills, D.P.; et al. Influence of Pressure on a Dysprosocenium Single-Molecule Magnet. *Chem. Commun.* **2023**, *59*, 2656–2659. [CrossRef]
37. Pointillart, F.; Le Guennic, B.; Cador, O. Pressure-Induced Structural, Optical and Magnetic Modifications in Lanthanide Single-Molecule Magnets. *Chem. Eur. J.* **2024**, *30*, e202400610. [CrossRef] [PubMed]
38. Norre, M.S.; Gao, C.; Dey, S.; Gupta, S.K.; Borah, A.; Murugavel, R.; Rajaraman, G.; Overgaard, J. High-Pressure Crystallographic and Magnetic Studies of Pseudo-D5h Symmetric Dy(III) and Ho(III) Single-Molecule Magnets. *Inorg. Chem.* **2020**, *59*, 717–729. [CrossRef]
39. Parmar, V.S.; Thiel, A.M.; Norre, M.S.; Overgaard, J. High-Pressure Investigation of a Five-Coordinate Dy Single-Molecule Magnet. *Cryst. Growth Des.* **2023**, *23*, 6410–6417. [CrossRef]
40. Pointillart, F.; Flores Gonzalez, J.; Douib, H.; Montigaud, V.; McMonagle, C.J.; Le Guennic, B.; Cador, O.; Pinkowicz, D.; Probert, M.R. Reversible Pressure-Magnetic Modulation in a Tetrathiafulvalene-Based Dyad Piezochromic Dysprosium Single-Molecule Magnet. *Chem. Eur. J.* **2023**, *29*, e202300445. [CrossRef] [PubMed]
41. Chen, W.-B.; Chen, Y.-C.; Liu, J.-L.; Jia, J.-H.; Wang, L.-F.; Li, Q.-W.; Tong, M.-L. A Piezochromic Dysprosium(III) Single-Molecule Magnet Based on an Aggregation-Induced-Emission-Active Tetraphenylethene Derivative Ligand. *Inorg. Chem.* **2017**, *56*, 8730–8734. [CrossRef] [PubMed]

42. Chen, K.; Wang, J.; Feng, Y.; Liu, H.; Zhang, X.; Hao, Y.; Wang, T.; Huang, X.; Hao, H. Multiple Stimuli-Responsive Flexible Crystal with 2D Elastic Bending, Plastic Twisting and Photoinduced Bending Capabilities. *J. Mater. Chem. C* **2021**, *9*, 16762–16770. [CrossRef]
43. Koehne, I.; Pietschnig, R. Synthesis of Geminal Bis- and Tetrakisphosphonate Ester Derivatives and Their Coordination Behavior Towards Ca(II) Ions. *Eur. J. Inorg. Chem.* **2022**, *2022*, e202200194. [CrossRef]
44. SAINT. *Program for Data Extraction and Reduction, Version 8.40A*; Bruker Nano. Inc.: Billerica, MA, USA, 2019.
45. Sheldrick, G.M. Crystal structure refinement with SHELXL. *Acta Cryst. C* **2015**, *71*, 3–8. [CrossRef]
46. Casanova, D.; Llunell, M.; Alemany, P.; Alvarez, S. The Rich Stereochemistry of Eight-Vertex Polyhedra: A Continuous Shape Measures Study. *Chem. Eur. J.* **2005**, *11*, 1479–1494. [CrossRef]
47. Schmidt, G.M.J. Photodimerization in the Solid State. *Pure Appl. Chem.* **1971**, *27*, 647–678. [CrossRef]
48. Pandiyan, B.V.; Deepa, P.; Kolandaivel, P. Do Resonance-Assisted Intramolecular Halogen Bonds Exist without a Charge Transfer and a σ-Hole? *Phys. Chem. Chem. Phys.* **2015**, *17*, 27496–27508. [CrossRef]
49. Wu, X.; Guo, J.; Cao, Y.; Zhao, J.; Jia, W.; Chen, Y.; Jia, D. Mechanically Triggered Reversible Stepwise Tricolor Switching and Thermochromism of Anthracene-o-Carborane Dyad. *Chem. Sci.* **2018**, *9*, 5270–5277. [CrossRef]
50. Suresh, S.K.; Raju, P.D.; Krishnan, A.; Ramachandran, L.M.; Suneesh, C.V. Supramolecular Interactions-Assisted Polymorphism and Unique Mechanofluorochromism in 9,10-Bis((E)-4-(Trifluoromethyl)Styryl)Anthracene. *Chem. Eur. J.* **2023**, *29*, e202204030. [CrossRef] [PubMed]
51. Dong, Y.; Xu, B.; Zhang, J.; Tan, X.; Wang, L.; Chen, J.; Lv, H.; Wen, S.; Li, B.; Ye, L.; et al. Piezochromic Luminescence Based on the Molecular Aggregation of 9,10-Bis((E)-2-(Pyrid-2-Yl)Vinyl)Anthracene. *Angew. Chem. Int. Ed.* **2012**, *51*, 10782–10785. [CrossRef]
52. Kannen, F.; Adachi, T.; Nishimura, M.; Yoza, K.; Kusukawa, T. Mechanofluorochromic Properties of 1,4-Diphenylanthracene Derivatives with Hypsochromic Shift. *Molecules* **2024**, *29*, 407. [CrossRef]
53. Yu, H.; Li, M.; Chen, X.; Han, N.; Xu, Z.; Wang, M. Synthesis of Anthracene-Based Mechanofluorochromic Molecules and Their Differential Mechanochromic Effects Triggered by Nitrogen Atoms. *Eur. J. Org. Chem.* **2024**, *27*, e202400876. [CrossRef]
54. Xu, W.-J.; Luo, Q.-C.; Li, Z.-H.; Zhai, Y.-Q.; Zheng, Y.-Z. Bis-Alkoxide Dysprosium(III) Crown Ether Complexes Exhibit Tunable Air Stability and Record Energy Barrier. *Adv. Sci.* **2024**, *11*, 2308548. [CrossRef]
55. Li, D.-P.; Zhang, X.-P.; Wang, T.-W.; Ma, B.-B.; Li, C.-H.; Li, Y.-Z.; You, X.-Z. Distinct magnetic dynamic behavior for two polymorphs of the same Dy(III) complex. *Chem. Commun.* **2011**, *47*, 6867–6869. [CrossRef] [PubMed]
56. Horii, Y.; Katoh, K.; Sugimoto, K.; Nakanishi, R.; Breedlove, B.K.; Yamashita, M. Detailed Analysis of the Crystal Structures and Magnetic Properties of a Dysprosium(III) Phthalocyaninato Sextuple-Decker Complex: Weak f–f Interactions Suppress Magnetic Relaxation. *Chem. Eur. J.* **2019**, *25*, 3098–3104. [CrossRef] [PubMed]
57. Long, J.; Tolpygin, A.O.; Mamontova, E.; Lyssenko, K.A.; Liu, D.; Albaqami, M.D.; Chibotaru, L.F.; Guari, Y.; Larionova, J.; Trifonov, A.A. An Unusual Mechanism of Building up of a High Magnetization Blocking Barrier in an Octahedral Alkoxide Dy^{3+}-Based Single-Molecule Magnet. *Inorg. Chem. Front.* **2021**, *8*, 1166–1174. [CrossRef]
58. Cole, K.S.; Cole, R.H. Dispersion and Absorption in Dielectrics I. Alternating Current Characteristics. *J. Chem. Phys.* **1941**, *9*, 341–351. [CrossRef]
59. Gu, L.; Wu, R. Origin of the Anomalously Low Raman Exponents in Single Molecule Magnets. *Phys. Rev. B* **2021**, *103*, 014401. [CrossRef]
60. Ma, Y.; Zhai, Y.-Q.; Luo, Q.-C.; Ding, Y.-S.; Zheng, Y.-Z. Ligand Fluorination to Mitigate the Raman Relaxation of Dy^{III} Single-Molecule Magnets: A Combined Terahertz, Far-IR and Vibronic Barrier Model Study. *Angew. Chem. Int. Ed.* **2022**, *61*, e202206022. [CrossRef]

MDPI AG
Grosspeteranlage 5
4052 Basel
Switzerland
Tel.: +41 61 683 77 34

Magnetochemistry Editorial Office
E-mail: magnetochemistry@mdpi.com
www.mdpi.com/journal/magnetochemistry

www.ingramcontent.com/pod-product-compliance
Lightning Source LLC
LaVergne TN
LVHW070452061126
851858LV00011B/225

9783725871605